(1994) 17.99

How Soon Is Now?

HOW SOON IS NOW?

THE TRUTH ABOUT THE OZONE HOLE

Nicholas Booth

SIMON & SCHUSTER
LONDON · SYDNEY · NEW YORK · TOKYO · SINGAPORE · TORONTO

First published in Great Britain by Simon & Schuster Ltd, 1994
A Paramount Communications Company

Copyright © Nicholas Booth, 1994

This book is copyright under the Berne Convention
No reproduction without permission
All rights reserved

The right of Nicholas Booth to be identified as author of this work has been asserted in accordance with sections 77 and 78 of the Copyright Designs and Patents Act 1988

Simon & Schuster Ltd
West Garden Place
Kendal Street
London W2 2AQ

Simon & Schuster of Australia Pty Ltd
Sydney

A CIP catalogue record for this book is available from the British Library.

0–671–71180–6

Typeset in Sabon 11/14 by
Hewer Text Composition Services, Edinburgh
Printed and bound in Great Britain by
Butler & Tanner Ltd, Frome and London

For my parents, who've bravely fought the spectre of cancer

Then I saw that there was a way to hell, even from the gates of heaven.

John Bunyan, *Pilgrim's Progress*

Contents

Acknowledgements	ix
1. Introduction: Whatever Happened to the Ozone Hole?	1
2. Polar Twilight	11
3. Crisis! What Crisis?	52
4. Eight Miles High	103
5. Global Perspectives	141
6. Pieces in a Puzzle	190
7. Instant Science	231
8. Deadly Springtime	271
9. Dr Watson Goes to Washington	316
10. Planet in Peril?	361
Acronyms	388
Glossary	391
Selected Bibliography	394
Interviewees	397
Index	402

Acknowledgements

The germ of the idea to do this book came in the summer of 1990 when I was asked to profile Joe Farman for the late, lamented *Green* magazine. Like a great many people, I suspect, I blithely assumed that the Antarctic ozone hole was all figured out. Five minutes in the company of Joe made me realise it wasn't and indeed, increasingly thereafter, I started to think that here was an untold story worthy of detailed examination in book form. If nothing else, I wanted to strip bare the basic facts which had remained perpetually concealed by the environmental hysteria of the late 1980s and early 1990s.

The end result is this book to which I subsequently devoted my energies and writing talents from October 1991 to October 1993. As will be clear from the text, I travelled widely – across the United States (twice), throughout Britain, to Sweden and France and down to the southernmost tip of Chile. I sought the expertise, perspective and guidance of many scientists, engineers, political advisers, industrial chemists and environmentalists. Without their readiness to be interviewed this book would not have been possible. I appreciated their help and kindness in explaining the more arcane aspects of a bewilderingly complex subject, particularly as new developments took place as the text took shape. Any errors of interpretation or fact are my own.

The scientists whom I interviewed are listed in the bibliography. In particular, some of them put themselves out in many ways and at the risk of invidiousnes, I should wish to thank those who went to the trouble of reading and re-reading drafts, as well as answering my

How Soon Is Now?

persistent, increasingly esoteric, questions: Sherry Rowland, Adrian Tuck, Joe Waters, Susan Solomon and Bob and Elizabeth Festa Watson in the United States as well as Fred Taylor, John Harries, Joe Farman, Brian Gardiner, John Pyle and Neil Harris in Britain.

Perhaps the most daunting task of all was my travel to the world's most southerly city. Thankfully, the remarkable Bedrich Magas ensured that I was able to piece together a true picture of the health effects in and around Punta Arenas. He also took time and trouble to help set up interviews and often served as a translator.

The task of collating material, as any writer knows, can be smoothed along by public relations officials who, by the very nature of their task, often have a difficult time of it. This book could not have been written without the cooperation of many public information officials within the US National Aeronautics and Space Administration. In recent years as NASA's prestige has fallen, so too has the service provided by its public information officers. It isn't what it once was, but there remains a core of excellent people who provide invaluable help and assistance: at JPL, Jim Wilson and Bob McMillin; at NASA Ames, Peter Waller; at NASA Goddard, Dolores Beasley, Allen Kenitzer and Jim Elliott; and at NASA Headquarters, Debbie Rahn and Brian Dunbar.

Other public information people bent over backwards far beyond professional necessity to aid me on my travels. Joan Vandiver Frisch at NCAR not only arranged me somewhere to stay in Boulder, she also made sure that I was restored to health after a particularly nasty bout of flu. My time in Boulder was helped by Bill Brennan at NOAA, Peter Caughey and Jim Scott at the University of Colorado. The indefatigable Don Bane, now with Lockheed in Palo Alto, saw to it that, in the words of the song, I found my way to San José. Other valued sources of information were: Diana Steele at the University of Chicago; Barbara Lubinksi at the University of Denver; Barry Parker at the Met Office; Jackie Hutchinson at the Rutherford Appleton Laboratory; Neil Pattie at the National Remote Sensing Centre; plus Corin Milais at Greenpeace and Dr Doug Parr at Friends of the Earth.

I owe heartfelt gratitude to my agent, Vivienne Schuster of Curtis Brown, for her wise advice and constant encouragement. This project has also been blessed by the counsel, interest and good naturedness of Nick Webb, Managing Director of Simon & Schuster, who provided a voice of objective sanity as multiple drafts swirled around – and

ACKNOWLEDGEMENTS

threatened to engulf – me. Ingrid von Essen copy-edited the text with precision and aplomb which improved the material immeasurably. Thanks also to everyone else at Simon & Schuster who nursed this book to press, particularly Jenny Olivier, Jacquie Clare and Aruna Mathur.

During the protracted gestation of this book, my friends had to endure my ramblings and witterings as I tested ideas and chapters upon them. In particular, Dominique Torode and Julian Parker allowed me to invade their spare room to finish off the final chapter. I would like to thank the following for help above and beyond the call of friendship: Steph Bruder and Paddy Merryfield, Niki Boyce and Kevin Rodgers, Penny Finlay, Kate FitzGerald, Simon Savage, Stuart Eaton, Helen Eaton and Lee Garner, Laura McGee, Mandy Wheeler, Marion Morrell, Katie Normington, and Alex Sutton-Vane. I owe a particular debt to Roz Lowrie who is unique in exerting a calming influence with characteristic creativity, humour and care.

Finally, I would like to thank my oldest, wisest friends of all. Al and Marka Hibbs have put up with my irregular appearances in Pasadena and disturbances to their routine for over a decade now. Frank Miles has also been a tireless supporter and cast his critical eye over some early drafts of material. Finally, and fundamentally, I owe a great debt to my parents for allowing me to disturb their lives for the extended periods that I spent in Cheshire coaxing a book out of multiple drafts and logistical panics throughout 1992–93. Somehow I did, and without them, this book simply would not have been possible.

NICHOLAS BOOTH
May 1994

1
Introduction

Whatever Happened to the Ozone Hole?

We simply had to understand the Earth's atmosphere, to chart both its resiliency and its fragility, and to come to terms with our own, largely unwitting, ability to damage it. This was a formidable task, and the ozone scare forced us to make a start on it, but the job is far from done. Large global systems like the Earth's atmosphere are characterised by exceedingly complex chemical and physical processes that frequently defy an adequate understanding, much less control. Chemicals like the spray can propellants never existed in nature. In his increasing use of such chemicals, man seems at times to be engaged in some bizarre game of environmental Russian roulette.

It may be that, on a global scale, it is a game in which he is out of his league.

Lydia Dotto and Harold Schiff, *The Ozone War*

[Although] ozone destruction would continue for decades, the controls now put in place would eventually halt it. The worst years are still to come [from] 1995 to 2005, with the amount of ozone destroying chlorine in the atmosphere expected to peak at around the turn of the century at about four parts per billion. It would be the middle of the next century before the level returned to that of the early 1980s.

News item, *The Times*, 26 November 1992

* * *

How Soon Is Now?

On his first day back at work after a three-month sabbatical, Dr Jaime Abarca finally became convinced by the threat posed by the Antarctic ozone hole. What surprised him, though, was the severity of the symptoms exhibited by the youths who had sought his medical advice. As the only dermatologist in the city of Punta Arenas at the southern tip of Chile, Abarca had long ago become used to people reporting rashes, headaches and all manner of perplexing scrofulas which they almost inevitably blamed on El Agujero – that hung, however uncertainly, like a modern-day Sword of Damocles above their heads. It was the price to pay for living in the world's most southerly city, a quirk of geographical proximity to the South Pole, and the opening-up, each spring, of a hole in the ozone layer above it.

Abarca had always maintained a professional detachment and remained sceptical. Although, as he well knew, ozone depletion would bathe the Earth's surface with increased levels of dangerous ultraviolet radiation, it was believed that it would be years before the effects were discernible. A deleterious legacy of increased skin cancers and cataracts in humans, as well as damage to plants and animals, could possibly result. Already reports of sheep and cattle going blind in the farmlands surrounding the Straits of Magellan had increased in number, but these were largely apocryphal. If they indicated anything, it was perhaps the capacity of the population at large to undergo a mild form of psychosis in blaming each and every unusual occurrence upon the ozone hole.

On Monday 5 October 1992, Abarca changed his mind. He had been away in the United States to learn the latest developments in dealing with skin cancer. But nothing had prepared him for odd symptoms exhibited by a backlog of patients. That afternoon, two teenagers reported an intense and painful redness of the face which had resulted from their playing football for a couple of hours outside the day before. Several of their footballing friends exhibited similar, but less intense, skin problems. One of the youths was given painkillers and told to report back in two days' time. By then, a nurse at the local hospital had told Abarca of other cases of severe sunburn. The teenager reported back with a face that was already peeling.

Abarca was puzzled until he learned a disturbing fact. That

weekend an orbiting NASA satellite had watched the northerly extremities of the ozone hole brush across South America. Seven years after the phenomenon had first been announced to a shocked world, that year's hole had already been the quickest forming and deepest ever seen. Ozone levels over Tierra del Fuego were the lowest ever recorded by the Nimbus 7 satellite in nearly fourteen years of operation. And, as the hole wobbled uncertainly around the Antarctic landmass, the areal extent of the ozone loss was the largest ever observed, covering sizeable portions of southern Chile and Argentina. To Abarca the coincidence was unequivocally disturbing. 'What I saw that day is not normal for here in Punta Arenas,' he told me two weeks later. 'These are the first biological impacts directly on human beings.'

A few weeks later, however, a team of medical researchers from Johns Hopkins University in Baltimore, intrigued by fanciful newspaper reports of 'bug-eyed bunnies' and oddly flowering plants, visited southern Chile as part of a pilot study to quantify the health effects of ozone depletion. 'It is clear that the citizens of Punta Arenas are not experiencing a burden of illness due to ozone depletion,' said the project director, Dr Oliver Schein, after their visit. Though he expressed concern that long-term, cumulative exposure to the ozone hole could be significant, he suggested it had had little effect so far on the flora and fauna of southern Chile.

This was the conclusion of the report which his team submitted to the US Environmental Protection Agency six months later, after carefully considering all the available evidence for increased levels of the most dangerous form of ultraviolet radiation, known as UV-B. Dr Schein and his colleagues concluded:

> Approximately a doubling of UV-B exposure occurred on isolated days in September and October 1992 in Punta Arenas as a result of the ozone hole.
>
> The evidence for acute, immediate health effects of this UV-B exposure is sparse. Such effects were not supported by medical chart review from ophthalmologists or dermatologists in the region. Similarly, acute diseases known to be related to UV-B exposure in animals were not found.

How Soon Is Now?

So who is to be believed? Not for the first time would the science related to the ozone issue cause researchers to come to opposite conclusions from looking at roughly the same sort of information. Ambivalent conclusions are common in scientific enquiry. The ozone issue has particularly polarised discussions of the science involved and the uncertainties and inconsistencies within it have often been glossed over. It would be foolish to suggest that the very fate of the Earth is threatened by ozone depletion but that is not to say it will not present any problems in the future. The purpose of this book is to give a balanced assessment of the dangers.

Perhaps the most curious fact about the ozone layer is just how unlikely is its existence. Ultraviolet radiation forms just one small part of the constant stream of light from the Sun which allows life to flourish on Earth. High above our heads, a diffuse layer of ozone is formed by the action of sunlight upon oxygen, and in the process this absorbs the most dangerous forms of ultraviolet radiation. It is no exaggeration to say that the continued well-being of life on Earth depends on what would otherwise be a very obscure form of high-altitude chemistry. And, indeed, the relationship between life and the ozone layer has been delicately poised for millennia.

Life could never have evolved on Earth had the ozone layer been part of our primordial atmosphere. And yet, without it, life could not exist today. When the Sun and planets formed 4.5 billion years ago out of a swirling, amorphous disc of dust and gas, the nascent Earth was ideally placed for life to evolve. The basic biochemical building blocks were present and conditions were conducive for life to form. As in the fable of Goldilocks on an unimaginably vast, interplanetary scale, Venus was too warm, Mars was too cold but Earth was just right.

Ironically, the complex biochemistry out of which simple amino acids evolved was triggered by the same ultraviolet radiation that now threatens life on Earth. Beyond those wavelengths to which our eyes are sensitive, ultraviolet radiation is more energetic and dangerous than visible light. Constantly radiated from the Sun, ultraviolet light streamed down to the Earth's surface unimpeded in past epochs and destroyed many of the chemical reactions that could trigger the formation of life from the 'primeval soup' of organic material. Deep within the oceans or mudlands, however, these nascent life forms were protected from the harshest ultraviolet radiation and began to photosynthesise food production.

Over millennia, these basic, cellular life forms then produced oxygen which gradually accumulated and pervaded the whole of the Earth's atmosphere. Ultraviolet light broke down this oxygen into an unstable form high above the Earth to create a layer of ozone. Thereafter life on Earth was protected from the harshest ultraviolet radiation and began to oxidise organic material directly. The significant biochemical energy which was released enabled life to move onto the land and, eventually, evolve into higher forms of multicellular animals and plants. And thanks to the protection afforded by the ozone layer, life was able to flourish on planet Earth.

Only since the 1930s have scientists had unequivocal proof of the existence of this protective layer of ozone and a sense of just how precarious it truly is. Roughly 350,000 kilos of this pungent, poisonous and colourless gas is produced every second high above the equator and then continuously circulates around the Earth to diffuse into a permanent layer on the edge of space. The balance between the natural creation and destruction of ozone is precarious for if all the ozone in the stratosphere was brought down to ground level, it would amount to less than the thickness of the cover of this book.

Sadly, mankind's unparalleled profligacy in polluting our environment has threatened the continued well-being of life on Earth. Just before Christmas 1973, two research chemists in California, Sherwood Rowland and Mario Molina, stumbled across the astounding fact that artificial chlorine compounds – specifically, the CFCs employed in aerosols and the halons found in fire-fighting equipment – would inexorably diffuse up into the stratosphere and remain there for a century and a half. During that time a cascade of chemical reactions would ensue, resulting in the steady stripping away of the ozone layer right across the globe. Their findings were, as might be expected, highly controversial, and all the more so for the fact that it took over a decade for unequivocal proof of ozone depletion to be found.

As recently as 1984, the US National Academy of Sciences

produced a report which said that although ozone would indeed be depleted by CFCs, it would be decades before that depletion would actually make itself apparent. That same August, the US magazine *Science Digest* carried a story entitled 'Ozone: The Crisis That Wasn't'. The Environmental Protection Agency estimated that a ban on CFCs would cost 9,000 jobs and add $1 billion to the burgeoning federal deficit. Industrial manufacturers of CFCs expressed concern but procrastinated, preferring to see evidence of damage before anything was done to curb their production. All too typical of the perceived wisdom of the time was the assertion by the Chemical Manufacturers' Association that 'we pay too much attention to those who go around proclaiming the sky is falling'.

A quiet, unassuming British scientist called Joe Farman was the unlikeliest of Chicken Lickens. But in the early 1980s, his research team at the British Antarctic Survey was the first to see that ozone was indeed depleting – but not as predicted. Instead of the slow stripping-away of the ozone layer across the globe, Farman had found that each spring an area the size of the United States was depleted of more than half its ozone in the skies above the South Pole. When his results were published in May 1985, they amounted to a scientific bombshell, for, as Farman himself has often reflected, 'Antarctica was the last place on Earth to look'.

In the years since then, ozone depletion has worsened over Antarctica each spring as the atmosphere continues to be loaded with chlorine from chlorofluorocarbons. The ozone hole in the austral spring of 1993 registered as the most depletion yet seen, with three-quarters of the ozone above the South Pole removed on 5 October. For Sherry Rowland, it underlined the seriousness of the situation: 'If we have more chlorine in the stratosphere in the future we'll lose ozone faster. But you obviously can't lose more than all of it. Twice as much chlorine, say, and you'll lose it in three weeks rather than six. In that sense, we've reached saturation point.'

Slowly but surely our protective ozone shield is weakening around the globe. In the late 1980s, it had become abundantly clear that the slow stripping-away of ozone was most pronounced over the northern hemisphere. Throughout that decade, 8 per cent of the total ozone above northern mid-latitudes had been lost, as opposed to the 4 per cent loss which had been predicted. Despite international controls which will phase CFCs out by the end of 1995, increased 'chlorine loading' will continue until at least the year 2005. The prognosis for the 1990s is that we are fast approaching the most dangerous point of all.

Any sense of relief at the phasing-out of ozone-depleting chemicals has to be tempered by the knowledge that ozone depletion will continue and gather pace throughout the 1990s and well into the twenty-first century. Twenty million tonnes of CFCs have been pumped into the atmosphere since they were first manufactured. The ozone depletion that is seen today is caused in part by the first chlorofluorocarbons which were injected into the atmosphere as long ago as the 1930s and 1940s. Even with today's accelerated phase-outs, 5 million tonnes will remain in the stratosphere until the end of the twenty-first century. As NASA's Michael Kurylo has pointed out: 'Even if we don't add another ounce of chlorine to the atmosphere, it will take about a century for their levels to return to their pre-1985 levels before the hole occurred.'

On the other hand, nobody has yet died from ozone depletion and the direst predictions about its effects have yet to come true. It is still by no means certain how perceived changes in the ozone layer will affect the inhabitants of planet Earth and the flora and fauna upon its surface. The most worrying developments so far are not those which some environmentalists would have us believe.

Take, for example, the issue of skin cancers. Throughout the world, incidences of normally rare cancers of the skin have been rising alarmingly in recent years. In the spring of 1992, for example, Greenpeace began a very emotive awareness campaign to highlight the dangers which may accumulate from ozone depletion. A child inside a plastic protection suit (bearing a bizarre resemblance to Macaulay Culkin) accompanied the caption 'You'll need more than a brave face when the ozone goes'. What the leaflets failed to point out is that ozone depletion hasn't been round for long enough to cause such damage. In fact, summer holidays in sunnier climes and the perception that a suntan is healthy underpin the frightening skin-cancer statistics.

It is for reasons like this, perhaps, that worldwide concerns about the environment have taken a back seat to the problems of economic recession. During the writing of this book, the pendulum of popular perception most definitely swung the opposite way. The fact that many of the promised horrors from ozone depletion have failed to be realised has been taken as evidence that we have somehow been saved from the peril. Diagnosed as the 'Ozone Backlash' by *Science* magazine in 1993, it 'graphically demonstrates the problems of doing research on a politically charged issue where there are still many scientific uncertainties'.

The starting point for the Backlash came early in 1992, when it appeared that an ozone hole might form over the North Pole. Over the winter of 1991–92 both American and European scientists had large investigations of the conditions over the North Pole. In early February 1992, NASA held a press conference to alert the world to that possibility, though in the end the stratosphere warmed and it was averted. In many ways, the press conference acted like a lightning rod for many of the criticisms of current scientific knowledge. Most of them can be dismissed as half-truths, misinterpretations or deliberate lies.

If in the 1980s each and every development in ozone science was taken as a portent of doom, they now seem to reinforce the impression that it might all have been irresponsible scaremongering. 'After nearly a decade of headlines and hand-wringing about erosion of the Earth's protective ozone layer,' the *Washington Post* declared in April 1993, 'the problem appears to be well on the way to solution.' Alas, the paper had totally misread some of the latest developments in the science, a conclusion which other, less respectable journals were keen to exploit.

The truth is that the problems remain, as do the dangers which may accumulate in the future. To suggest anything less would be to perpetrate a new set of myths as ridiculous as many of those environmentally correct exaggerations from the late 1980s.

Ozone scientists find themselves working in a constant glare of media interest. They endeavour to make sense of their data against a backdrop of politics, pressure from the environmental lobby and prophecies of doom. The atmosphere is a dynamic, constantly shifting laboratory. Unlike the ones in which scientists work, it is totally beyond their control, as is the prevailing political climate of the day. Scientists find that even their most cautious statements, hedged in caveats and tentative terms, can be misrepresented, both unknowingly and deliberately. They have learned as best they can how to perform their work in the fishbowl of media attention.

Scientists are not infallible and need leeway in the normal course of their work. Yet scientific fallibility is harder to accommodate under intense media scrutiny than it would normally be in the corridors of the conference centre or the learned pages of academic journals. The Ozone Backlash, like extreme examples of environmental evangelism, is a farrago of half-truths and insinuations to which ozone researchers have had to inure themselves. Unwittingly, they have found that their words have been amplified and distorted by politicians, polemicists and environmentalists alike. Cautious speculation can become tomorrow's headlines or the cornerstone of a governmental bargaining stance. It is, for scientists at least, the most disagreeable aspect of living in the age of the sound bite. They have had to learn the hard way a lesson about which Bob Dylan once sang: you don't need a weatherman to tell you which way the wind is blowing.

One prominent researcher offered this perspective:

> For twenty or more years, stratospheric science was a backwater. And then all of a sudden ozone became 'trendy'. Our research takes in all sorts of complex factors which will ordinarily take many years to resolve. But now we're forced out of the quiet of the lab and into the limelight. All of a sudden – almost overnight – we're thrust into the position of having to comfort the public and enter the political arena as well. If you go to your GP he can't tell you when you're going to die. I can't tell you what the future climate will be. All I can say with any degree of accuracy is that the Earth a few decades from now will be a different planet and that's all.

The ozone story is full of strange episodes, odd characters, unlikely adventures and bizarre coincidences. At times, it reads like a cross between a Greek tragedy and a nineteenth-century farce. The narrative could be thought of as a detective story bringing the sorry tale of ozone depletion up to date. There is, however, no dénouement for developments continue apace. The totally unexpected discovery of the Antarctic ozone hole also suggests that there may be other unpleasant surprises in store. The text is also part travelogue, underscored by a unique mix of politics, science and human nature. It presents the latest scientific research and the worldwide reaction to it by politicians, environmentalists and the general public. It is my hope that I will provide the average non-technical reader with enough insight to make informed decisions about a bewilderingly complex subject.

Throughout 1992 and 1993, I travelled extensively to monitor the work of scientists and engineers intimately involved with the ozone issue across the globe. From Punta Arenas in Chile to Kiruna in northern Sweden, I conducted nearly a hundred interviews with prominent researchers in the field. In France and Sweden, I talked with experimenters who use balloons to hoist their equipment into the stratosphere; in California, I talked at length with the men who have for the past few years routinely flown the ER-2 aircraft over the polar regions; in Colorado and New York, as well as Oxfordshire and Cambridge, scientists shared their experiences in modelling the atmosphere using supercomputers; and in and around the US capital, I interviewed at length NASA scientists who keep a constant vigil on the ozone layer from space or advise world governments on the state of the stratosphere. This book, then, represents the fullest popular account of global concerns about ozone depletion yet assembled.

We can only untie the Gordian knot of ozone depletion if we are fully aware of its intricacies. The price to be paid for ignorance is great: if we fail to comprehend its Byzantine complexities and contradictions we will surely be consigned to a fate aptly described by T. S. Eliot – 'We had the experience but did not heed the lesson.' That would be the ultimate folly, the total abrogation of responsibility and the handing-down to the next generation of a problem which by then would undoubtedly prove to be intractable.

2

Polar Twilight

> It is astonishing and even terrifying to contemplate the narrow margin of safety on which our lives thus depend. Were this trifling quantity of atmospheric ozone removed we should all perish.
>
> Dr Charles Greeley Abbot of the Smithsonian Institution on the apparent thinness of the ozone layer, quoted in the *New York Times*, 30 October 1933

It was Friday the thirteenth and it showed on the faces of the scientists sitting around the large conference table. Outside, slow flurries of snow made a suitably depressing backdrop to their discussion of the weather over northern Sweden. Inside, the air of despondency was heightened as all the participants were feeling tired, cold and more than a little homesick. It had been a long campaign – the longest yet – and the assembled group were almost battle-fatigued after the concentration over the past five months. The novelty of being able to ski every day had long since worn off. Worse still was the chronic indigestion resulting from the food provided by the canteen at Europe's only privately owned space base.

The state of the weather is of direct concern to scientists involved with one of the most intense investigations of the ozone layer over the North Pole. For the primary tool of the European Arctic Stratospheric Ozone Experiment is large scientific balloons which can be launched only under the calmest of conditions. Here, well within the Arctic Circle in the bleak tundra north of the city of Kiruna, the weather is not exactly at its most clement in March. But the geographical proximity to the North Pole means that balloons will rise up into that part of the stratosphere where the most interesting chemistry will be taking place. The multipurpose research centre known as ESRANGE (which launched Sweden's first satellite in the 1970s) is the only place for hundreds of kilometres around with the requisite facilities to mount a large, coordinated balloon campaign.

Today's met briefing is typical of the five-month experiment. In the conference room, ESRANGE officials sit at the far end of the table and lead the discussion through its preordained course. By the window are the scientists whose experiments are being prepared for flight. Opposite them are the French technicians who build, fly and operate the balloons. Caught somewhere in the middle is Neil Harris, a young atmospheric chemist from Cambridge University. Though he has no experiment of his own, he acts as an informal coordinator for the European Arctic Stratospheric Ozone Experiment.

Scandinavia is covered by a nearly solid, swirling block of cloud. The assembled group know that this doesn't augur well for one of the last flights in the EASOE campaign. Their tiredness is all the more acute because of the endless waiting for the weather to oblige. It is frustrating to be at the mercy of conditions illustrated by a profusion of weather charts. The balding French meteorologist who is displaying them on an overhead projector seems to be taking the bad weather personally. He is thus a mumbling apologist for the state of the lower atmosphere: 'Very cloudy skies, no rain, winds from east-northeast. There are no storms.'

Fortunately, there is better news in the lower stratosphere. Neil Harris shows some overheads faxed from the European Centre for Medium Range Forecasting in Reading. 'We're under the main part of the vortex,' he says with a smile, referring to the winds which form a 'collar' around the North Pole. In Antarctica, cold temperatures and the vortex coincide, but the Arctic vortex wobbles around like an ungainly amoeba at the behest of the wind patterns. Since balloons are at the mercy of these circumpolar winds, they will either rise inside the vortex or remain resolutely outside. So if there is a balloon launch over the next few hours it will ascend inside the vortex, where ozone depletion is most likely to occur.

It is clear, however, that no flight will occur today, a consensus at which the French technicians and experimenters have little trouble arriving. The assembled group then try to second-guess weather conditions across northern Sweden for the next day, right up to the stratosphere.

'An educated guess?' asks a French scientist wearing oversize boots and the demeanour of one who has been through this more times than he'd care to remember. His name is Jean-Pierre Pommereau and, as head of polar research at the Centre National de Recherche Scientifique, he is probably the doyen of European balloon experimenters. Up to this point he has stood quietly by the window chainsmoking and occasionally grimacing at the charts displayed by his colleagues.

'I'm educated so I won't guess,' Harris replies, with a knowing smile.

'Both will go tomorrow?' asks another Frenchman. Claude Camy-Peyret has a large experiment package due for launch.

'Maybe,' says Harris. 'Just possibly,' he adds to inspire a little more confidence.

A stark question presents itself for the decade of the 1990s. Will an ozone hole form over the North Pole? That, in essence, was what the European Arctic Stratospheric Ozone Experiment set out to answer over the winter of 1991–92. Previous experiments in the late 1980s had revealed that the same chemical processes which occur over Antarctica were now taking place in the stratosphere above the Arctic. When CFCs and halons drift into the stratosphere they absorb ultraviolet light from the Sun and release their chlorine. In the freezing temperatures above Antarctica, these chlorine atoms devour oxygen from ozone molecules to form chlorine monoxide. This further reacts with ozone to form more oxygen and chlorine, thereby repeating the process and accelerating it. One chlorine atom can devour 100,000 molecules of ozone. The result is a chain reaction set off by a lethal cocktail of human-made chemicals which result in massive ozone destruction.

Hitherto, there had been an ostrich-like comfort in thinking that the most severe ozone depletion was far removed from the everyday concerns of the population at large, restricted, or so it seemed, to the strange ice desert at the bottom of the world. But if similar springtime ozone depletion took place over the North Pole it could be catastrophic for the more populated northern hemisphere below. Yet unequivocal evidence for ozone destruction over the North Pole would be that much more difficult to identify. Whereas meteorological conditions over the South Pole are the same from year to year, the North Pole is – statistically speaking – a crap shoot. During the winter season, the Arctic is on average 10°C warmer than its southern counterpart. Air masses constantly move to and fro across northerly latitudes, mixing air more effectively between the equator and pole. This depositing of warmer air over the Arctic regions generally rules out marked depletion of ozone which might otherwise take place.

Over Antarctica during the springtime, the air is largely cut off from the rest of the atmosphere owing to the formation of a large vortex which is unique to that part of the world. This means that ozone cannot be replenished from outside the polar regions. A faint vortex may be found over the North Pole in spring, but it is smaller, leakier and less well defined. As a result, atmospheric circulation around the North Pole tends to obfuscate the underlying picture. Generally speaking, the flow of air spreads in a clockwise direction outwards from the pole and descends gradually into the lower stratosphere. This circulation naturally varies the amount of ozone present and can effectively thin out its distribution.

So, before it could be ascertained that ozone had been destroyed by chemical means – specifically as a result of the breakdown of man-made chlorofluorocarbons in the stratosphere – natural fluctuations in ozone levels due to atmospheric dynamics have to be determined. The easiest way to do this is by measuring the presence of long-lived gases – both natural and artificial – in the stratosphere, whose abundancies in the presence of ozone is known before it is destroyed. This in itself is not easy as they are present in extremely minute quantities, at most a handful of molecules in many billions. Small wonder then that they are known as trace gases, and their detection has been likened to searching for a drop of vermouth in an Olympic-standard swimming pool full of gin. Large scientific balloons offer the possibility of detecting them with remarkable precision through a wide cross-section of the upper atmosphere, either remotely by highly sensitive detectors, or else directly by canisters which sample the air as they are lowered on the end of parachutes.

Even with today's full panoply of space technology, balloons are still useful in atmospheric research. They are relatively cheap, easy to operate and provide reasonable access to the upper reaches of the stratosphere. ESRANGE's contribution to the overall experiment was to provide a launch site from which scientists had a good opportunity to access the Arctic vortex. The experiment campaign as a whole had taken twenty months to organise and would take five months to complete. Coordinated by the European Community, some 250 scientists from sixty research groups in seventeen countries participated. Such a large-scale, logistically demanding enterprise is needed to unravel the complexities of the upper atmosphere over the Arctic.

Because balloons cannot be steered, launches and recoveries are at the mercy of the prevailing air currents. For this reason, the thirty-seventh and thirty-eighth balloon launches in the EASEO campaign on Friday 13 March 1992 were definitely on hold until meteorologists could be certain of the special conditions required to launch them.

Kiruna and its surrounding areas are inhabited by around 27,000 people. Despite the ceaseless march of technology, which has brought with it central heating, satellite television and snow bikes, Kiruna still has the unmistakable feel of a frontier town. Without rich deposits of iron ore in the nearby foothills, it is doubtful the barren countryside would ever have been developed in the first place. Kiruna blossomed from a series of mines at the end of the nineteenth century, and now boasts that the underground mine in nearby Kirunavaara is the largest in the world. Its 42 kilometres of galleries will one day, the city fathers hope, make an ideal setting for an underground marathon. Though the city's name is derived from the Lapp word for snow grouse, the small bird is almost inevitably contained within the chemical symbol for iron in Kiruna's coat-of-arms. Even its most luxurious hotel is called the Ferrum. This holds a dance each week, and some people drive for two hours to get there. The group playing rock-and-roll standards in a country-and-western style, with lyrics shouted in Swedish, is sophistication itself compared to the regular stripteases arranged for the Finnish mine workers who also congregate at the Ferrum during the long winter darkness.

The astounding natural beauty of the countryside is probably reason enough to visit Kiruna during the summer. For the gloomy winter months, however, a visitor's guide does its best to entice you by declaring in the most reverential tones: 'The City Hall is a must, not only because it was named by the Swedish Architects' Association as the most beautiful public building in 1964, but also because it houses Kiruna Municipality's justly renowned art collection.' It is a treat probably ignored by most visitors to Kiruna, for rapidly flourishing space and satellite industries act as a bigger magnet than even iron ore today. Quite simply, Kiruna is an ideal place from which to gather data from polar-orbiting satellites: virtually all the satellites that provide information for weather maps on European television pass overhead many times a day. Large satellite dishes pepper the landscape for many miles around.

Because of ozone-related work, Kiruna also trumpets the Arena Arctica, which is a large new hangar out at the airport. During the EASOE campaign, it housed the Fokkers and Transalls that made up to 400 hours of flights in and around the polar vortex to complement the balloon flights. Publicity brochures for the Arena Arctica boast that 'it can house a Boeing 747', or, pointedly, 'a DC-8, a Transall, a Fokker 50 and four Falcons' – all of which are used in ozone research. One brochure concludes with a cryptic statement: 'Looking at the map, it may seem Kiruna is at the end of the world, but nothing could be more incorrect. If you fly from Tokyo, you reach Kiruna in 10 hours, from New York it takes 8 hours, and from Bahrain your flight takes 6½ hours. The end of the world is a lot further away.'

Anyone leaving Kiruna and heading for ESRANGE could be well forgiven for thinking that they really were on their way to the end of the world. The taxi ride seems to take far longer than the forty-five minutes by the clock. En route it can be disconcerting to see cars being driven at normal speeds in the ice and snow, but the local drivers are used to it. The gently undulating landscape of tundra is still heavily covered by snow even in March. Reindeer can be seen on the approach to ESRANGE, which adds to the sense of impossible remoteness.

At the base's outer limits is a fence and a security gate beyond which the space centre beckons, located downhill, its grounds extending for over 20 square kilometres. The eye is drawn to an L-shaped building with what appears to be two giant golf balls atop it. It could easily pass for Blofeld's headquarters in an early James Bond film. This is the main ESRANGE building, where the centre's hundred or so staff work. The golf balls are geodesic domes housing radar installations which monitor small sounding rockets that are intermittently launched to investigate the mysteries of polar physics, such as the aurora borealis or northern lights. About a kilometre beyond is a fluted building next to a red and white tower. The former appears like a giant, upturned concrete flask, from which sounding rockets emerge – not always successfully. In March 1993, sad to relate, a sounding rocket exploded while it was being prepared for launch and killed an ESRANGE technician.

Halfway between both is a smaller building with a pointed roof known as 'the Church'. Inside are the facilities needed to prepare the experiment packages for balloon flights during the European Arctic Stratosphere Ozone Experiment. It can take anything from a day to a fortnight to prepare a payload, depending on their level of complexity. During the EASOE campaign the Church was home to the two dozen or so technicians from the Centre National d'Études Spatiales, France's space organisation. Normally they are based at Aire-sur-l'Adour, just outside Toulouse. They have travelled all over the world to launch *les ballons stratosphériques* for scientists who require their services.

From the entrance to ESRANGE, a large clearing to the right of the main building is immediately obvious. Drive closer and you'll see a sign for the balloon-launching field, whose Swedish translation can be found in large letters: ballonguppsläppningsplan. Above it, and rather more prominently, are two red circles: one contains a cigarette, the other a naked flame. Both represent the most dangerous hazards for the French balloon teams because they still use hydrogen. This afternoon, though, there is little need for caution as the launch field is empty in its entirety, and although temperatures often rise above freezing point in March, it seems much colder as the Sun drops towards the western horizon. The winds pick up as darkness encroaches upon the desolate, icy landscape, whipping up flakes of snow which desultorily float around the *ballonguppsläppningsplan*. It is clear that the decision not to tempt fate on Friday 13 was a good one.

Saturday dawns brighter and the met briefing at 11.30 a.m. is relaxed by comparison. More faces crowd into the ESRANGE conference room and their number includes a German team from the Max Planck Institute for Chemistry in Lindau and Russian scientists from St Petersburg. The most agitated man in the room is probably Claude Camy-Peyret, director of an atmospheric research laboratory at the Université Pierre et Marie Curie in Paris. A small man, exhibiting all the classic signs of nervous energy, he wears large-framed glasses which magnify his eyes so that he has the unfortunate appearance of an agitated owl.

Camy-Peyret is principal investigator for one of the largest EASOE balloon experiments, known as the Limb Profile Monitor of the Atmosphere or LPMA Scientifically speaking, the 'limb' of the planet is the thin crescent which it presents to an observer. In this case, as the LPMA rises it will home in on the thin ribbon of the stratosphere and scrutinise the 'heat signature' of interesting trace molecules to be found there. At over 400 kilogrammes in weight, the LPMA will have to be launched by one of the largest balloons CNES can provide, some 100,000 cubic metres in volume. Small wonder that Camy-Peyret is agitated, particularly as a previous flight of his experiment was not totally successful.

Camy-Peyret's laboratory has provided the LPMA gondola, the platform upon which the scientific experiments are housed. There are, in fact, two separate experiments aboard, one of which is an absorption spectrometer developed by his group in Paris. It detects trace molecules by observing how they absorb sunlight directly and therefore has to point towards the Sun. On the other side of the gondola is a radiometer provided by physicists at the University of Denver, which tunes into the emission from the molecules themselves. Normally balloon experiments navigate by automatically charting a course via the Earth's magnetic field lines. At this latitude (68°N) however, magnetic disturbances (which give rise to the northern lights) are too commonplace to tempt fate. So the Denver physicists have provided a mechanical tracking system which orients the gondola by locking on to the Sun.

Sunlight being at something of a premium in the polar regions in spring sets an operational constraint on the Limb Profile Monitor of the Atmosphere instrument. It obviously has to be launched in sunlight, and, unlike most balloon flights, has to make measurements all the while it is ascending (the normal practice is to wait until it reaches 'float altitude'). This means the gondola has to be steadied as its ascends, which will result in mechanical imbalances while the Denver orientation system is in operation. On its first flight in February, the LPMA gondola also carried a German-built ultraviolet photometer and the whole gondola rocked and jittered all the way up to the stratosphere. This time around, Camy-Peyret decides to play safe and the German instrument is unceremoniously left on the ground.

He is still understandably anxious and sits near to the front of the conference table in an almost conscious effort to will good news from the launch officials. 'There is a chance of launch?' he asks. '*Un petit chance?*' he adds to his French colleagues. He has a scheduling problem: he must leave Kiruna by early Tuesday to fly back to Paris, where he will catch a 747 flight to Miami to be on hand for the forty-fifth launch of the space shuttle, which is carrying a more complicated version of LPMA into orbit.

The bald French meteorologist presents more charts and, once again, seems to be taking the barest hint of bad weather as a personal affront. Things look better, however, for although the rest of the weekend will still be very cloudy, winds will be weaker than he had originally expected and it will be colder. Neil Harris reports that the latest forecast from the ECMWF suggests that any launches over the following week will take place inside the vortex. This good news is weighed against another, more immediate scheduling problem. The balloon bearing the LPMA package has to be launched at least three hours before sunset to ensure the Sun will still be visible even from the stratosphere.

As the LPMA gondola essentially watches the Sun, the timing of its launch is critical: the longer the launch is delayed that afternoon, the lower the Sun will be in the sky. By 5 p.m. it will be below the horizon as seen from Kiruna, and, an hour later, will be completely gone from the perspective of the stratosphere. The earliest time for launch is 3:25 p.m. local time and a nominal three-hour mission would be expected. Given that it takes about four hours to prepare another balloon for launch, this will delay the Max Planck instrument until after 10 p.m.

This second flight will carry an emission spectrometer known as MIPAS which requires total darkness to detect the infrared radiation emitted by trace molecules. MIPAS requires a much longer flight – five hours – and the delay to 10 p.m. means that by the time it has reached float altitude, just after midnight, there will only be a couple of hours of darkness left before the Sun rises, shortly after 2 p.m. This leaves precious little time for the Max Planck team to obtain data on the concentrations of different molecular fragments which are implicated in ozone depletion.

The Germans are understandably not very happy. Sitting opposite Camy-Peyret, one experimenter is heard to say 'Ve are making compromises' while his colleagues mumble darkly among themselves. It would be unfair to characterise these discussions in the style of some awful sitcom where national stereotypes attack each other. In reality, all participants are extremely reasonable, though the pressures are intense. Some of them have been working for months, if not years, on the planning for their experiments. To lose the chance of obtaining data because of something as ludicrous as the equivalent of a stratospheric traffic jam is not to be taken lightly.

'We can bring forward the first launch,' says another CNES man. Some of his compatriots are puzzled: they think he said 'lunch'. The MPI team seem happier at this prospect and there follows an involved discussion about the weather conditions. Neil Harris makes his decision: 'My personal feeling is that we ought to go and go today.' There is broad agreement. The meeting breaks up in, if not elation, then something approaching relief.

Out on the *ballonguppsläppningsplan*, the dozen or so members of the CNES team are working up to the launch of LPMA. The gondola is a large metallic frame filled with a series of boxes. It is taller than the technicians who are loading it onto a guide rope which is attached to two small balloons, the structure as a whole anchored to a large winch. Those two balloons will take the strain of launch when the instrument gondola is jerked upwards and cushion its delicate electronics from the 'kick' which results. The larger balloon itself is laid flat against the ice after being carefully unfurled by the French technicians and is connected to the gondola by a large 'train' made of reinforced plastic-fibre rope. At various points along the ladder-like train are navigation beacons, radar reflectors and the all-important parachute which will be used to lower the hefty gondola gently to the ground at the end of the flight.

The gondola as a whole weighs nearly 430 kilogrammes and is heavily insulated for good reason. The great enemy of the infrared spectroscopist is stray thermal radiation which is emitted by objects as diverse as the instrument itself and the surface of the Earth far below. Therefore, the instrument is cooled by a flask of liquid helium. The boxes within it appear as though they've been put together haphazardly, but their structure is precisely ordained. Protected within the structures are the infrared instruments, the heliostat to lock the gondola on to the Sun and the transmission and recording systems for the precious data. Engineering data on the state of the experiments will automatically be transmitted back to ESRANGE to enable Camy-Peyret and his colleagues to see that they are at least working properly. The data collected will be recorded on a magnetic cartridge, slightly smaller than a domestic video cassette.

The LPMA gondola is unusual in the EASOE campaign in that it carries US-built hardware. As the French and American governments cannot exchange funds directly, the Denver scientists get a flight for free in exchange for providing a proven gondola-orientation system. Claude Camy-Peyret and John Starkey of Denver University are now giving the LPMA gondola a last-minute inspection. Starkey is responsible for the communications system which links both experiment and control systems. The transmission and recording electronics are contained on a special plate attached to the gondola by metal-sprung mounts. They will absorb any shocks during the ascent and particularly on landing. More crash pads are attached to the base of the gondola to reduce any damage when it lands by parachute. Stickers with the label 'Reward Given' are liberally sprinkled about the external structure, along with the address of ESRANGE, should it be found by an unsuspecting Lapp far away in the unseen tundra, who – like aliens in sci-fi films – naturally speak English.

Away from the gondola, the balloon canopy is starting to expand as it is filled with hydrogen. It is held in place by a large drum to which a spring mount that will actually release it is attached. Two CNES technicians stand either side of the prostrate balloon and use large 'air guns' attached to a compressor which chugs quite contentedly to itself. Slowly the transparent envelope of the balloon starts to fill out with hydrogen and lurches uncertainly upwards. After about twenty minutes, it has swelled into a large tear-shaped object. Over the Tannoy comes a warning that it is ten minutes to launch and all onlookers are asked to retreat to within the boundaries of a garage-like building at the edge of the *ballonguppsläppningsplan*. Although a mishap is unlikely, 400 kilos of gondola veering and swerving across the ice could be lethal.

Although things become frenetic up to the moment of launch, the events are hypnotic and seem to take place in slow motion. The drum mechanism swings back and the large balloon jerks upwards, momentarily appearing to deflate from the force of the launch. It makes a whooshing noise and moves upwards to the sound of flapping. At the other end of the balloon train, the auxiliary balloons are released and the experiment gondola moves upwards. When the balloon train tautens, the auxiliaries are automatically released and soon they separate, floating uncertainly in the bright sky like giant plastic scrota.

The balloon as a whole moves upwards and appears to brighten as it rises, catching more rays from the Sun with increasing altitude. It will still be visible for another twenty minutes, and many of the experimenters stay to watch their LPMA instrument for as long as possible. Shortly they will be transported to a small hill overlooking the balloon field which houses the control centre for the flight and is crewed by the CNES technicians. Known as Radar

Hill, it keeps tabs on the balloon flight and the health of the LPMA gondola.

Neil Harris stops looking upwards, smiles and joins colleagues in a car which speeds off towards Radar Hill. The thirty-seventh balloon of the EASOE campaign has just been launched.

Back inside the main ESRANGE building, a monitor announces that LPMA has been launched and automatically records the minutes and seconds which have subsequently elapsed. Next to it is a cryptic message scrawled on a noticeboard:

LAUNCHES		RECOVERIES	
France	3	France	2
Germany/USA	2	Germany/USA	2
England	0	England	0

It is an in-joke hinting that although Cambridge houses the European Community's Ozone Research Coordinating Unit, the UK plays no active part in the balloon launches.

Next door is the computer centre, in which all the scientific and meteorological data from the EASOE campaign will be archived. 'Housekeeping' data used by engineers to monitor the balloon's health are also relayed from Radar Hill to enable scientists working at terminals to follow the progress of their experiments. An hour into the flight, the telemetry indicates that all is going well. The balloon is heading in a northeasterly track taking it towards the border with Finland. After the excitement of the launch, things are quieter and many experimenters start to trickle back to the main ESRANGE building. There is a murmur of excitement when a bright balloon is seen passing overhead, but it is self-evidently a weather balloon, carried northwards by lower-level air currents which mimic those affecting the stratosphere.

A computer terminal is dedicated to radar data indicating range, altitude and speed. The gradual procession of numbers attests to the fact that the balloon is heading on much the course it should. But just short of one and a half hours into the flight, something strange happens. The altitude figures start to decrease rapidly and the other data seem inexplicably frozen on the screens. In a matter of minutes, the balloon appears to have descended from its operational altitude of 23 kilometres to the ground. This should not occur for another four hours, and, even then, should take over half an hour as it descends by parachute.

ESRANGE technicians gather around the screen, looking extremely worried. Someone whispers the dreaded word: 'Freefall!' The radar tracking and range show that it has reached 68°10'N, 21°44'E, somewhere in the tundra of eastern Sweden. Anxious phone calls are made to a helicopter pad which ESRANGE uses at Kiruna Airport.

Outside, the Sun has dissolved into a menacing orange glow delineated by grey streaks of high-altitude cloud. Out towards the east, the skies are clear. Somewhere out in the gathering darkness are the remains of the balloon and its precious scientific cargo. Sitting dejectedly in the ESRANGE canteen is John Starkey of Denver University. 'There's the remains of a quarter of a million dollars out there,' he says to no one in particular.

Over the Tannoy comes a call for Claude Camy-Peyret to report to the ESRANGE helipad to await the arrival of a helicopter from Kiruna Airport. It will head off towards the Finnish border to search for the free-fallen gondola and Camy-Peyret will try to retrieve whatever is left of his experiment. A harassed ESRANGE official appears in the canteen to tell Starkey there's room on the helicopter for an extra person. He declines for the simple reason he has no cold-protection clothes. Though the balloon is only 30 kilometres distant, should anything happen to the helicopter en route there would be a very real possibility of lingering death.

At 6 p.m., the helicopter arrives and heads off with Claude Camy-Peyret aboard. In the canteen Neil Harris, John Starkey and a handful of the CNES balloon team earnestly discuss what could have happened. At 7 p.m, word comes through that the helicopter has located the remains and they will make a cursory inspection. It is immediately apparent that half a million dollars of scientific gondola free-fell into the Swedish tundra because the balloon train snapped below the parachute. The feeling is that there will be little trouble retrieving the equipment tomorrow at first light.

'This should have happened yesterday,' says one of the more

superstitious CNES team members. He and his colleagues eat their meal after the LPMA mishap in a dejected silence, preparing for the next balloon flight with a distinct morning-after feeling. The premature loss of the LPMA mission is a slight blessing for the MIPAS team. The radio telemetry channels are free sooner than had been expected and its launch can thus be brought forward. On an earlier flight, the Max Planck team was bitterly frustrated with telemetry problems: intermittently, beautiful spectra would be displayed on their screens, but would then disappear as if on a badly flickering TV set. Its equipment transmits data in real time because it is much less complicated. The MIPAS team have spent the subsequent weeks checking and double-checking their data transmission system.

It is bitterly cold and dark when MIPAS lifts into the sky at 8.52 p.m., illuminated by powerful searchlights. 'Psychologically we need this baby to go,' Jean-Pierre Pommereau says to me with grim determination. The launch was delayed slightly because of suspected pressurisation problems with the main balloon. All seems to be well for the first two hours of the five-hour flight. But in the wee small hours of Sunday, another catastrophic failure occurs: the balloon train carrying the MIPAS gondola also snaps below the parachute.

The next morning, even more dejected faces can be found in the ESRANGE canteen. For the truly superstitious, significance is attached to the fact this latest accident took place on 15 March – the Ides of March. But all is not lost, for this time around, the Max Planck Institute researchers console themselves by the fact that at least they have obtained some useful data.

'If you'd told me there'd have been thirty-seven launches before a failure, I wouldn't have believed it,' Neil Harris told me the next day as the mangled electronics of both LPMA and MIPAS were retrieved by keen-eyed helicopter pilots. The odds were that the balloon losses should have been even greater than they actually were. A quick look at the previous experience of the French launch teams in the Arctic shows just how perilous an exercise it could be. They had problems similar to those NASA encountered with scientific balloons a few years before.

How Soon Is Now?

In the late 1980s, CNES had mounted four balloon campaigns, three of which were known by the acronym of CHEOPS – CHEmistry of Ozone in the Polar Stratosphere. The first two CHEOPS campaigns, in the Arctic winters of 1987 and 1988, were only moderately successful. On CHEOPS I, the first flight was lost; the team waited for warmer conditions to launch the second, which was successful. The following year, only four out of seven balloon flights were successful. The freezing temperatures of the Arctic stratosphere caused the polyethylene skin of the balloons to disintegrate and shatter.

Manufacturing scientific balloons is so tricky that it has been likened to a black art. The choice of material used for the balloon's skin is critical. The material has to extend up to 100 per cent its normal length as the balloon expands, yet it should not deform in the process. Large scientific balloons may increase in volume by 300 times during ascent. One French balloon engineer provides a graphic metaphor: at launch they look like giant condoms which could fit over the Eiffel Tower and by the time they reach the stratosphere they could fill the Parc de Princes rugby ground.

On some flights, once a float altitude is reached the balloon should stay there. So the balloon skin must not be susceptible to changes in sunlight or reflected radiation from the ground which might alter its altitude. Most balloons would not encounter temperatures colder than $-90°C$ in the lower stratosphere. But in the Arctic and Antarctic it can get very much colder than that – as the experimenters on CHEOPS I and II had found to their chagrin.

So in 1989 a new campaign, known as TECHNOPS, was carried out with a new balloon skin which, its manufacturers said, would work in temperatures as low as $-95°C$. During daytime flights the temperature was not so much of a problem as at night when it could plummet to below $-100°C$. As a result, TECHNOPS technicians attached sensors to the outside of the balloons to watch how the temperature altered during ascent. By monitoring the skin temperature, it was then possible to predict the ascent rate and alter it if necessary to avoid any anticipated problems. Four totally successful flights went some way to assuage fears of scientists who had watched their experiments fall out of the sky on previous campaigns. The next year, a third CHEOPS campaign launched eleven balloons successfully out of Kiruna. French and German scientists obtained a wealth of data about conditions in the Arctic and found extensive evidence for chemical processes similar to those at work over Antarctica.

As head of research at the CNRS Service d'Aéronomie, Jean-Pierre Pommereau was the French coordinator for CHEOPS III. He has good reason to recall the first week of February 1990, when severe storms lashed northern Europe and caused the circulation of air in and around the Arctic to be distorted. In particular, air at the edge of the storm systems was forced up into the stratosphere from mountainous regions in Norway and Greenland. The prevailing wisdom was that the deposition of this warmer air over the Arctic would rule out large-scale ozone depletion. Yet the skies above Kiruna were filled with polar stratospheric clouds. Conditions were too windy to launch balloons, but ground-based instruments at ESRANGE recorded stratospheric temperatures as low as −94°C. Within days, other instruments watched ozone levels drop to 60 per cent of their normal values, giving the lowest values ever seen over the Arctic.

Subsequently, when balloon flights began again after the winds had dropped sufficiently, the instruments detected large diurnal dips in ozone levels in and around the vortex. To Pommereau it presented a clear picture of 'mini-holes' – hundreds of kilometres across and tens of kilometres deep – opening up in the ozone layer. Their cause seemed to be the upward passage of atmospheric waves generated by the storms below, disturbing temperatures in the stratosphere to the record lows measured from the ground. Within a few days these mini-holes had filled up with ozone again, the natural result of stratospheric dynamics. 'We were using the same sort of instruments that are commonly used in the Antarctic,' Pommereau told me. 'And basically with CHEOPS III we saw the same chemistry. That was an important result because the dynamics of the atmosphere over the north and south poles are so different. What we found was that at the same temperature, we got the same results. So we could say that chemistry was responsible.'

To some of the French researchers it looked like the unsteady genesis of an Arctic ozone hole. But because there were only a handful of flights, separated in time and entry point to the Arctic vortex, it was difficult to obtain an overall picture of the extent of this enhanced chemical processing. It was out of a desire for such a wider view that the larger, more extensive EASOE campaign was born.

The European Arctic Stratosphere Ozone Experiment drew to a close ten days later, and, despite the two balloon failures, was deemed a success overall. The ambitiousness of the enterprise was matched in extent only by NASA's second Airborne Arctic Stratospheric Experiment, which had finished at roughly the same time. This had been centred on Bangor, Maine, and predominantly used aircraft in its investigations. Both experiments revealed that the northern hemisphere was saved from the worrisome prospect of an Arctic ozone hole by a quirk of meteorology. Though the ozone layer was cold enough for polar stratospheric clouds to have formed during January, no large-scale ozone loss occurred because the Sun was largely absent from the stratosphere at that time. When it returned in the middle of the month, it did not fire chemical reactions on anything like the scale of those that have stripped away ozone molecules over the South Pole.

By February, temperatures in the upper atmosphere had increased so that polar stratospheric clouds could no longer form. A large ridge of high pressure had sat over the north Atlantic, serving to warm the stratosphere. Partly as a result of this there had not been the atmospheric wave activity which had generated the mini-holes seen in previous years. Paradoxically, however, the high pressure had also served to thin out the ozone layer, in much the same way as water in a stream passing over a rock. The net result was that somewhere between 10 and 20 per cent of the ozone layer normally seen over the Arctic regions had been lost during the EASOE campaign.

This observed loss could not be blamed on the dynamics alone. Chemistry was responsible, partly in the guise of the lingering dust cloud from Mount Pinatubo, the Philippine volcano which had erupted the previous summer. Sulphur dioxide particles in the volcanic plume had subsequently spread across the globe and provided surfaces upon which ozone-destroying reactions could occur at warmer temperatures than are routinely found over the

Arctic. The sulphur dioxide cloud from Mount Pinatubo had also served to repress the levels of oxides of nitrogen which serve to act as a check on the build-up of chlorine. Normally oxides of nitrogen will react with the chlorine resulting from the breakdown of CFCs to form chlorine nitrate and hydrogen chloride, which leave no chlorine available to destroy ozone. But – as the results from the LPMA and MIPAS experiments, among others, would ultimately show – the oxides of nitrogen had been depleted by reacting with the sulphur dioxide from Mount Pinatubo.

Early in April, the initial, end-of-experiment results from EASOE were released to the media. Accompanied by a press release entitled 'Future in Balance for Arctic Ozone Layer', it stated outright that 'although a major localised ozone loss (an ozone hole) was not observed this winter over the Arctic, scientists have seen conditions which could lead to an ozone hole if a future winter was longer and colder and especially as stratospheric chlorine is expected to increase into the next century'. The loss of 10–20 per cent of the ozone layer over the Arctic over the winter of 1991–92 did not seem as dramatic as the 50 per cent or more depletion now routinely observed over the South Pole each spring. But, as Neil Harris pointed out when questioned by journalists, it was hardly something to celebrate. 'For some reason these results have been seen as good news,' he said quite forcefully. 'I don't see why a 10 to 20 per cent loss is good news.'

The scientists were keen to point out these were the initial quick-look results. The full explanation of what had happened would require many more months of meticulous analysis.

Twelve months later to the day after the loss of his LPMA gondola, Claude Camy-Peyret smiles grimly at the memory. He has spent many waking moments since pondering why his experiment was destined to suffer such an apparently ignominious demise. We are sitting in his office in a tower of nightmarish aspect among the hideous concrete landscape of the Université Pierre et Marie Curie in the fifteenth arrondissement of Paris. His desk is cluttered with print-outs, computer disks and electrical connectors, the normal detritus of the balloon experimenter.

Camy-Peyret bears the mien of one who has been triumphant in the face of adversity. On the wall directly in his line of sight are a handful of spectra which he has been able to retrieve from his LPMA instrument. To the untrained eye they appear similar to the doodlings of a child who has experienced difficulty in completing a dot-to-dot puzzle. But to Camy-Peyret and other cognoscenti they represent the first detailed observations of one of the more elusive molecules involved in the complex chemistry of ozone depletion.

On that bleak March morning when he inspected the remains of the gondola, he had thought that retrieving scientifically valuable information 'was a lost cause'. Among the mangled electronics, however, one of his keen-eyed engineers saw what looked like a crushed data cartridge. Somehow, he managed to remove it without causing any further damage and proudly handed it to his boss. 'It wasn't in what you would call very good shape,' Camy-Peyret says. When the unfurled magnetic tape was brought back to Paris, Camy-Peyret contacted its manufacturers in Germany to attempt a mechanical resurrection. The company meticulously respooled the tape in a magnetically 'clean' laboratory, and now, a few months later, Camy-Peyret has been able to assemble a collection of nearly forty spectra in all.

They profile conditions in the atmosphere from 3,000 metres up to 23,000 metres, at which point the balloon train snapped. 'I can see lines of hydrogen chloride,' he says with pride. 'Very few of the other EASOE experiments were able to measure that.' This elusive chemical species acts as a reservoir for chlorine activated inside the polar vortex and is thus a measure of the stratosphere's natural safety mechanism protecting the ozone layer. Both the French and American instruments aboard revealed fascinating information on the way in which chlorine returns to its 'reservoir' forms at the end of the winter. They showed that they do so far quicker and more efficiently than had previously been thought. Some of the other spectra lie hidden on his desk by the rather more obvious result of LPMA's fateful flight – insurance forms. They are in triplicate and have to be meticulously filled in to enable Camy-Peyret to claim insurance from CNES to build a replacement instrument.

In the year since, Camy-Peyret and his team have had ample time to carry out an extended postmortem on the 14 March failure, as well as the problems which preceded it on 5 February. That first flight had also carried a German-built photometer which had attempted to monitor the flux of ultraviolet light streaming in from the Sun. 'We'd never tested this configuration before in flight,' Camy-Peyret says. 'A strong rocking motion resulted because the experiment package was not properly balanced.' As the Denver heliostat tried to lock on to the Sun, the gondola kept rocking in the opposite direction to compensate for the loss of balance. The French spectrometer saw the Sun wobble in and out of its field of view, thereby rendering the spectra it was trying to obtain useless.

An extensive investigation by both the CNES launch team and ESRANGE officials has failed to diagnose exactly why the LPMA flight on 14 March failed. The most obvious contributory factor was the unusual flight procedure dictated by the need to lock on to the Sun early on in the ascent. Normally, flights out of the CNES balloon centre in Aire-sur-l'Adour in southern France do not commence observations until they reach float altitude. The balloons sit in the stratosphere and watch the changing illumination of the atmosphere as the Sun rises and sets. The gondola does not have to stabilise itself until it is in the least dense reaches of the upper atmosphere. From Kiruna, however, the Sun effectively stays still as the balloon rises upwards and measurements are made from the resultant change in perspective. So the LPMA has to use the heliostat and its motors in the densest layers of the atmosphere, thereby inducing greater stresses than usual. These extra forces probably twisted the balloon train and snapped it below the parachute package. The gondola as a whole was thus resigned to destruction many kilometres below.

Camy-Peyret does not wish to assign blame for the accident. It was, he reflects, one of those things that are sent to vex the balloon experimenter. Once the insurance forms are completed, he expects to get the money to build a replacement gondola for the next European balloon campaign, planned for the winter of 1994. After the problems he encountered last time, Camy-Peyret could be forgiven for never wanting to attempt further flights or even participating in another Arctic campaign. He smiles as he tells me he will gladly go through it all once more, though the next time he would be perhaps a little wiser about the gondola loading and operations of the balloons. 'For simple measurements, we can use simple instruments,' he says. 'For those species which are not so abundant or do not have strong spectral signatures, you have to have powerful, complicated instruments.'

In short, then, these observations are just one small piece in the complex, three-dimensional jigsaw puzzle that is ozone depletion. The apparent failure, yet ultimate triumph of Camy-Peyret and his team, is a perfect encapsulation of scientific progress. Behind each and every hard-earned addition to the body of scientific knowledge, there are countless other stories which remain largely unknown. And was it ever thus: if there is one theme linking today's ozone research with past episodes throughout its protracted and convoluted history, it is surely the human stories involved, more often as unlikely as they are entertaining. They encompass a broad sweep of human history with some extraordinary characters and bizarre happenings.

The annals of nineteenth-century chemistry contain many references to an 'electrical odour' associated with the first tentative experiments involving electrostatic plates. But it would not be until the year 1840 that a Swiss chemist identified the noisome gas in question and found that it could be created both by the electrolysis of water in oxygen and when phosphorus glowed unhindered in air. Christian Schonbein named his discovery after the Greek verb *ozein*, to smell, as he related in a letter to Michael Faraday a few years later: 'It will perhaps interest you to learn on this occasion that my friend, being an excellent Greek scholar, acted the part of god-father, when I christened my child "Ozone".'

Ozone is simply an unstable form of oxygen, formed by the action of sunlight on the air we breathe. Whereas oxygen molecules comprise two atoms, ozone has three: it has a marked propensity to give up that extra oxygen atom whenever it can. Schonbein came up with a simple chemical test to determine its presence by use of papers doused with starch and potassium iodide. Ozone would oxidise the iodine content of the paper and turn the starch a blue colour. Even today this basic test is still used to determine the presence of ozone, and a slightly fancier version is employed by ozonesondes to detect its concentrations as they ascend upwards.

Ozone's discovery portended a myriad uses in keeping with the Victorian mania for novelty and invention. These were heady times: hardly a month went by without some new discovery in science and technology which promised to change the world. Ozone was no exception and in some circles it became all the rage as it was widely believed to have life-enhancing properties. The well-to-do would go to inordinate lengths to 'take in' naturally occuring ozone, which was known to be found at the tops of mountains and near the seaside. Lamentably, it did not appear to exist in towns and cities, which moved the librettist W. S. Gilbert to pen a poem in its honour in 1860. Not one of his better works, 'Ozone' ran in part:

> I don't quite see how it can act – in fact
> In a room where a hundred are packed, it's lacked;
>> In a tenanted place
>> Not a ghost of a trace
> Of the gas that is known as ozone is shown
> Not a trace of this useful ozone!
>
> But if on Ben Nevis's top you stop,
> You will find of this gas there's a crop – but drop
>> To the regions below,
>> And experiments show
> Not a trace of this useful ozone is known
> Not a trace of this useful ozone!
>
> In a desert 'twill cover the ground, all round,
> And up in the clouds I'll be bound it's found;
>> But oh, it's a pity
>> That here in the city
> The divvle of a drop of ozone is blown,
> Not a drop of this useful ozone!
>
> It's because I'm an ignorant chap, mayhap,
> And I daresay I merit a slap or rap,
>> But it's never, you see,
>> Where it's wanted to be,
> So I call it Policeman ozone – it's known
> By my friends as Policeman Ozone!

By the time Gilbert wrote his poem – twenty years after the discovery of ozone – it had been found that the gas possessed excellent disinfectant properties. As a result, ozone was produced industrially in vast quantities for this purpose from controlled electrical discharges. This led Gilbert clumsily to pun:

> Did you hear of the use of ozone, ohone?
> It's the best disinfectant that's known, they've shown

And indeed it was: later in the nineteenth century, ozone was pumped into hospitals, theatres, schools and even into the murky depths of the London Underground. Other uses were not nearly as prosaic: chronicles of wine-making from the turn of the century refer to its properties in helping to mature brandies and fine wines.

Nearly a century later, the use of ozone as a disinfectant is about to make a comeback as remarkable as that of Lazarus. Some industrial chemists predict nothing short of a renaissance so far as ozone is concerned in the 1990s. As a detergent not only can it destroy bacteria and viruses, but ozone kills the microbes responsible for dysentery, which normally survive chlorine-based disinfection. It can also be used as an efficient water purifier and scrubber to deodorise the exhaust emissions from sewage works. A company called Triogen uses ozone water purification in more than 200 public swimming pools in Great Britain. In France there are already more than 600 drinking-water-purifying plants which employ ozone.

Many prisons and hotels in the United States are using ozone in the laundry. They employ an electric-discharge system which generates ozone out of ordinary air rather than pure oxygen, thereby slashing costs. According to its manufacturers, O_3 Tech of Fort Pierce, Florida, ozone laundries reduce shrinking of clothes, remove more organic stains, use far less water and do not release potentially harmful detergents into the environment. Doubtless it will complete industrial chemistry's rehabilitation in the eyes of its many detractors when industrialised ozone itself largely comes to replace chlorine-based compounds as a common detergent around the world.

This resurgence in the use of 'policeman ozone' is ironic, for by the middle of the twentieth century it had come to be regarded as more of a criminal. Its propensity for strong oxidation meant that whenever it came into contact with any organic matter it reacted violently. Public opinion shifted when it became apparent that ozone was a severe irritant and could cause protracted respiratory problems. But more than anything, ozone's unwanted presence in the natural home of the pulp detective story in the 1950s led to widespread acknowledgement of its undesirability.

After the Second World War, smog levels in Los Angeles escalated. City leaders realised that the unpleasant pall which hung over the City of Angels could create serious health problems. The most obvious culprit was industrial pollution but, after the introduction of strictly enforced controls of industrial emissions, smog levels persisted. Jan-Arie Haagen-Smit at the California Institute of Technology was asked to investigate the problem. As a plant physiologist, he was curious why crops seemed to be damaged by the citywide smogs. After fumigating plants with nitrous oxides and hydrocarbons – the usual culprits in smog pollution – and simulating levels of Californian sunshine, they exhibited all the signs of smog damage normally found in LA. Haagen-Smit realised that ozone was involved because photochemical reactions within the smog produced it. The chemistry involved is a variation on the high-altitude reactions which would preoccupy stratospheric researchers two decades later.

The blame lay with motor vehicles. Nitrous oxides in car exhausts can be affected by ultraviolet radiation during the summertime. Ultraviolet light absorbed by nitrogen dioxide will produce nitric acid, leaving a stray oxygen atom to combine with diatomic oxygen. The end result is an ozone molecule, helped along by the hydrocarbons burned in car fuel. By monitoring ozone at ground levels, the smog damage was found to peak during daytime when traffic swelled the city's freeways. In 1960, California introduced the first legislation to minimise car pollution and Haagen-Smit became known as 'Dr Smog' in the popular press.

Haagen-Smit also showed that ground-level ozone could have serious health effects. At Caltech he observed that rubber bands would show signs of damage when exposed to smog and sunlight because of the ozone which was produced as a by-product. Subsequent medical studies have implicated ozone inhalation in causing headaches, nausea and damage to the respiratory system and emphysema. As a result, the World Health Organisation has set a recommended limit of exposure to ground-level ozone of 0.1 parts per million (ppm) by volume over the space of an hour, after which time eye irritations will start.

At the Earth's surface, ozone naturally occurs at about 0.02 parts per million: during the summer months, when the unpleasant combination of sunshine and car pollution is at its most pronounced, levels can often surpass the WHO standard. In Britain, the record is 0.26 ppm, established during the drought summer of 1976. Cities that are prone to photochemical smogs because of geographical circumstances are susceptible to increasingly dangerous levels of ozone production. Perennially smog-bound cities like Mexico City, Tokyo and Santiago often exceed WHO standards. In August 1992, for example, ozone levels in Mexico City nudged 4 parts per million, at which point the government shut factories and schools. It is a problem that will only be alleviated if draconian clean air legislation is enacted.

Prolonged exposure to ozone can also lead to impotence. Related concerns inspired perhaps the weirdest experiment ever attempted. It is a variation of Haagen-Smit's rubber-band experiment on a similar material in the same city. Doubtless quite a few doctors did a double take as they browsed through the 9 September 1988 issue of the *Journal of the American Medical Association*. In its normally po-faced pages, Dr Richard Baker and colleagues at UCLA reported that ozone can damage condoms.

When twenty unravelled condoms were placed in proximity to 300 parts per billion of ozone for seventy-two hours, only two remained intact. Electron microscopes revealed the exposed condoms had been pitted with small craters. Given that Los Angeles experiences such levels of ozone during the summer, Baker's work could have come as a timely warning indeed, but it was written with the Indian subcontinent in mind. The complete title of Baker's letter was 'Precautions when lightning strikes during the monsoon: the effect of ozone on condoms'. The monsoon season brings with it lightning which generates pronounced increases in ground-level ozone and Baker's research was addressing the pitfalls of inadequate packaging in India. It has serious implications for health programmes and, in particular, the spread of the HIV virus.

* * *

The detection of the whereabouts of most ozone in the atmosphere came about by a circuitous route entirely typical of the scientific process. The first steps were taken by balloon experimenters in the nineteenth century who slowly reached the stratosphere – not always successfully. The first recorded scientific balloon flight took place in August 1784. Guyton de Moreau and Abbé Bertrand reached 3,000 metres, and used a thermometer and barometer to prove that both temperature and air pressure decreased with altitude. Twenty years later, Joseph Gay-Lussac flew as high as 7,000 metres, where he started to feel light-headed and fainted, but not before he had collected an air sample. When later analysed, the sample revealed that atmospheric gases remained in the same proportions as at ground level.

In 1862, an aeronaut named Henry Coxwell and the physicist James Glaisher attempted to go higher than ever before in the name of science. They used a hygrometer to determine how much water vapour was present and found that as air cooled it held less moisture. On their very first flight, however, Glaisher lost consciousness at 8,800 metres above Wolverhampton. Coxwell lasted longer, but could not manipulate the valves because the cold had given him frostbite. Somehow, Coxwell managed to activate a valve cord with his teeth and they descended to tell the tale. They made more than twenty-six flights in the name of science and were fêted as fine, upstanding heroes in the Victorian mould.

Without knowing it, balloonists were on the threshold of the ozone layer. But to reach it would exact a terrifying toll. In 1875, three Frenchmen ascended aboard a balloon christened the *Zenith* and reached 8,500 metres. All three fainted; when the balloon descended the pilot awoke to find his companions dead. This all but ended high-altitude ballooning by aeronauts until pressurised compartments could be constructed in the 1920s. Eventually, crewed balloons were abandoned and since the Second World War, scientists have largely used hydrogen-filled balloons similar to those employed in the European Arctic Stratospheric Ozone Experiment in 1992.

Exactly a century before, however, another French scientist came up with a far less dangerous way of using balloons to probe upwards. Léon Philippe Teisserenc de Bort was something of a scientific gadfly. He had resigned as chief meteorologist to the French weather service in 1892, thereafter setting up his own observatory in Trappes. He decided to use kites connected by hundreds of metres of piano wire to reveal how atmospheric conditions changed at increasing altitude. Self-contained instruments attached to the kites could profile the pressure, temperature and humidity as they rose. But after many accidents – such as breaking telephone wires, ensnaring steamboats and electric cables, thereby shorting out whole areas of Paris – he wisely turned to small hydrogen-filled balloons below which mechanised instruments automatically recorded data.

After three years and 236 flights, Teisserenc de Bort presented his findings to the French Academy of Science in 1902. They defied not only perceived wisdom but also plain common sense. With increasing altitude, manned balloon flights in the previous century had established that temperatures progressively dropped. But Teisserenc de Bort found that at about 11 kilometres above the Earth, the atmosphere started to warm up. Temperatures reached a minimum of −50°C, thereafter rising to about −40°C at an altitude of 14 kilometres. In the absence of any further data, de Bort presumed temperatures would remain constant all the way up to the edge of space. His readings suggested that weather would also stop there since it is caused for the most part by convection of air masses, a process which would no longer be relevant if temperatures remained constant.

Teisserenc de Bort used the Greek word *tropos*, meaning to turn over, to characterise this curious change in the structure of the atmosphere. The lower atmosphere thus became christened the troposphere and the boundary at which the temperature started to increase again was named the tropopause. He assumed that above it there would be layers of very thin air which would separate out in density because the temperature was constant. He used the Greek word *stratos* to describe the atmosphere at altitude.

Because his readings took him no higher than 14 kilometres, Teisserenc de Bort was not to know that the air density is not stratified. He was essentially correct in his assumption that there would be very little convective motion, for, although there is a form of weather in the stratosphere, it is a very anaemic imitation of the storms below. Air motions within the stratosphere are controlled directly by the transfer of radiation from the Sun and radiation reflected back from the Earth. It is a measure of Teisserenc de Bort's integrity and standing that he was taken seriously.

The obvious question was why is the stratosphere so warm? The answer came from two of de Bort's countrymen who approached the problem from another perspective. Instead of probing upwards, Charles Fabry and Henri Buisson, both physicists at the University of Marseilles, considered what would happen as sunlight progressively descended through the tenuous reaches of the atmosphere towards the surface. To this end they were helped by one of the great triumphs of nineteenth-century science: spectroscopy. It would enable Fabry and Buisson to investigate the composition of the atmosphere from the ground without risking their lives in a balloon.

Isaac Newton in the seventeenth century had used a prism to split visible light into its component rainbow spread of colours, each corresponding to a different wavelength of light. The splitting of light in this way revealed a great deal about the atmospheric molecules through which the light had passed. Each and every molecule absorbs light at a characteristic set of wavelengths. As unique as fingerprints, these absorptions gave scientists the key to learning about the chemical composition of the atmosphere remotely. In time, as optical instrumentation improved, scientists became familiar with each and every telltale sign of spectral absorption. Most spectacularly, absorptions of an unknown gaseous element had been found when spectroscopes had been attached to telescopes pointed towards the Sun. At this time, the gas in question had yet to be discovered in the laboratory and in honour of its obviously solar origins, it was christened helium.

By the end of the nineteenth century, spectroscopy had established that visible light formed just a small portion of the electromagnetic spectrum, the name given to the full gamut of radiations which are wholly invisible to the naked eye. It was clear that these invisible radiations adjacent to those which can be seen optically contained interesting information about the atmosphere. But it was not just a case of building better equipment, for both infrared and ultraviolet spectroscopy presented their own difficulties. Most infrared spectroscopy was compromised from the ground because atmospheric water vapour absorbed most of the infrared signatures from molecules higher up in the atmosphere. In the 1950s this obvious limitation would be circumvented by sending infrared spectrometers aloft. The infrared portion of the spectrum is especially valuable in spectroscopy because many molecules in the stratosphere vibrate with energies which correspond to individual infrared wavelengths.

Beyond the blue part of the visible spectrum, further complications would come into play. Fabry and Buisson's work would have its genesis in an informal competition between astronomers to see who could find the shortest wavelength of ultraviolet light that reached the ground. It seemed to be cut off at around 300 nanometres (three thousandths of a millimetre), for which there was no good theoretical explanation. Surely the Sun didn't automatically stop shining below those wavelengths?

By the 1870s there was some speculation that the explanation could lie in the systematic absorption of shorter ultraviolet wavelengths higher up in the atmosphere. An Irish chemist by the name of W. N. Hartley had looked at how ozone absorbed ultraviolet in the laboratory around 1880. He suggested that it had the right properties to explain the cut-off, but spectrometers were not then sensitive enough to measure the faint spectral signal which would prove it beyond doubt. There the matter lay, though Hartley was to lend his surname to that part of the ultraviolet spectrum where ozone absorbs ultraviolet most strongly: today they are known as the Hartley bands.

Twenty years later, Charles Fabry and Henri Buisson performed more advanced experiments with the better equipment which had become available. In 1912, they found in laboratory experiments that artificial sources of ultraviolet could transmit at shorter wavelengths than those measured from the sky. 'It was evident that there was a strong atmospheric absorption,' Fabry would recall in a memoir of his work, *L'Ozone atmosphérique*, published half a century later, 'but the attribution of this absorption to ozone was still very hypothetical. On the other hand, was atmospheric absorption, whatever its cause, sufficient to explain all the facts?' To investigate, Fabry and Buisson built a spectrometer from scratch to observe how exactly the ultraviolet dropped off in the stratosphere at frequencies characteristic of the presence of ozone. Though primitive by today's standards – it has been described as being of the 'string and sealing wax variety' – their spectrometer revealed, beyond doubt, the existence of ultraviolet absorption due to ozone.

Intriguingly, they went one stage further. By determining very carefully how ozone absorbed the corresponding amounts of artificial ultraviolet produced in the laboratory, Fabry and Buisson could estimate how much of it there would have to be in the upper atmosphere to cause the cut-off measured at the ground. The answer seemed to indicate that there would be a thin, diffuse layer of ozone located somewhere 25–30 kilometres above the Earth. If it were compressed and brought down to ground levels and pressures, they speculated, then the ozone would amount to only about half a centimetre.

Subsequent measurements showed that Fabry and Buisson were spot on in their estimates. Had the First World War not intervened, they could have carried out further experiments to prove it conclusively.

To generations of physics students at Oxford University from the 1920s onwards, Professor Gordon Dobson was a familiar, well-liked figure. A large, jovial man, he was, according to one colleague from later years, 'the last specimen of a generation of the English country gentleman scientist', and 'a nice old Victorian gentleman' in the phrase of another. Though he spent a proportion of his time demonstrating experimental physics at the Clarendon Laboratory, 'Dobby' liked nothing better than to work at the laboratory he had assiduously built up at his home, thereby allowing him time to spend in his beloved orchard and garden. His singular contribution to science was as the founding father of modern ozone studies.

Gordon Miller Bourne Dobson was born in 1889 in the Lake District, and, after showing an early aptitude for science, studied natural sciences at Cambridge University. In his own phrase he 'amused himself by reading geophysics' in his last year and experimented with measuring waves in Lake Windermere, which led to his first scientific paper, published in *Nature* in 1911. After serving in various branches of the Meteorological Office, he ended the First World War at the RAF base in Farnborough, where he had met the physicist Frederick Lindemann. Together they had performed the first ever experiments to see how aircraft could help improve weather forecasts. When Lindemann subsequently moved to the Clarendon Laboratory in Oxford after the cessation of hostilities, Dobson followed him a year later and was appointed lecturer in meteorology.

Almost immediately, Dobson was to collaborate with Lindemann on an enterprise which combined the latter's theoretical skills with his own considerable experience in experimentation. It involved the study of the uppermost reaches of the stratosphere by the ingenious method of observing meteors, the fragmentary remnants of interplanetary dust which burn up as shooting stars in the night sky. The ancient Greeks presumed they were atmospheric phenomena and called them after their word for lofty. They were wrong, but *meteoros* was adopted as a label by modern-day scientists who study the atmosphere.

Lindemann and Dobson both knew that meteors burned up in the highest reaches of stratosphere, far beyond the reach of even the largest balloons. By accurately measuring the length and brightness of the incandescent trails which meteors fleetingly left behind, Dobson and Lindemann reasoned that it would be possible to apply the laws of thermodynamics to work out the density of the air through which they had passed. If teams of observers observed the same meteor trails from different sites, the altitude at which the meteors burned up could be worked out by simple triangulation. Once the density of the atmosphere at a certain altitude was known, its temperature could also be worked out.

In the first two years of the 1920s, they performed a series of triangulation experiments whose results were nothing less than astonishing. At 50 kilometres above the Earth, the temperature of the upper stratosphere was 21°C. If temperatures had remained constant above the tropopause, as suggested by Teisserenc de Bort, −50°C could have been expected. When Lindemann and Dobson presented their findings to the Royal Society in 1922, the press picked the story up straight away. There was some speculation that instead of travelling 800 kilometres to the Riviera,

Britons could ascend 50 kilometres to experience the same tropical temperatures.

Media hype aside, they had corroborated the earlier findings of Fabry and Buisson by increasing the certainty that there was a layer of ozone which would absorb ultraviolet light from the Sun and heat the stratosphere in consequence. It was directly as a result of this work that Dobson was elected a Fellow of the Royal Society, the highest accolade which could be bestowed on a British scientist by his peers. To his chagrin, Dobson never collaborated much further with Lindemann, who was destined for even greater things, most notably as Churchill's scientific adviser during the Second World War. Among other almost apochryphal legends, Lord Cherwell (as Lindemann soon became known) dropped a block of ice while the great man was in the bath to prove that icebergs could be used as aircraft carriers.

The only heights to which Gordon Dobson aspired were literally in the stratosphere and so he decided to further the work of Fabry and Buisson which had been interrupted by the First World War. He soon hit upon the idea of making routine ozone measurements from various sites around the European continent. He also recognised that a simpler spectrometer would be needed than the one the Frenchmen had used. It would have to be portable and easy to furnish in a laboratory like the one he had built up at his house at Boar's Hill outside Oxford. And he decided to improve on their work by making photometric measurements, which would reveal far more easily how much ozone was present in the stratosphere.

By choosing two ultraviolet wavelengths – one of which was strongly absorbed by ozone, the other weakly – it was possible to calculate the amount of ozone in the stratosphere from their relative intensities. To begin with, Dobson constructed instruments which employed a photographic plate to record these photometric measurements. His very first prototype (made entirely of wood) is preserved at the Science Museum in Kensington.

Further spectrophotometers of this type would be employed in a putative ozone-measuring network. In 1925, Dobson obtained financial backing from the Royal Society to establish a total of six instruments in Scotland, Ireland, Sweden, Germany, Switzerland and one at his home in Oxford. The Swiss instrument was placed in Arosa, from where a simpler spectrograph had been employed in making ozone measurements from time to time as early as 1922.

Since Victorian times, the Swiss Alps had been a fashionable place among the well-to-do 'take in' ozone. The rarefied air, together with the appreciably higher levels of ultraviolet radiation to be experienced there, was believed to be vaguely energising. Even stranger to posterity, Arosa was also advertised as a spa, though today it is known its waters contain salts of uranium, not exactly beneficial for health. Nevertheless, as we will see, Arosa was to play an important role in the story of ozone depletion decades hence.

In later years, Dobson was to recall two minor hiccups in setting up his network. The first was the General Strike of 1926, which delayed its implementation until July 1926. The second was the German Foreign Ministry, whose diplomatic bag was used to ferry photographic plates to London from the instrument stationed at Lindenberg. Often the plates were opened in the post between London and Oxford, thereby becoming ruined and, to add insult to injury, Dobson himself had to pay the carriage since the German embassy refused. Nevertheless, by the late 1920s, more spectrophotometers of Dobson's basic design were incorporated at additional sites around the world.

As the network spread, so did Dobson's reputation. One day a letter arrived from Peking which, to his surprise, announced: 'I enclose cheque for £100, please arrange to send me one ozone spectrograph. If you want more money I will send it.' In the interim, the first photoelectric cells had been developed, which meant spectrophotometers could be built without recourse to cumbersome photographic plates. Dobson used the first photoelectric version of his spectrophotometer in the summer of 1930, the basic design of which is still in use today. Though even modern-day versions seem antiquated (they look like props from *Young Mr Edison*), they are the simplest way of making routine ozone measurements at regular intervals from the ground.

At the Clarendon Laboratory, Dobson meticulously recorded the data from his network. The record of observations showed that ozone levels seemed to be correlated with weather fronts. There was a strong seasonal variation and most ozone was produced in the springtime. There were also noticeable variations with latitude.

Belatedly, Dobson realised that the ozone layer also explained the riddle of why the sound of battles on the Somme could be heard in London but not in the Channel ports. Its densest region acted as a reflecting barrier for the sound of gunfire: soundwaves would literally bounce off the ozone layer and thus miss whole swathes of the countryside. This was shown by a series of remarkable experiments involving cannons carried out at Dobson's behest in the late 1920s by the superintendent of armaments at the Woolwich Arsenal.

A few years later the full explanation of why Dobson and Lindemann had measured such relative warmth at altitude was provided by their Clarendon colleague Sydney Chapman. Ordinary oxygen molecules absorb the shortest wavelengths of ultraviolet and in the process would be heated: this would break the molecular bonds and result in two oxygen atoms. If an oxygen atom attached itself to an oxygen molecule, ozone would be formed. But if that stray oxygen atom reacted with the ozone which had been formed, then two oxygen molecules would result. The upshot was that in chemical terms there was equilibrium: ozone would be formed at the same rate as it was destroyed. It was like water filling up a bathtub as quickly as it was escaping through the plughole.

The ozone layer was no less delicately poised for although it was a natural consequence of the action of sunlight, it was hardly solid, highly variable in nature and precariously thin. In 1933, Chapman confirmed Fabry and Buisson's calculation that if all the ozone above our heads were brought down to ground level, it would exist as a tissue-thin layer. One leading writer of the day was percipient enough to note that 'this layer of ozone, no thicker than a wafer biscuit, is all that stands between us and speedy death'.

Gordon Dobson was immortalised not only by his ozone-measuring device but also by the unit of measurement which was bequeathed to stratospheric science. It is so complex to define that suffice to say here that an average value for ozone at any one time, 300 Dobson units, is equivalent to 3 millimetres at ground level. For his part, Chapman termed the study of the upper atmosphere *aeronomy*, a label which has been used ever since.

The Second World War interrupted the steadfast work of the Dobson network. During the war, Dobson himself was to take part in the first ever measurements within the stratosphere from Boston bombers, a story which will be told in greater detail in Chapter Four. To his surprise, the stratosphere was much drier than had been expected, and he later proposed, with his Oxford colleague Alan Brewer, that the ozone produced at the equator was effectively 'dried out' because the tropopause there was higher and colder than anywhere else on Earth. The ozone would spread to the poles by the natural dynamic circulation of the atmosphere, as his network of instruments below had begun to show beyond all doubt.

In 1948 the Dobson network was put on a surer footing with the formation of the International Ozone Commission, which, to begin with, was based at the Clarendon Laboratory. Its first head was Sir Charles Normand, who had recently retired as the head of the Indian Met Service. With his formidable administrative skills – finely honed in that last outpost of the Empire – Normand ensured that by the mid-1950s there were over forty stations returning daily measurements of ozone. The task of collating the material was given to a handful of secretaries whose importance was recorded by Dobson himself in a summary of his four decades of research in 1968: 'Miss Trollope, Miss Robinson (Mrs Cole), Miss Luck and finally Miss Trollope again (now Mrs Sykes) helped successively in this work.'

And so during the intervening years these good ladies with the unlikely names uncomplainingly recorded ozone measurements from around the world. Their nuances and subtleties were faithfully pored over and analysed by Gordon Dobson. Whenever he was asked why he was interested in such an obscure topic, 'Dobby' would good-naturedly shrug and say that it was just another trace gas in the atmosphere which required monitoring for no other reason than that it was interesting. He was quite happy to make daily observations, knowing full well that many of his colleagues thought it a strange pursuit, a sort of scientific equivalent of train spotting or collecting matchboxes. In the years after his death, the work he had set in motion would take on a new meaning and urgency that he could hardly have anticipated.

Although Dobson retired from the Clarendon Laboratory in the late 1950s, he still went into the department and enjoyed following the trends of atmospheric research. There were always visitors and a buzz of activity at his new home on Watch Hill near Shotover, from where he continued to make ozone observations each day. Desmond Walshaw, a postgraduate researcher in the early 1950s, describes a typical visit there as a perfect encapsulation of English domesticity:

> There would be coffee or tea by the fire of home-grown and home-sawn wood, with the family portrait, the grand piano and the fine bracket clock with its penetrating strike, or in good weather on the terrace. Then the visit to the laboratory to see the latest results or try the new device, or, more frequently in later years, discussion of work by the fire.

Dobson remained active until the very end. Another colleague, John Houghton, recalled visiting him one day when he was eighty-five to find him replacing apple trees in his orchard. He was to make his last observation of atmospheric ozone on 30 January 1976, the day before he suffered a massive stroke. He died six weeks later.

The name Antarctica derives from the Greek words for 'against' (*anti*) and 'bear' (*arktos*). As cosmos opposed chaos and hubris contrasted nemesis, the ancient Greeks expected symmetry in geography. At the zenith above the Arctic was the constellation of Ursa Major, the Great Bear, which navigators used to chart their way across the northern hemisphere. At the Earth's invisible nadir, the Greeks reasoned, there would be a landmass whose formal name derived from this supposed astronomical symmetry, though for centuries afterwards it often just appeared as *terra incognita australis* on woefully inadequate maps. The Greeks also had a notion of four elements – earth, water, air and fire – which Sydney Chapman later remarked were involved in the greatest scientific endeavour ever planned. That enterprise would place Antarctica in the canon of modern science and elevate it to a central role in the drama of ozone depletion in the decades which followed.

It has been said that the Second World War was a triumph for physics and engineering. The technology developed for waging war would have many applications and lead to a postwar frenzy of activity and advance in virtually all the scientific disciplines.

It was soon clear that large-scale scientific collaborations could play a healing role by replacing former enmities with a sense of common purpose. International collaboration would provide a priceless opportunity to learn more about the world as a whole. With this philosophy in mind came the notion of the International Geophysical Year, whose express purpose was to transform our understanding of the planet upon which we live.

According to legend, the IGY originated at a dinner hosted by a pioneering (and then largely unknown) physicist called James Van Allen at his home in Maryland in 1952. It was held in honour of Sydney Chapman on one of his regular visits to the United States, and during the course of the evening an earnest discussion developed about the merits of creating a concentrated effort to learn more about the physics of the Earth in all its manifestations – the weather, the core and its magnetic interactions with the Sun. The last-named requirement meant that if there was going to be a concerted effort it should coincide with maximum solar activity. It would focus on Antarctica, where much remained to be learned. Every eleven years the Sun's disturbing influence is at its greatest and this would enable scientists to obtain the clearest picture of how the Earth's atmosphere responded. The next solar maximum would occur in 1956–57.

From today's jaundiced perspective the International Geophysical Year seems to have been more a metaphor for geopolitics than geophysics, for the greatest impact of a year that lasted eighteen months was to reverberate around the globe in shrill beeps against a crackle of shortwave static. The IGY gave the United States and the Soviet Union the excuse to launch rockets out beyond the stratosphere and into the brave new world of the space age. The Soviets got there first with Sputnik 1 in October 1957, an accomplishment which tended to overshadow all else. US national pride was swiftly restored when a geiger counter designed by James Van Allen was flown on America's first satellite, Explorer 1, in January 1958 to reveal the belts of solar radiation trapped by the Earth's magnetic field which now bear his name.

Scientifically, the IGY resulted in advances in many areas of geophysics, most spectacularly in the understanding of what happens when the Earth's magnetic field interacts with the Sun's and how that influences the weather. The IGY also set the tenor for subsequent international collaborations in science that would largely become routine. Sydney Chapman wrote in 1959 that 'the planning and execution of the IGY was marked by a most cooperative and harmonious spirit among the scientists of the sixty-seven nations associated with it. Their common interest in its subject and purposes made it possible for them to work together despite differences of race, creed, or political organisation.' (Almost – China pulled out because Taiwan participated.)

Developments in atmospheric research from the IGY were to have far-reaching consequences. Air-sample measurements instigated by a young Caltech graduate named Charles David Keeling would alert the world to the inexorable rise of levels of carbon dioxide and the increasing certainty of global warming. Another discovery of similar import began with the arrival of a recent Cambridge graduate as part of an expedition mounted by what was then known as the Falkland Islands and Dependencies Survey. As part of the British contribution to the IGY, Joseph Charles Farman's task was to establish a Dobson instrument near Antarctica. Little did he suspect that it would become a better part of his work for the rest of his career.

Joe Farman was remarkably cultured by comparison to most of the Young Turks who arrived on the ice. Most remarkable was his endless fund of knowledge about wines, a topic about which the larger British public was wholly ignorant. And many IGY veterans recall that the earnest young man with a distinct Norfolk accent was then, as now, more often than not surrounded by a small plume of tobacco smoke from his beloved pipe.

After graduating from Cambridge, Farman had been working on designs for the Blue Streak missile within De Havilland Propellers ('for some reason that was the high-tech part of the company,' he said casually) when he saw a recruitment advertisement for the International Geophysical Year. He applied, and within two years found himself on Argentinian Island off the coast of the Antarctic Peninsula. With characteristic single-mindedness, Farman had visited Gordon Dobson at his Oxford home to learn about the intricacies of his famous spectrophotometers. Before he left Britain he had made practice measurements from Lerwick in Scotland and developed a lasting friendship with the grand old man of ozone science.

Those first few years of ozone observations from Antarctica showed something that at first seemed to be very odd. The values of springtime ozone were decidedly lower than had been expected, certainly compared to similar observations in the northern hemisphere. At Spitsbergen in the springtime, for example, ozone abundances established a world record of 480 Dobson units. It was blithely assumed that similar values would be mirrored in the austral spring, but by comparison, Antarctica was covered by around 300 Dobson units. Why was Antarctic ozone comparatively so low? Farman himself was puzzled while Gordon Dobson, a hemisphere away, was, colleagues recall, excited by these curious observations.

Atmospheric scientists later realised that they had made a fundamental mistake by assuming that conditions in the southern hemisphere would replicate those in the north exactly six months later. The Earth's atmosphere did not exhibit the symmetry beloved by the Greeks because of the peculiar phenomenon of the Antarctic vortex. Gordon Dobson himself acknowledged the fact in his seminal work *Exploring the Atmosphere*, published in 1962, by noting that it seemed 'as if in winter the south polar stratosphere is cut off from the general worldwide circulation of air by the very intense vortex of strong westerly winds which blow round the Antarctic continent, enclosing very cold air which is rather weak in ozone'.

This strange vortex was just one of a handful of strange phenomena which served as a reminder of how little was really known about our planet before the IGY. At the time, these observations from in and around the South Pole seemed like a small point, but thirty-five years later they would stir up a peculiar controversy around whether scientists had somehow stumbled on an ozone hole even then. It is a measure of how politicised stratospheric research had become in the intervening years.

Joe Farman continued to oversee ozone measurements and made several more trips south in the early 1960s, by which time his pay cheques were issued by the British Antarctic Survey. Eventually, he became responsible for the Dobson instruments at the main survey base on the ice at Halley Bay, and was one of the more dedicated of a handful of scientists who were making systematic ozone studies in the strange ice desert at the bottom of the Earth. As the world changed beyond all recognition in the years that followed, the Dobson spectrophotometers at Halley Bay and elsewhere in Antarctica provided some measure of continuity. Out of sight, and virtually out of mind, they ceaselessly probed the cold, clear twilit skies high above 'the last place on Earth'.

3

Crisis! What Crisis?

> Like children with a new toy, we are inclined to play with our science before we properly understand it. There has been much pressure in recent years to concentrate our scientific efforts on the applications at the expense of fundamentals. The lesson of the CFCs is clear and should be learned now – it is that we must not use the power of the knowledge that we now possess to manipulate the world without ensuring that we also have the knowledge to understand and mitigate unforeseen errors and disasters.
>
> The unforeseen consequence of the otherwise safe and excellent CFCs cannot be blamed on the individual chemists or their managers, who did all reasonable testing that was expected of them. But it was left to the individuals, freely following hunches that were in the first place totally unconnected with the ozone layer ... to say things that had never been said or thought before. It is they who saved us from what might have been disaster.
>
> Sir George Porter, presidential address to the Royal Society,
> 30 November 1989

IF we wanted a culprit to blame for forcing two of the most perplexing and seemingly intractable environmental problems on an unsuspecting world, we might look no further than Thomas Edison Midgley. For not only did he create the family of chlorinated fluorocarbons, or CFCs, which cause ozone depletion, he also put lead into petrol to stop engine knock. At first glance it would be easy to characterise Midgley as an irresponsible industrial inventor who created a dreadful legacy for subsequent generations. But if he were alive today, Midgley would doubtless justify his work on the grounds that CFCs have a myriad uses, and he could not have known about the drawbacks. 'Midge', as he was known to one and all, was an indefatigable defender of the faith in the progress of science to solve problems and make life easier. He was as much a representative of the first half of the twentieth century as the environmental activist or media consultant of the latter half, upon which he unwillingly wrought so much damage.

The dawn of the twentieth century glinted on freshly minted technological marvels. Within a few years, the aeroplane and the motorcar would transform transportation, particularly across the vast expanses of the United States. Chemistry, too, would bring about unprecedented technical change – particularly as a result of its remarkable successes during the First World War – and thereafter become a rapidly expanding arena in which scientists and entrepreneurs could make fortunes. It was yet another manifestation of the American dream and something to which Thomas Midgley would make a unique contribution. So perhaps it was apt that he had been born to parents who had left Victorian Britain for the opportunities afforded by the New World. His father was an inventor who had come up with the idea of a detachable rim tyre made of rubber. Midgley senior also knew the more illustrious inventor after whom their son gained a middle name – Thomas Alva Edison. Shortly after Thomas junior was born, in May 1889, the Midgleys moved to Columbus, Ohio, where Thomas senior worked as a machine designer and created the sensation of the moment, the 'safety bicycle'.

His son followed in his father's creative footsteps and studied engineering at Cornell University. His genius seems to have been recognised early on, for his was the only name proffered to the National Cash Register company, who were actively seeking the brightest students. In 1911, Midgley reported to work for the company's Inventions Department, a forerunner of today's research and development departments, where he could do more or less what he wanted. Even then, they were relentlessly commercial, very much in the image of Edison after whose tireless self-promotion the myth of the scientific inventor had grown.

In time, Midgley too became an astute populariser and pre-eminent self-publiciser with a panache which even today causes astonishment. While he was experimenting with lead additives in petrol, an engine blew up in his face, splattering his right eye with metallic fragments. Though an optician removed most of the larger pieces, his vision was seriously impaired by the smaller fragments. Midgley knew that mercury could amalgamate with other metals, so he decided to administer droplets of a purified form of the metal from a pipette. As predicted, the smaller pieces of metal adhered to the drops and his vision was cleared.

This work had come about when Midgley had joined the Domestic Engineering Company in the summer of 1916. It had been founded by an NCR alumnus, Thomas Kettering, who had amassed a considerable fortune as a result of inventing the electrical starter which had banished the handcrank forever. But one problem still remained: knocking. Early American automobiles had the propensity to make the strange noises that Mack Sennett captured for comedic effect on celluloid. Knocking was caused by fuel being prematuraly ignited in the engine, and it seriously impaired the performance of the car engine.

Kettering believed Midgley was just the sort of person to figure out how to cure knocking, so he hired him. Midgley experimented for over three years with over 3,000 additives that would stop knocking. An idea of Kettering's was to focus on metallic compounds, and here they found encouraging results. Five years after, their search had begun they homed in on tetraethyl lead, which seemed to offer the best combination of fuel efficiency and ease of manufacture.

Tetraethyl lead was dyed red and marketed as Ethyl. The chemical company Du Pont manufactured it and distributed it in highly concentrated small bottles which Midgley christened 'Ethylizers'. But just as sales started to mushroom, the great 'loony gas' scare of 1924 occurred: workers at an ethyl-manufacturing plant in

New Jersey developed deliria, paranoid psychoses and eventually succumbed to convulsively painful deaths. Media hysteria fanned the story. Something dramatic was needed – which Midgley duly provided. At a press conference a week later, Midgley explained that the workers had refused to wear protective masks against the power of concentrated ethyl. He proceeded to wash his hands and face in the diluted form commercially available and then inhaled the vapours for over a minute. This powerful demonstration alleviated most fears about its toxicity and within two years nearly a billion litres of ethylised petrol had been sold.

Midgley's singular vision also came to bear on another technological miracle of the age – refrigeration. Refrigerator sales had skyrocketed throughout the 1920s, despite the fact that the ammonia and sulphur dioxide, commonly used as the refrigerant, often leaked and spoiled the food. Worse still were fridges which used methyl chloride as a refrigerant which, being odourless, caused many fatalities when it leaked. So Midgley was loaned to Frigidaire, the leading fridge manufacturer of its day, to find something that was safe to use. He hit upon fluorine as having the right combination of stability and low boiling point: the only problem was that it was toxic and quite possibly corrosive. But experiments in combining fluorine, chlorine and carbon seemed to work because they would vaporise at the low temperatures needed for extremely energy-efficient coolants. Different combinations of chlorine, fluorine and carbon resulted in a family of inert chemicals which were both stable and non-flammable. Neither did they appear to be toxic nor corrosive and produced no discernible aftereffects when inhaled by guinea pigs or (briefly) human beings. This was the extent of safety testing when Midgley begat chlorofluorocarbons on an unsuspecting world.

He was to score another remarkable PR triumph when he announced their existence at the annual American Chemical Society meeting in Atlanta in April 1930. Midgley poured liquid CFCs into a glass and inhaled the vapours; he then exhaled and blew out the flame of a candle in front of him. The implication was clear enough: refrigerators using CFCs would never again claim lives if they inadvertently leaked. Under the slogan of 'End Cube Struggle!', fridges using CFCs became an overnight sensation and accounted for virtually all the refrigerators on the market by the end of the 1930s. Frigidaire, General Motor and Du Pont joined together to market the new wonder chemicals under the trade name Freon. Du Pont owned the trademark and subsequently mushroomed into a giant corporation, licensing manufacture of CFCs around the world.

Quite simply, CFCs were not only miraculous in their many desirable properties but they were also amazingly cheap to produce and soon their uses multiplied. Because of their energy efficiency, CFCs were ideally suited to be used in air conditioning and also as propellants for spray cans. When the newly invented insecticide DDT was distributed in 'bug bombs', CFCs were employed as a propellant. Agricultural production soared throughout the United States. In time CFCs would also be used in spray cans for cosmetic products, during the manufacture of plastic-based foams and insulators and later as cleaning solvents in the electronics industries. In short, CFCs transformed the world and lived up to one of Du Pont's most famous advertising slogans: 'Better Living Through Chemistry'.

In many ways Thomas Edison Midgley became its most obvious living embodiment, for after his morning's work in December 1928, he became unimaginably rich. He moved into an enormous mansion in Columbus, Ohio, with what he wistfully called the most beautiful garden in the world, and constructed a giant tunnel into the hillside behind the house. It opened into a ravine and he lit it with electric candles: from photographs it bears an uncanny resemblance to the Batcave in which Bruce Wayne could be found in the TV series *Batman*. Midgley's personal fame multiplied as he liked to travel and expound his theories. Many contemporaries noted that he was also fond of his drink. Many an evening in the company of Midge would involve discussions of how the onward march of technology would lead to genetic engineering, intelligent computers and, ironically, climate modification. There were apparently no boundaries to scientific and technical innovation.

By the time of America's entry into the Second World War, Midgley realised that his most productive days were over. Viewed as the grand old man of industrial chemistry, he soon expounded the value of youth, citing the fact that most discoveries are made in the early years of a career. He suggested that beyond the age of forty-five, scientists should become administrators instead. Midgley's own sense of preternatural aging was underscored by polio, which struck him down in 1940. He weakened and became increasingly despairing. An inventor to the last, he created a lifting mechanism to get himself out of bed and into his wheelchair. And it was in this strange contraption that Midgley met his death by his own hand in November 1944. Although newspaper accounts tried to play off his death as some sort of bizarre accident, Midgley's death certificate records a verdict of suicide.

On the way home from Midgley's funeral – in a car whose engine didn't knock and which was air-conditioned using CFCs – Thomas Kettering was moved to think of a suitable epitaph for his most illustrious employee. At the funeral the priest had quoted from the Bible that 'we brought nothing into this world and it is certain that we can carry nothing out'. 'It struck me then,' Kettering was later to recall, 'that in Midge's case it would have been so appropriate to have added, "but we can leave a lot behind for the good of the world."'

Unwittingly, Midgley himself had provided an all too apposite comment on the legacy he had created in CFCs. Just after he was struck down with polio, Midgley calculated the chance of catching it was equal to drawing a particular card 'from a stack of playing cards as high as the Empire State Building'. And doubtless he would have given similar odds against having, in the family of chlorofluorocarbons, created an almost intractable problem for future generations.

Within thirty years of Midgley's death, the social and environmental impact of technological innovation had struck the inhabitants of planet Earth with a resounding thud. Whereas the First World War had been a triumph for chemistry, the second produced a renaissance in physics, most noticeably in atomic power. By the 1960s, however, cheap, endless electricity from 'our friend the atom' had proved a chimera and the first warnings about standards of engineering at nuclear power plants were raised. The illusion of the miracle of pesticides had reached prominence in 1962 in an eminently reasonable, sustained piece in the *New Yorker* by a former US government biologist called Rachel Carson. Her words were later published in book form as *Silent Spring*, which acted as clarion call for the nascent environmental movement in the United States.

The ozone layer first began to concern environmentalists in the context of aviation. Jet airliners had transformed the economics of tourism: air travel was no longer the preserve of the well-to-do, and had flourished in the postwar consumer and population booms. To transport ever greater numbers of people, the creation of a supersonic airliner was mooted. The Anglo-French Concorde, developed in the 'white heat of technological revolution' forged by the Wilson government, was the first. Its existence threatened the wellbeing of the large US aerospace industry, so in the last year of his life, President John F. Kennedy sanctioned a Super Sonic Transport (SST) programme to be carried out by Boeing. Soon magazines and newspapers were filled with articles predicting hundreds of supersonic airliners filling the skies in the brave new world of the 1980s.

Initial opposition to SSTs centred on their attendant sonic booms, which many felt would make life unbearable for urban communities close to airports. It was also suggested that water vapour from SST exhausts would modify climate in an equally unaesthetic way: in the dry stratosphere, it was feared, the skies could fill with a continuous patchwork of vapour trails. By radiating heat back to space before it reached the ground, the vapour would inadvertently change the Earth's climate. A study conducted by James McDonald of the University of Arizona for the National Academy of Science reported in 1966 that in reality this posed little threat.

Subsequently, however, it was realised that a variation on the chemical reactions involved in water vapour production could upset the chemistry of the ozone layer. By the early 1960s, researchers had been puzzled by regular measurements of ozone in the lower stratosphere from balloons, sounding rockets and high-flying aircraft. There seemed to be one-third less ozone than there should have been according to Sydney Chapman's broad-based mechanisms. There was obviously some sort of chemical reaction going on which was significantly reducing the stratospheric ozone layer. It was known that the main constituents of air didn't destroy ozone, so oxygen, nitrogen, carbon dioxide or water vapour itself could be discounted. That left trace gases in the atmosphere, present in minute quantities, which would work in some unknown way at eating ozone.

The answer to this conundrum involved chain reactions in which unstable molecular fragments played a central role. Known as free radicals, they are present but short-lived in many chemical reactions as they are left with a 'rogue' election which will strip away relatively unstable molecules like ozone. They are particularly voracious because they themselves remain unscathed and keep on eating ozone like a particularly hungry Pac Man in a peculiarly demented computer game. The first to be investigated were oxides of hydrogen, formed during fuel combustion in aircraft and generically known as HOx. Further free radicals were later studied until the right culprit could be identified. From a scientific perspective, the late 1960s and 1970s could be viewed as narrowing down the search for the catalytic chain most likely to destroy ozone.

It is doubtful whether discussions of this esoteric chemistry would have burst into the American media had it not been for James McDonald's involvement. He looked at the effects of HOx from a catalytic chain point of view and in March 1971 reported to a Congressional committee the result of his findings. Eight hundred SSTs, he calculated, could provide enough HOx to cause a 4 per cent decrease in ozone worldwide. This in itself was not particularly apocalyptic: it was when he made the first public link with ozone depletion and skin cancer that things went critical. Based on statistics relating increases in skin cancer with increasing proximity to the equator, McDonald suggested that about 5,000–10,000 deaths from this would result from the additional ultraviolet light let in.

By this point, the SST debate had become very involved and vitriolic. Cost overruns and political opposition to the US SST programme meant that Boeing would be funded only to build two prototypes. Its allies were fighting tooth and nail to keep the programme alive, and unwittingly McDonald had played right into their hands. He was one of the few scientists who had looked into the question of unidentified flying objects and had suggested they were worthy of further, serious investigation. McDonald's Congressional testimony on the ozone issue was verbose and, in consequence, difficult to follow. At its conclusion, one representative went straight for the jugular and asked if there was a relationship between UFOs and skin cancer. This unfair juxtaposition was enough to give McDonald a credibility problem, particularly when the *New York Times* later reported his earlier Congressional testimony that UFOs could be causing power cuts in Manhattan.

McDonald repeated his ideas at a scientific meeting organised by the Department of Commerce a few weeks later at its environmental research laboratories in Boulder, Colorado. Its purpose was to discuss the general environmental effects of SSTs on the stratosphere, but the first day saw McDonald being roundly criticised by scientific colleagues for linking ozone depletion with skin cancer. Some Boeing officials openly referred to him as 'that UFO nut' and refused to believe any of his calculations. McDonald would commit suicide later that summer (for reasons unconnected with the ozone issue). After his death, McDonald's methodologies were recognised to be essentially correct. He remains the only unequivocal victim of the ozone debate.

McDonald's vilification was a counterpoint to the important outcome of the Boulder meeting: the step from HOx to NOx. It was taken by Harold Johnston, an internationally renowned chemist at the University of California at Berkeley. He was an expert in the chemistry of nitrous oxides, having studied at Caltech in Pasadena in the 1950s under Jan Haagen-Smit, where he had worked on those pioneering studies of smog. Johnston was astonished to hear a presentation that claimed that the oxides of nitrogen would present little problem in the stratosphere. As he well knew, nitrogen oxides are involved in photochemical production of ozone at ground level. One presentation showed that in the mesosphere, above the stratosphere, NOx would be involved in photochemically driven catalytic cycles. The meteorologist who had done these calculations suggested that similar cycles would be unimportant in the stratosphere. But, Johnston thought, could that blithe assumption be made? He suspected that NOx would be important and could be involved in the destruction of stratospheric ozone. Overnight, Johnston calculated that if a fleet of 500 SSTs regularly plied the stratosphere for two years, then global ozone levels would be reduced by 10 per cent.

Many of the participants – cheered on by Boeing – viewed these calculations as suspicious. The outcome of the Boulder meeting was the first of many bureaucratic buck-passings: the Secretary of Commerce was advised that not enough information on the environmental impact of SSTs was available at that time upon which a reasonable decision could be made. By this point it was really academic because Congress had voted against building the two prototypes, and a few weeks later, the Senate followed suit. But the papering-over was too much for Johnston, who refined his calculations and published them in *Science* in August 1971. Five hundred SSTs flying seven hours a day would strip the ozone layer by 10 per cent within a year.

Although the US SST was effectively dead, the European Concorde was ready for takeoff. Under the auspices of the US Department of Transport another group of scientists was brought together to investigate their influence on the stratosphere. The Climatic Impact Assessment Program (CIAP) cost some $60 million, involved a thousand scientists from ten countries and reported its findings in January 1975. It concluded that Johnston was essentially correct but that better modelling was needed to assess the problem accurately; and, indeed, a few years later, improved calculations showed that stratospheric ozone might actually increase by a few per cent from SST exhausts. One scientist was later famously to quip that it seemed an expensive way to maintain the ozone layer.

By then, a handful of Concordes had landed in the United States, but they hardly posed much of a threat to the ozone layer or, for that matter, American national pride. In the end the US SST programme had been killed off by financial and political rather than environmental concerns. And by this time, another threat to the ozone layer had been identified: the chlorofluorocarbons, which were Thomas Edison Midgley's contribution to the world.

The way in which the world became aware that CFCs could damage ozone was, like so many scientific breakthroughs, accidental and, even with the benefit of hindsight, bizarre. Had a British scientist working from the seclusion of his home not been intrigued by pollution on holiday, it is entirely possible there would have been no ozone wars in the 1970 and, thus, no idea why the ozone hole would appear a decade later.

As the icy wastes of the Weddell Sea came into view, James Lovelock had little notion of how far-reaching his work would soon become. His reputation as a scientist was based on his invention of a device known as the electron capture detector in the 1950s. It had revolutionised many branches of science by enabling researchers to measure minute traces of atmospheric gases. The patent for his instrument enabled Lovelock to work in seclusion in the English countryside and study more or less what he wanted.

Jim Lovelock has at various times been called a natural philosopher, genius, guru, inventor and – to his surprise – maverick, although he would prefer 'independent scientist'. As a schoolboy in Brixton in the 1930s, Lovelock was told by his headmaster that he was a fool to take up science. 'There is no place there except for those of genius or with private means,' the teacher imperiously declared. Despite this discouragement, Lovelock was to amass the latter as a result of the former.

After twenty years as a civil servant at the National Institute of Medical Research, Lovelock was happy to be out of the index-linked rat race, which he felt stifled creativity. In so doing he courted the risk of being mistaken by some as a loopy Professor Branestawm character, though his peers had no such doubts, electing him to a fellowship of the Royal Society. Today, he is most associated with the Gaia theory, developed while working as a consultant to NASA, for which he has become a darling of New Age hippies (it chimes with their holistic view of Mother Earth) and free-marketeers (to whom it means the Earth will take care of as much pollution as can be released into the environment). In fact it says neither – and Lovelock is careful not to espouse any one cause, a standpoint derived from his experiences during the CFC debate of the 1970s.

Working aboard the British research ship *Shackleton* in the austral summer of 1971–72, Lovelock was to show the presence of chlorinated fluorocarbons throughout the atmosphere. He had come to the Antarctic in order to measure background levels of CFCs using the unparalleled 'sniffing' capacity of an advanced electron-capture device, whose sensitivity he had especially enhanced a millionfold. The relative inertness of CFCs made them ideal as tracers – chemical species whose motions would provide interesting clues to how the atmosphere circulates. As he often built prototypes at home, the Lovelock family was one of the first to give up aerosol spray cans: they interfered too much with his delicate atmospheric measurements.

His interest in CFCs came about when his family took a holiday in Southern Ireland during the gloriously hot summer of 1966 (which moved Lennon and McCartney to write 'Good Day Sunshine'). Lovelock was puzzled by the presence of haze which made it difficult to see a lighthouse 205 kilometres away. With the electron-capture technique, Lovelock had at his disposal the means to see if the unseasonal haze was caused by manmade pollution, and where it came from. Two years later he systematically set out to answer the question. He could monitor minute traces of pollutants in the atmosphere such as ozone or oxides of nitrogen – and CFCs. The last was the clincher because they could only come from industrial sources. Lovelock made a startling discovery: when the wind blew from the Atlantic, he found 15 parts per trillion of pollution; when from Europe, it was 50 parts. The greater numbers of tourists who descended upon southern Europe with spray cans were the cause of the pollution.

This discovery prompted Lovelock to think about the long-lived presence of CFCs in the atmosphere. Were they accumulating because they were so stable? An obvious starting point would be to try to measure them across the whole of the globe. To get a background reading, Lovelock realised that a trip to the last great unpolluted continent at the bottom of the world was needed and dutifully applied to the Natural Environment Research Council for funding for a trip to Antarctica. As is customary, his application was deliberated upon by other scientists and Lovelock later recalled: 'One of the reviewers said: 'This application is clearly bogus. Every schoolboy knows that CFCs are the most stable compounds and it will be impossible to measure them at parts per million, let alone trillion, because of their lack of reactivity.'' The civil servants who managed NERC's meagre coffers were more amenable to his proposal. Although they could not provide him with funds, he was welcome to cadge a lift aboard the *Shackleton*.

So Lovelock sailed south and made the first of his measurements from Antarctica. He found that CFC-11 and -12 (the most common of the CFC family) were present in minute quantities, a handful of parts per trillion. But when those numbers were extrapolated throughout the atmosphere it was clear that CFCs had been steadily accumulating since Midgley had invented them. In time, Lovelock would make other measurements from the western Pacific and help set up a network of measuring stations which would record the inexorable rise of CFCs in the years that followed.

When the results of his Antarctic voyage were published in *Nature* in December 1973, Lovelock, by his own admission, made 'a bit of a boo-boo'. He concluded that CFCs would pose no conceivable hazard to humanity by the long-lived nature of their build-up. At that time, it had become fashionable to decry the benefits of science and industry. This baffled Lovelock, who is quick to point out that Rachel Carson's chronicle of the effects of pesticides, *Silent Spring*, could not have been written without electron-capture detection. Two decades later he said: 'The last thing I had in mind was that these measurements would cause environmental problems – which I was happy to say. I had a fear that some kook of a green would go around saying we're all going to die.'

And indeed by the early 1970s, the only indication that CFCs did any harm came from the counter-culture. For those who couldn't quite manage the trip out to Haight-Ashbury, nor the expense of LSD tabs, inhaling CFCs from spray cans caused quite a modestly priced sensation. In the drippy, trippy kaleidoscopic world which was America start the 1970s, cerebral aneurisms caused by CFC inhalation were yet another form of distraction for youths looking for ever more pointless thrills. To the forces of social order it was proof of what things were coming to rather than any threat to the Earth as a whole. In a similar vein, it was perhaps appropriate that the scientist who determined exactly how ozone was at risk was based in California – that unique depository of way-out thinking which best characterises the time – and, even then, was only prompted to look at the problem because of a conversation resulting from a chance meeting on an Austrian train.

Just after the Second World War, a young American scientist at the start of his or her career could find no better mecca than the University of Chicago. Its formidable reputation had been bolstered by the first coaxing of nuclear fission from a curious collection of graphite piles in what had been a students' squash court. This was the start of the Manhattan Project, which would culminate in two mushroom-shaped explosions that would cast their pall over the postwar world and change the basis of international relations for ever. After the war – the very timing of whose conclusion they had dictated – many of the Manhattan alumni remained in Chicago, providing an unparalleled pool of intellectual talent to be drawn on by students. Their number included the Nobel laureates Enrico Fermi and Harold Urey and a future Nobel prizewinner, Willard Libby, whose carbon-14 radiodating technique would transform many sciences, most obviously archaeology.

When asked about the prospects for the atomic age, Libby said at the time: 'If we treat the atom well and handle her wisely she can give us material benefit almost without end, and the awful aspect of the warlike atom will be softened by the great power for good that she has brought about.' And indeed this use of atoms for peaceful purposes would play a crucial role in the ozone story. Not only did the scientist who first recognised the threat posed by CFCs, F. Sherwood Rowland, study under Libby at that crucial time, his later work was also funded out of the munificence of the Atomic Energy Commission.

Sherry Rowland was a bright child and had entered Ohio Wesleyan University at sixteen in 1943. His father was a professor there, and Sherry continued to live at home. After serving in the US Navy at the close of the war, Rowland left Ohio Wesleyan in 1948 and started his graduate studies in earnest in Chicago. 'My choice of graduate school was a very simple one,' Rowland admitted. 'It was the same one my parents had been to.' Before graduation, he was still unsure which of the sciences he had majored in (physics, mathematics and chemistry) he would pursue: with lecturers of the calibre of Urey, Fermi and Libby it was not the easiest of decisions. In the end, Sherry Rowland plumped for chemistry, although he was distracted by an overwhelming interest in sports.

At nearly 2 metres in height, Rowland had the physique to become a basketball player, and he also had aptitude. He played semi-professional basketball in and around Chicago as well as baseball in Canada during the summer vacations. On one occasion he ended up coaching a team when their manager quit, and, for good measure, won the Canadian semi-pro championship. But Rowland realised he had to leave these sports behind. 'At some level sheer physical talent takes over,' he said. 'I had it at a good level, but not outstanding.' Even today his office is decorated by basketball pennants and awards, including one which testifies that he was one of the first Western scientists to play against a Russian team in 1967 (it doesn't mention that he ended up playing barefoot).

Basketball's loss was chemistry's gain. After gaining a PhD in the hot atom chemistry of bromine, Rowland went to Princeton to continue his work on another element – the radioactive tracing of the hydrogen isotope tritium. In the summer of 1953, he measured the amount of tritium occurring in the atmosphere. Months previously another group had made similar measurements, from which it was obvious there had then been appreciably smaller amounts present. 'If you compared the results you might have concluded that a substantial amount of tritium had been released into the atmosphere,' Rowland noted. Its source was the first US hydrogen bomb test. His first scientific paper on the subject was not published until 1961, by which time the H-bomb connection was well known.

In the intervening years, Rowland became a jobbing radiochemist, first at the University of Kansas, then moving to the new University of California campus at Irvine in Orange County in 1964. He built up a chemistry department from scratch, starting laboratory studies of radioactive fluorine and chlorine. Ironically, they used a chlorofluorocarbon as an inert source for both gases in a neutron accelerator. Though these studies had some medical benefits, it was for the most part abstract work involving the investigation of complex chemical reactions for their own sake. And so it might have continued, had Rowland not decided to change the direction of his career. 'By the 1970s, I was ready for a change, 'he said. 'My own view is that as a scientist you need to do something different every so often. So in 1970 I went to a conference on the application of radioactivity on the environment.'

He had heard about the Salzburg conference while working at the International Atomic Energy Agency in Vienna the previous year. To his chagrin, it was to be held in October 1970, and not during the summer, when the music festival is held. Rowland was an avid opera buff and, even today, sports pennants vie with opera posters from his limited office wall space. But since his radioactivity work was partly funded by the US Atomic Energy Commission, he could justify his attendance. Had he not shared a train compartment from Salzburg to Vienna with the AEC's Bill Marlow, he might never have become involved in the ozone issue at all. Marlow was planning a meeting in Fort Lauderdale, Florida, where chemists and meteorologists could get together and learn from each other. Rowland said he would attend.

Hitherto chemists and meteorologists had had little need to consult each other. The ozone issue changed all that and, even worse, blurred the demarcations between the two disciplines. It generated perpetual conflicts, at the heart of which was a sense of intellectual ownership of the stratosphere. To meteorologists, chemists were laboratory experimenters who knew nothing about the real world: how could lab chemistry play any part in an atmosphere dominated by large-scale motions? Chemists viewed meteorologists as using those same large-scale motions to explain away any problems or inconsistencies.

Scientific conferences can sometimes be staggeringly dull, even to those who are interested in the debates. The gossip-sharing outside the formal presentations usually makes the most impact, as it did in Fort Lauderdale. The corridors were abuzz with discussions of the significance of Lovelock's CFC measurements after they had been formally presented by Lester Machta of the National Oceanic and Atmospheric Administration. Rowland was intrigued and wondered if they would drift into the stratosphere, which seemed the only place they could go. It was just the sort of new direction in which to move his career.

Although he was not intimately acquainted with the science of atmospheric chemistry, Rowland had some considerable advantages in looking at the problem. For a start, he had been using a neutron accelerator to experiment with chlorine and fluorine and was well acquainted with the difficulties of radiochemistry. He began to wonder if trace gases in the atmosphere would have analogies with the trace-gas measurements he had been overseeing in the laboratory. The answer would be yes – up to a point. 'In the lab molecule A often reacts with molecule A,' Rowland explained. 'In the atmosphere, trace-gas molecule A never sees another A as it will always react with something different. It's an interesting problem.'

By the time Rowland started to investigate what would happen to CFCs when they reached the stratosphere, word of Jim Lovelock's measurements had spread through the scientific community and they were about to be published in *Nature*. It was also known that chlorine could attack ozone, as laboratory experiments carried out as far back as the 1950s showed. Early in 1973, researchers at the University of Michigan had also looked at how chlorine from volcanoes and exhausts from the space shuttle, then under construction, might affect the ozone layer. They concluded that they would provide little threat, if any, ironically identifying a catalytic chain reaction involving chlorine. Curiously, they had not considered the most obvious source, a family of chemicals applied at the start of the day under most people's armpits.

Connecting the right chain reaction with the right source would be carried out by the arrival in Irvine of Mario Molina as Sherry Rowland's postdoctoral associate in October 1973. They had met a few months previously when Molina had expressed an interest in the CFC problem which Rowland was about to investigate. At the time, Molina was busy writing up a PhD on chemical lasers from the University of California at Berkeley. For him, too, it was a nice change of direction. He was a bright student who hailed from a well-to-do Mexican family; his father had been ambassador to the Philippines. Though Molina had studied in Europe, he had preferred the stricter regimentation of the US postgraduate system and had achieved excellent results at Berkeley.

Anticipating Molina's arrival, Rowland had asked the AEC for additional funding to investigate the CFC question but it wasn't forthcoming. However, the commission did say that he could spend part of his renewed annual radiochemistry grant on CFC-related work if he so wished. Molina's first weeks under Sherry Rowland's tutelage involved searching out unfamiliar equations and textbooks in trying to address the CFC issue. As he told Lydia Dotto and Harold Schiff in their excellent chronicle of the early years, *The Ozone War*: 'We took it as a challenge [to] see what happens to these things. We didn't even think about the environmental effects.'

The first part of the problem was to consider what happened to CFCs as they rose upwards and whether there were any physical or chemical processes which would remove them. Their inertness and their very ubiquity, as demonstrated by Jim Lovelock, strongly suggested there weren't. Methodically, Molina came up with possible mechanisms to remove CFCs in the lower atmosphere, only to find they were untenable. They wouldn't rain out, nor did they seem to dissolve in the oceans. They determined that the CFCs would mix in throughout the atmosphere, taking five years or so to drift up to the stratosphere. 'The ozone layer not only protects us,' Sherry Rowland said, 'it also protects CFCs in the troposphere. Once we realised they would go up into the stratosphere, we then wondered what would happen to the chlorine.'

They determined that once CFCs reached the stratosphere they could remain there for up to 150 years. This meant, in effect, that 90 per cent of all the CFCs produced since Midgley's time would still be in the atmosphere. Rowland already knew that chlorine could destroy ozone in photolytic reactions, and when they found the correct catalytic chain, they were surprised at its length: one chlorine atom could effectively destroy 100,000 ozone molecules. It was only when Molina checked on the volume of CFCs produced annually (800,000 tonnes in 1974), that they realised the magnitude of the problem. Those CFCs would accumulate in the stratosphere for over a century, even if production levels were halted overnight. Plugging those enormous numbers into the catalytic chain, Molina then calculated that between 7 and 13 per cent of the total ozone would be eaten away over that hundred-year period. And production levels of CFCs were on the increase year by year, so the depletion would increase apace.

Molina came up with this staggering possibility just before Christmas 1973 and managed to contact Rowland just as he was preparing to leave for Vienna on another six-month sabbatical at the International Atomic Energy Authority. Rowland had been planning to use the sabbatical to ponder new areas of research, but ended up spending most of his time in Austria looking at the CFC issue. After Rowland had meticulously cross-checked Molina's calculations, they both agreed the results had to be published no matter how disturbing they were. That Christmas, when his wife,

Joan, asked him how the work with Mario was going, Rowland startled her by the starkness of his answer: 'It's going well but it looks like the end of the world.'

What was soon to be christened the Rowland-Molina hypothesis took shape in a scientific paper which they submitted to *Nature*. Yet, when it was published in June 1974 it elicited little reaction from the press. The Watergate hearings were in full swing and, besides, there was a feeling that there had been a surfeit of environmental doom and gloom stories. It was only when Rowland and Molina presented their findings at that year's meeting of the American Chemical Society in September, that they made a splash. It was the opening of hostilities in the Ozone Wars of the 1970s.

They had refined their calculations and found things to be even more serious than they had first thought. If CFC production increased at an average of 10 per cent per annum until 1990 (as it had from 1964 to 1974) and stabilised thereafter, between 5 and 7 per cent of the ozone would be depleted by 1995 and up to 10 per cent by the year 2050.

Spray cans accounted for about three-quarters of CFC emissions in the United States, whose chemical manufacturers accounted for about half the world production. Those numbers translated into a multimillion-dollar industry which was now at risk from what it perceived as irresponsible environmentalists. Certainly that was how Sherry Rowland viewed the debate, finding himself thrust into the role of scientific prophet of doom as chief defender of his own theory. 'That attitude was so prevalent that one simply couldn't win against it,' he said later. 'In a sense you were lucky to win against aerosol propellants.' It was, to be sure, an unequal battle, for industry spent vast sums on advertising, appealing to base common emotions. One advertisement declared in 1974: 'Don't give up on fluorocarbon aerosols YET! We believe in what US law holds clearly and we cherish dearly: you are innocent until proven guilty!' And that was to be the leitmotif of the years that followed: innocent until proven guilty.

In time, the debate polarised: in one corner were scientists, trying to learn more about a subject which was bewilderingly complex. In the other, manufacturers trying to play down the significance of the threat. The industrialists cast the atmospheric chemists as so many unrepentant Henny Pennies expecting the sky to fall. But the industrialists were prevaricating and procrastinating in the face of danger. As Sherry Rowland was to ask, with increasing desperation, over the next few years: 'What's the use of having developed a science well enough to make predictions if in the end all we're willing to do is stand around and wait for them to come true?'

Part of the problem was that there was very little consensus in the scientific community on the intricacies of the stratosphere. Computer models of the atmosphere could not reproduce its current workings, so how could they be gauged as reliable for predicting the future? Worse still, officially sanctioned reports convened by committees of one sort or another invariably tried to appease the scientific factions within them. As a result, their conclusions were of the bet-hedging variety, giving the appearance (sometimes erroneous, more often deliberate) of passing the buck. Accordingly, in late 1974, a committee with the grandiloquent title of the Committee on the Inadvertent Modification of the Stratosphere (which became known as IMOS) was charged as a presidential task force to look at the Rowland–Molina hypothesis. By June 1975, it had reported that CFCs were a 'legitimate cause for concern' but concluded that the federal government should defer action until the publication of a report then being carried out by another committee of twelve scientists convened by the US National Academy of Sciences.

The National Academy study showed just how complicated the science of ozone depletion was, best exemplified by what came to be known as the chlorine nitrate débâcle. In the merry-go-round of chemical reactions photolysed by ultraviolet radiation, chlorine nitrate had been thought of as a short-lived reservoir for active chlorine. But laboratory studies by Rowland's group at Irvine showed that it could exist for many hours, as opposed to the seconds which had been previously thought. The net effect was that not so much active chlorine would be available to eat at the ozone layer. When this data was plugged into the computer models of the stratosphere considered by the NAS study, it implied that a third less ozone than predicted by Rowland and Molina would be destroyed.

In early 1976, lurid rumours circulated that the Academy had had to rewrite the report as a result. In fact, an emergency meeting was called in Boulder to look at the issue. Different models produced different results depending on the assumptions built into them. Those that minimised the ozone loss had a mathematically convenient Sun which shone continuously at half strength. As chlorine nitrate is produced by the direct action of the Sun, the net result was to overproduce it. Other models didn't take into account that ultraviolet radiation is reflected from clouds back up to space, thereby doubling the flux passing through the stratosphere. This served to destroy ozone more efficiently exactly as Rowland and Molina had suggested. When factored in, the 7–13 per cent depletion remained.

The chlorine nitrate issue delayed the NAS report's publication until September 1976. It concluded that Rowland and Molina were essentially correct, but that government action should be postponed because there was not enough experimental evidence to warrant it. The curious result was that both environmentalists and industry could claim the NAS report as a victory. This, too, would be a recurring feature of the ozone story. Further National Academy reports, commissioned in 1979, 1982 and 1984, varied the total predicted ozone loss from 3 per cent to 16 per cent and all points between. 'After 1979 that 16 per cent kept shrinking as rate constants were re-measured,' Sherry Rowland remembered.

If anything, the public impression of stratospheric science was a morass of conflicting statements awash with contradictory evidence. The simple fact was stratospheric science was constrained by a data bottleneck: no new theories could be approved or demolished since very little new data was available to test them. For Jim Lovelock – who viewed the brouhaha in the United States with bemused contempt – the ultimate absurdity of the issue was when some of his data were rejected 'because they didn't fit into the models'. It was, he reflected, part and parcel of the 'vicious, competitive world of American science', the main motivation for which, in Lovelock's eyes, was 'prizes, glory and grandeur'.

The aerosol industry was the first to break ranks from other users of CFCs, partly because some of its number didn't use CFCs as propellants. In 1975, the Johnson Wax company declared its intention to stop using CFC propellants completely. Precipitate action by the Food and Drug Administration in October 1976, followed a week later by the Environmental Protection Agency, tipped the balance by decreeing that all CFC-propelled aerosols should be banned within two years. But early in 1978, the US government decided not to proceed with further bans in refrigeration, air conditioning, solvents and other industrial processes. The second NAS report a year later effectively forced their hand by saying that there should not be a wait-and-see policy. So in April 1980, the US Environmental Protection Agency proposed limiting CFC production to 1979 levels.

General distrust of the government in the wake of Watergate went hand in hand with a newfound eco-awareness in some parts of the United States. In the late 1970s, energy conservation and environmentalism had started to gain a political platform. A number of states enforced controls to reduce pollution. Oregon became the first state to ban the aerosols, quickly followed by New York State. By the end of the 1970s, however, a political sea change of far greater significance had taken place in America. If the first part of the Rowland–Molina saga had coincided with Nixon's disgrace and Watergate, its concluding years were crowned with the election of a former B-movie actor, swept to victory on a tide of nationalism and a reaffirmation of traditional values. In short, Ronald Wilson Reagan's ascendancy into the White House ensured that the EPA proposals were totally ignored. An administration that was elected on a platform of reducing governmental interference could hardly have done otherwise.

The incoming Reagan administration, however, went much further, containing as it did any number of free-marketeers to whom environmental legislation was anathema. Most vociferous was the secretary of the Environment himself, James G. Watt Jr, who seriously suggested that restrictions on national park development wouldn't be necessary as any damage would be swept away by the Second Coming. His first administrator at the EPA, Anne M. Burford, was less apocalyptic but nevertheless refused to meet with environmentalists for at least a year. At her confirmation hearings, she stated outright that in her opinion the CFC theory was controversial and that there was a 'need for additional scientific data before the international community would be willing to accept it as a basis for additional government action'.

Indeed, subsequent scientific assessments seemed to prove her suspicions were not entirely unfounded. A third National Academy report in 1982 lowered ozone depletion to 5–9 per cent and by February 1984, when its last report on the subject appeared, the NAS concluded that the figure would be 2–4 per cent. The consensus was that the distribution of ozone worldwide would change, rather than wholesale depletion taking place across the globe. By then, levels of CFC production worldwide had risen beyond those at which they had levelled out when the US aerosol ban came into force.

By the mid-1970s, balloon experiments had shown that CFCs were reaching the stratosphere, but was there any evidence for the end product of the Rowland–Molina hypothesis? It would be in the form of chlorine monoxide, a free radical that would eat something like 100,000 ozone molecules before the chain would be broken. The problem was that chlorine monoxide had no atomic transitions to be seen spectroscopically, so to determine its presence would require sampling in the stratosphere itself, not an easy task with the technology then available. 'The lower stratosphere is hard to reach,' Rowland said. 'Balloons on the whole didn't give you the detailed chemistry unless they were big payloads. We needed not just the stable molecules, but the radicals.'

One person who was very good at measuring those radicals was Jim Anderson, a tall, blond-haired scientist whose patience belied his tenacity and determination, often in the face of adversity. Like many of the key players in the unfolding story, he had started off doing something else. As an astrophysics student at the University of Colorado, he had been performing rocket-borne spectroscopy of the OH radical, sometimes called the Howard Hughes radical because of its elusiveness. 'In setting up the experiments I'd had to produce OH or hydroxyl in the laboratory,' he recalled. 'It ignited my interest in chemistry, which as an undergraduate I'd always found deadly!'

In 1970 he had moved to the University of Pittsburgh to pursue a PhD. There he impressed his supervisors with his ability to measure the OH radical in the laboratory. The technique he pioneered was the use of lasers to induce resonance within the electronic transitions of the radical he was investigating. By measuring how they fluoresced at X-ray wavelengths, he could determine one molecule in 10 billion in the laboratory. Unlike rocket or balloon experiments where 'you work for two years for forty seconds worth of data', Jim Anderson could quite contentedly perform experiments in the lab day after day. But he desperately wanted to use his new techniques in the field once he'd been granted his doctorate.

Like that of Lovelock in England, Anderson's work was not appreciated by his peers. When he applied to the Environmental Protection Agency for a grant for using the X-ray fluorescence technique in the field, he was turned down. 'They said it would be irresponsible to fund this sort of work as it had nothing to do with the real atmosphere.' Eventually, he was able to test it in the real atmosphere, armed with a NASA grant to measure radicals from balloons to characterise the detailed chemistry of the stratosphere in 1974. When chlorine monoxide became an issue owing to the interest in the Rowland–Molina hypothesis, Anderson's team adapted the X-ray fluorescence technique to measure it from balloons launched from the National Scientific Balloon Facility in Texas, which had been established in the dusty scrublands just outside Palestine in the 1960s for reasons not entirely scientific. LBJ was president and his Congressional cronies conspired to scrape one last offering from the political pork barrel.

After launch from Palestine, balloons would rise to an altitude of 45 kilometres, where a large instrument package would be lowered by parachute, taking *in situ* measurements. Five minutes' worth of observations would show whether free chlorine atoms and chlorine monoxide existed in the stratosphere. In five flights during 1976–78, Anderson showed that there were 2 parts per billion of chlorine monoxide, slightly higher than the 1 part per billion predicted from the breakdown of CFCs since Midgley's time. Other groups were finding the same range of numbers. But on Bastille Day 1977, Anderson made a startling observation: he measured 8 parts per billion. 'It caused me no end of anguish,' he said later. 'I didn't like what I saw, but I saw it. I knew I couldn't throw it out, though it would have made life easier.'

Instead of being taken as an obvious manifestation of human influence, this bizarre 14 July observation would come to be viewed as a red herring since it didn't fit the models of stratospheric chemistry. One of the myths that circulated around this incident was that his technicians had cleaned the instruments with a chlorinated solvent and didn't dare tell their boss. Anderson categorically denies this and today he believes it was probably as a result of additional chlorine deposition from meteors, a maximum of which occurs at that time of year.

If anything, the Bastille Day results showed the limitations of balloon measurements. Balloon experimenters knew that their observations were snapshots of small samples of the stratosphere. It was difficult to weave a seamless understanding of ozone variations from a patchwork quilt of balloon observations. It was also not the easiest task in the world to assess whether the observed ozone depletion resulted from naturally occurring dynamic variations or artificial chemicals. What was needed was nothing less than a balloon-led assault on the stratosphere, which NASA duly embarked upon in 1983. Known as the Balloon Intercomparison Campaign, it was planned to usher in a new generation of advanced instruments to measure chemical species – the generic description of the trace gases in all their multifarious guises – in the stratosphere and provide the most comprehensive data ever obtained on the chemistry of the ozone layer. Different techniques were used to measure the same chemical species to see which instruments were the most accurate.

In September 1982, Jim Anderson (by now transplanted to Harvard) had launched another technique which he hoped would become routine in quantifying ozone-related chemistry. From an altitude of 40 kilometres, a suite of instruments was lowered down by a tough Kevlar reel to take measurements at predetermined altitudes. Instead of grabbing molecules as the instruments dropped by, the new technique would measure them rather more sedately. During his first test flight, it worked remarkably well.

Fate, however, would ensure that it could not play a part in the ozone story. Unknown to Anderson, his balloon was one of the last built by reliable hands. A factory which made the polyethylene balloon skin changed ownership and many of the old employees left as well. The new employees hadn't mastered the tricky extrusion processes involved in the manufacturing process. It spelled disaster for the Balloon Intercomparison Campaign. One balloon failed to get off the ground, another flight was only partially successful and one free-fell from the stratosphere: an infrared spectrometer one storey high ended up as 30 centimetres of mangled metal in the Texas desert. Worse still, its lithium batteries had burst and dissolved most of the instrument's electronics.

By the mid-1980s, the confidence of American balloonists was, like the material upon which they depended, in tatters. For a three-year period no balloons with payloads greater than 700 kilos were launched. The result was that when no scientific balloons flew, metaphorically speaking, the balloon went up for ozone.

Towards the end of 1984, anyone passing the industrial estate at High Cross, on the northerly flanks of Cambridge, would have seen the lights burning late in an office in an otherwise nondescript modern block. Closer inspection would have revealed a solitary figure poring over print-outs and, as ever, surrounded by a small cloud of tobacco smoke. To the men and women who worked at the British Antarctic Survey it came as little surprise, for Joe Farman had always worked every hour that God sent. Now he rarely seemed to leave the office until very late – often after midnight – as he pondered a deeply puzzling phenomenon. The computer print-outs were unequivocal: if they were correct, something like half the ozone above Antarctica was disappearing each spring. In nearly three decades of work for the Survey, Farman had grown accustomed to the quirks of Dobson spectrophotometers. But nothing in his experience had prepared him for the substantial ozone losses that had been growing apace in the early 1980s. After checking and rechecking the data, it was time to publish. The year 1984 was one already loaded with emotional resonance, thanks to George Orwell, and the data in Farman's possession were Orwellian in their implications.

If anyone was going to find an anomaly in Antarctic ozone, the odds were that it would be Joe Farman. But he was surprised by the size of the anomaly he had stumbled across. He was well known as a meticulous observer and prided himself on only ever having passed a handful of the scientific papers out of the hundreds that had passed his desk for peer review. (A story that shows the way his mind worked comes from the late 1970s, when he worked at the BAS offices in Edinburgh. Each day he walked past a golf course on his way to work and idly wondered where lost golf balls would most likely end up. After thinking about it for a few days, he was rewarded one Sunday with 600 balls. 'If you have a curious mind you can work these things out,' he told me, with a grin, in 1990.)

Farman had devoted his life's work to the British Antarctic Survey, rising through the ranks until reaching the post of head of geophysics, where he became responsible for all atmospheric and geomagnetic observations carried out on 'the ice'. A small fraction of his time was to continue the work he had started way back in the IGY, collecting ozone observations from Halley Bay and, funds permitting, to analyse and archive the data. His perseverance would be rewarded with a deeply worrying discovery that would catapult the otherwise obscure Antarctic Survey into the forefront of research at a time when its very existence was threatened.

Farman had stayed aloof from the vitriolic arguments that had raged across the Atlantic over the issue of ozone depletion. Stratospheric research in Britain was still a pleasant, non-politicised backwater, compared to the media-fuelled controversies in the United States. Joe Farman, like his former mentor Gordon Dobson, was the keeper of the record. Twenty years after he had first gone down to 'the ice', Farman published a summary of results for the Royal Society in 1977. Ozone, he was at pains to point out, was naturally variable, and as yet there seemed to be little evidence for any large-scale changes. Farman concluded that 'it is natural to ask what changes could occur before they were identified in the observations. The answer depends to some extent on the nature of the change. One of, say, 10% over 10 days should be immediately obvious at low-latitude stations, yet could easily escape notice at a high-latitude station, especially in late winter or spring.'

Later he would realise that the ozone layer over Antarctica started to change in the austral spring of 1977. By the turn of the 1980s, stratospheric ozone started to show an appreciable loss each October. As ozone work was a minuscule part of his workload, Farman would often 'send my lads down to the ice', meaning the eager young recruits who worked under him at the Survey. In 1980, it was the turn of Jonathan Shanklin, who had recently graduated from Cambridge and was pursuing a PhD in atmospheric chemistry. The glamorous world of ozone spectrophotometry involved number crunching on a mind-numbing scale. 'We'd got lots of data but some of the numbers hadn't been converted into ozone values,' recalled Shanklin. In 1980, the ozone values in the spring looked suspiciously low. Shanklin enjoys telling the story that Farman's first reaction was: 'Don't worry about the low values.'

Maintaining the Dobson spectrophotometers was an arcane black art at the best of times, and out in Halley Bay, conditions made the antiquated equipment more temperamental. Farman decided the only course of action was the swift dispatch of a new instrument down to Halley Bay with the eager Shanklin to oversee its deployment. Swiftness, as Farman well knew, was of the essence. In 1981, the first major cutbacks from the Thatcher government were in the offing. There was an odds-on chance of no ship to go to the South Pole, let alone a new instrument, should he wait for another year. So towards the end of 1981 Shanklin tested the new instrument out on the ice at Halley Bay. He found that it was producing the same results as the older instrument. It was obvious something was happening in the stratosphere above his head.

Shanklin's return home was made unexpectedly hazardous by a plot hatched in the febrile minds of generals in Buenos Aires. By April 1982, the Falklands War had erupted and the British government declared an exclusion zone around the temporarily renamed Islas Malvinas. The Survey's ship *Bransfield* with Shanklin on board had to pass through that zone before the Task Force arrived. It could not refuel in Port Stanley as was customary. Shanklin and the precious ozone data made it. 'I've often wondered what would have happened had we suffered a similar fate to the *Belgrano*,' he said.

The 1982 data showed that ozone was indeed being stripped away with the return of sunlight. Shanklin wanted to publish the data as part of his PhD dissertation, but Farman persuaded him to hold fire for another two years. By 1984, it was abundantly clear that something was happening, yet Farman was worried because nobody else was seeing the springtime losses. With Shanklin and their BAS colleague Brian Gardiner, they went through the process of eliminating the various instrument malfunctions and other phenomena which might conspire to trick them. 'We didn't fully understand it,' Farman said, 'but we had to do something about it.'

One persistent criticism levelled at Farman is that he didn't publish sooner; he answers with the eminently sensible observation that they had to be confident of their results. 'Wolf had been cried so many times, we thought, Can another year matter?' A more serious criticism – in which one detects an element of sour grapes – is that he shouldn't have withheld his readings from the archives of the World Ozone Data Center, which had been established in Toronto in the 1970s. Worldwide observations from the Dobson stations were recorded in what were known as the 'red books'. The reason why he had not was that he wanted to analyse the data rigorously before he released them. 'Some of the critics have every right to complain,' he said ten years later. 'But they should have directed their complaints to the British government for not giving us the money to do the job at the speed necessary.'

By the end of 1984, it was clear that every possible error had been excluded. After a meeting with colleagues, Farman and his colleagues decided that they had to publish. A psychological boost had coincidentally come in the form of a paper in *Nature* by Michael Prather of NASA and others, arguing that ozone losses didn't necessarily have to be a uniform thinning of the global ozone shield. Distinct localised losses could occur, Prather *et al.* suggested. And Joe Farman had proof positive of a 'localised' phenomenon over at least Halley Bay and quite possibly the whole of the southern continent. Whereas Rowland and Molina maintained that most depletion would occur at 40 kilometres' altitude, his observations showed it was occurring at much lower altitudes. This implied there had to be a different, more efficient catalytic chain, which they duly suggested in a scientific paper submitted to *Nature*.

'Large losses of total ozone in Antarctica reveal seasonal ClOx/NOx interaction' chronicled Farman, Gardiner and Shanklin's deliberations in exhaustive detail. They were 'quite naughty', in Farman's phrase, to suggest that a chemical mechanism was entirely responsible. 'We were very dogmatic about it as there was no other way it could have been CFCs,' he explained. 'We were as certain as any scientists can be that here was a real, terrifying ozone change.' Farman dutifully travelled to London to drop their completed paper off at the offices of *Nature* on Christmas Eve 1984. It meant missing the Christmas party, which in the circumstances seemed like a small sacrifice. As he sat staring at the desolate countryside from the train on his way home, the wider implications of his observations struck him quite suddenly. 'We could see how the ozone hole could open up,' he said 'but what was most frightening, we couldn't see where it would end.'

An advance copy of Joe Farman's paper was sent to Susan Solomon at the National Oceanic and Atmospheric Administration's Aeronomy Laboratory in Boulder, Colorado, as part of the peer review process. Journals like *Nature* in the United Kingdom and *Science* in the United States are the most usual forum for reporting scientific advances. Before such reports are published, the counsel of other, supposedly dispassionate scientists is solicited to assess whether they truly represent new developments based on solid analysis. Then only twenty-nine years old, Susan Solomon had already established a reputation as one of the brightest atmospheric chemists of her generation. Farman's paper would elevate her to a key role in the ozone story. Instinctively, Susan Solomon could hardly believe that half the ozone over Antarctica disappeared each spring, but Joe Farman's work seemed to be 'solid'. She later recalled thinking: 'If this is true, it is the most important paper I've ever seen.' And she added: 'There wasn't any reason to suspect it wasn't true.'

Susan Solomon's interest in science began when she saw Jacques Cousteau's first appearance on American television in her native Chicago when she was nine. 'I thought it was the greatest thing,' she said. 'Science was something I *had* to do.' Almost immediately she declared to her parents that she wanted to be a marine biologist, later finding it was too qualitative for her liking. She pursued chemistry in high school and college. A decade later, while studying chemistry at the Illinois Institute of Technology, she learned that the measurements she was involved with would be important in theoretical studies of the atmosphere of Jupiter. This spurred her to become interested in atmospheric chemistry, a demanding area of study which was too Byzantine for most students. Susan Solomon simply adored it. 'Atmospheric chemistry is like organic chemistry in many ways,' she said. 'You have to have intuition at some level.

It's knowing that this reaction and that reaction are important but you have to forget the other five hundred.'

After studying at Berkeley under Harold Johnston, she completed her PhD at the National Center for Atmospheric Research in Boulder. In 1981, she joined the Aeronomy Lab operated by NOAA and has stayed ever since. Working with Rolando Garcia of NCAR, she had built up an impressive catalogue of two-dimensional models that had started to reproduce some of the workings of the atmosphere by combining dynamics and chemistry. But their work had not anticipated the evidence she had before her in February 1985 of a hole in the ozone layer over Antarctica. 'You can convince yourself it's something odd to stretch your imagination or disbelieve it,' she says. 'For me a much more positive attitude in science is if it talks like a duck, walks like a duck, then it probably is a duck.' She thought that something so glaring should also have been observed by others, and that the chemistry was not quite right, but the British Antarctic Survey had heard a duck quacking very loudly indeed. Farman's paper obviously had its basis in meticulously thorough analysis. With little hesitation, she recommended it should be published, and it appeared on 16 May 1985.

Farman's paper had the effect of making many of the scientists involved in stratospheric research feel that they were excluded from the field in which they were working. Jim Anderson's reaction was typical: 'Our sense of shock was profound. I cannot tell you how disturbing it was. We didn't have any idea of why it was so.' More than one American researcher was to mutter: 'What the hell is the British Antarctic Survey anyway?' Slowly but surely word spread that its work was faultless and steeped in meticulous analysis. Sherry Rowland met Brian Gardiner of the BAS at a conference in Hawaii, where the diffident Scotsman was amused to see that the American had a Xerox of their paper in his briefcase. Rowland suggested ways in which the Dobson instrument could have malfunctioned, and Gardiner explained that these possibilities had been eliminated. 'Clearly, this guy knew what he was doing,' Rowland said later.

Some of the suspicion exhibited by a number of American scientists, it must be said, was probably due to the simple fact that they hadn't picked the ozone hole up. Here they were receiving millions of dollars per year and yet they had been shocked to their collective core by a complete unknown who was having trouble scraping together the money needed to operate an antiquated instrument. Robert Watson of NASA Headquarters in Washington, DC, was particularly disconcerted, for not only had he orchestrated an international review of stratospheric ozone, it was his reponsibility to act as dispenser of funds to about three-quarters of the US community. 'Everyone was puzzled by Farman's results,' he admitted. 'But we didn't question the data.'

For Bob Watson, the British Antarctic Survey called up vague schoolboy associations with polar explorers, because he originally heralded from Romford in Essex. He is a short, loquacious man with a straggly beard and an engaging manner, who studied chemistry at Queen Mary College in the East End of London at the tail end of the 1960s. With bomb damage still apparent in the local environs from the wartime Blitz, his college days were barely touched by the influence of Swinging London. He too started off in laboratory chemistry, including the crucial gas-phase reactions involving bromine, chlorine and flourine, 'all of which would be central to the later ozone work'. But that wasn't his motivation for studying such a 'really esoteric field. At that time, I didn't really know what the ozone layer was.'

In the early 1970s, he crossed the Atlantic to become a postdoctoral researcher at Berkeley under Hal Johnston's tutelage, eventually ending up at the Jet Propulsion Laboratory in Pasadena in 1975. Operated by the California Institute of Technology for NASA, JPL was unusual in that, although it wasn't formally part of the civil service, its funding came direct from the federal government. JPL had a reputation for excellence in science and had a continuing programme of work in gas-phase chemistry. (Mario Molina had also moved there in the late 1970s.) 'If you didn't want to teach it was a great place,' Watson said, 'and our research group was really first class.' At the end of 1980, he moved to NASA Headquarters to head the space agency's research efforts into the stratosphere. 'My main task was to ensure that there was a critical mass of lab science, field work and theory,' he says. 'Ozone was a hot potato, politically, and we needed to really concentrate our efforts.' (Watson's singular contribution to the subject will be examined in Chapter 9.)

How Soon Is Now?

The Farman paper was disturbing to Watson because NASA supposedly had the wherewithal to spot large-scale depletions of ozone in the form of a dedicated ozone mapper aboard its Nimbus 7 satellite launched in 1978. Known as the Total Ozone Mapping Spectrometer, or TOMS, it should have seen the springtime dip in ozone over the South Pole. Watson soon learned that colleagues at the Goddard Space Flight Center on the northern outskirts of the capital (who were responsible for TOMS) had found something. There is a myth – often perpetuated by NASA itself – that space technology is miraculous and can automatically solve problems at the touch of a button. If there were problems in operating ground-based instruments, they were minuscule compared to those encountered in the day-to-day operations of a satellite.

TOMS worked much like the Dobson spectrophotometer, by determining how much ultraviolet was selectively absorbed at wavelengths characteristic of ozone. But from its perspective 1,000 kilometres above the Earth, it could see the whole of the Earth and returned some 196,000 observations each day. It would obviously not be possible for humans to crunch all those numbers, so a computer programme was developed to do the task. It flagged suspicious or spurious readings: one flag in particular automatically threw out readings below 180 Dobson units.

By the summer of 1985 it had become clear that the TOMS had been seeing suspiciously low readings in springtime over Antarctica, but that the computer had simply thrown the numbers out. 'We should have realised the ozone hole was there in the data by looking at the rejections,' Watson said. 'There were more and more suspicious data points at higher latitudes. In 1980, perhaps 1 per cent were being rejected, perhaps 5 per cent the next year, then by 1984, it was nearing 20 per cent.' When the TOMS data were suitably reanalysed, a global time-lapse movie was assembled which soon played on all the TV news shows. Soon the swirling, psychedelic amoeba – as large as the United States and as deep as Mount Everest is high – became ingrained in the public consciousness, helped by the simplistic verbal gimmickry attendant in the phrase 'ozone hole'. TOMS proved Farman to have been correct. For Watson, the fact that something as high-tech as a satellite had screwed up where ground-based observations hadn't was 'an important lesson and NASA has learned from it'.

Hindsight is a wonderful thing. When the various scientific records were re-examined, a clear picture emerged. It was later realised that the US-operated Dobson instrument at the South Pole had been miscalibrated and otherwise would have alerted the NASA Goddard people to double-check their TOMS data. Ozone sondes, small rubber balloons which measure the presence of ozone by chemical means, also revealed a decrease in ozone around the South Pole. In 1984, a Japanese researcher had also found evidence for suspiciously low ozone over Antarctica; he presented his findings at a meeting in Greece early the next year.

As a result of Farman's observations, it had struck Sherry Rowland just how useful Dobson instruments were, despite their reputation as being antiquated and unreliable. Since the mid-1970s, researchers had been trying to determine from the 'red books' if there were any changes in global ozone. The evidence was hard to find among the seasonal variations, and some observations seemed to be showing that ozone levels had been increasing in certain parts of the world. The fact that the ozone hole over Antarctica was seasonal had led Sherry Rowland to suspect that the Dobson record was worthy of re-examination. The obvious place to start was Arosa in Switzerland, from where measurements had been made continuously since the 1930s.

Neil Harris, who had joined Rowland as a graduate student from Oxford, set about looking at the record. Previous analyses had shown very little change, but, Harris realised, this was because people had been looking at yearly averages and not the monthly averages. 'From the monthly means it was clear that the first half of the year was markedly different from the second half,' he explained. 'There was a winter decrease relative to the summer months.' Though it was only of the order of a few per cent, 'there was a signal there,' said Sherry Rowland, 'and there was an obvious loss of ozone in the northern hemisphere'.

Clearly the Dobson record was valuable. At Bob Watson's behest, satellite and ground-based Dobson data would be combined to reveal unequivocally just how serious that problem was – and now it was happening over most people's heads. Exactly as Sherry

Rowland had said a decade before, the ozone layer around the world was slowly but surely starting to strip away.

If the ozone hole was the most frightening example of mankind's polluting influence, then mother-of-pearl clouds were a pleasing counterpoint – as beautiful and aesthetically appealing as anything to be found in nature. As far back as the 1800s, astronomers in Sweden had reported sightings of eerie, pink coloured clouds that could be seen against the dark of the long winter night. Though their explanation remained elusive, they were believed to be made of water-ice crystals. Their significance in the ozone story would far outshine their ephemeral beauty.

All stratospheric researchers agreed on one thing: Joe Farman's chemical mechanisms were wrong. The problem was that, as Susan Solomon realised, in the lower stratosphere the sort of photochemical mechanisms that he had proposed were not very active. She plugged Farman's data into her 2-D models to see if there was any way of getting the numbers right. 'I wrote to Joe and said: "I can produce ozone change higher up if I fool around with your mechanism,"' she remembered, '"but not at the lower region. What could cause it over the 10–25 kilometre height range?"'

The answer came to her at the annual meeting of the American Geophysical Union held in San Francisco in December 1985. By the mid-1980s, this had become a large jamboree with discussion of topics ranging from the internal workings of suboceanic volcanoes to weather observations on other planets. The aeronomy sessions were always lively and often spurred further research. When David Hofmann, a veteran balloon experimenter from the University of Wyoming, presented some balloon profiles on the existence of mother-of-pearl clouds, Solomon was astonished. In modern science, they were known rather more prosaically as polar stratospheric clouds (PSCs). What amazed Solomon was that their vertical profile was the inverse of the profile of the distribution of ozone loss. 'What happens if something happens on those clouds?' she thought.

Solomon also remembered an intriguing set of experiments initiated by Sherry Rowland's Irvine group in the early 1980s. Hydrogen chloride (HCl) and chlorine nitrate ($ClNO_3$) are two of the most obvious reservoirs for chlorine in the stratosphere, tying up the dangerous chlorine monoxide free radical, often termed the 'smoking gun' in ozone depletion. Could they react with each other? Laboratory experiments performed by Haruo Sato indicated that they did. When chlorine nitrate was placed in a vessel and HCl was released from a side arm, the chlorine nitrate disappeared within two seconds. When those numbers were plugged into a computer model of the Rowland–Molina hypothesis developed by Don Wuebbles at the Lawrence Livermore Laboratory, the ozone depletion went from 4 per cent to 25 per cent worldwide. But it had dawned on Rowland that this new reaction involving chlorine nitrate was taking place on the walls of the reaction vessel. 'We knew it would go on most surfaces,' Rowland said of the reaction, 'but there was nothing upon which to make it go.'

Except, as Solomon realised, polar stratospheric clouds could quite easily oblige over the poles. For modellers it meant opening up a new can of worms generically known as 'heterogeneous chemistry'. To date, stratospheric modelling had just considered gas-phase chemistry, involving reactions between individual gaseous molecules. Heterogeneous chemistry was something altogether more difficult to measure, involving reactions between gases and solid particles that would be near impossible to model in the laboratory. Yet it might explain why half the ozone over Antarctica disappeared in the springtime.

Accordingly, Susan Solomon and her partner in modelling, Rolando Garcia, pooled their experience with Rowland and Wuebbles to produce a paper that took into account heterogeneous chemistry. A similar set of calculations by a group at Harvard appeared in the same issue of *Nature*, thirteen months after Farman's original announcement. The Harvard group, led by Michael McElroy, a combative researcher who had once called science a 'contact sport' in an interview (and thereby aroused the suspicions of colleagues in perpetuity), suggested that something more than chlorine was needed to destroy so much ozone and added bromine into the picture.

In the end, neither team got the mechanisms exactly right, though heterogeneous chemistry was found to be responsible exactly as Susan Solomon had suggested. 'The contribution I made was to point out that this could be happening on PSCs,' she said. 'I didn't get the right catalytic cycle.' That would be determined later in 1986 by Mario Molina, who had gained a fearsome reputation for modelling ice in his lab at the Jet Propulsion Laboratory. He showed that the smoking gun was double-barrelled: chlorine monoxide reacted with itself to form what is known as a dimer, which was even more voracious at eating ozone. When this modification was plugged into the models, the predictions tallied with observations.

So within a year of Farman's discovery it was clear that polar stratospheric clouds could provide the surfaces upon which chlorine would be activated and from which ozone would be at risk. It was equally clear to chemists like Susan Solomon that the only way to find out more was to see for themselves.

Antarctica is as dangerous and forbidding as it is remote and mysterious. And to the group of fifteen US scientists huddled in a Hercules transporter in August 1986, the southern continent was certainly at its most perilous. It was a measure of the seriousness of the ozone problem that they were on a win-fly (winter fly-in) to try to see exactly what happened in the Antarctic stratosphere as the ozone hole opened up.

Their mission, known by the acronym of NOZE (which stood for National OZone Expedition), had only taken five months to put together. It had resulted from a meeting in Boulder the previous March, where ground-based observations of the ozone layer were under discussion. It had become clear to Bob Watson that the only way to get good, hard evidence on the ozone hole was to make measurements in Antarctica. When he first mentioned the idea in Boulder he realised he was coming across like Mickey Rooney in an Andy Hardy movie – 'I know, let's solve the ozone problem by going to the South Pole!' – but at least none of the attendees thought it was a ridiculous suggestion. And it was thanks to his determination to cut swathes through Washington bureaucracy that NOZE reached the ice that austral winter. Susan Solomon, to her own astonishment, found herself in charge. 'Prior to 1986 I had never touched an instrument other than a keyboard,' she says. And before the meeting in Boulder she had had little interest in visiting the South Pole at all.

CRISIS! WHAT CRISIS?

It is curious to speculate that, had she been a subject of Her Britannic Majesty, she never could have gone to the pole: even today women are expressly forbidden at Halley Bay in wintertime. In the world of American science, her sex was no hindrance, though it raised eyebrows en route to the pole in New Zealand, the normal springboard for American flights south. At the airfield in Christchurch, the pilot of the Hercules C-130 which would take them to the base at McMurdo Sound asked who was in charge. Much to Susan Solomon's amusement, his face registered disbelief when she put her hand up. It turned to acquiescence in the space of a few seconds, and he exclaimed: 'Good for you!'

As leader of NOZE, Solomon would be in charge of fourteen other scientists, only one of whom had ever been to the pole before. David Hofmann would, as he had been doing for over fifteen years, fly balloons to sample exactly what was happening inside the ozone layer. Barney Farmer of the Jet Propulsion Laboratory had adapted his design for a large infrared spectrometer which had flown the year before on the Space Shuttle, and built a new version which could sample many species associated with ozone loss from the ground. Nobody was at all sure if any of the hastily cobbled-together instruments would work in the freezing cold of an austral winter at the pole.

Ultimately, NOZE was looking for chlorine monoxide, which would directly implicate CFCs. Chlorine monoxide had no atomic transitions capable of spectroscopic detection, but it did react with lone oxygen atoms to form OClO, or chlorine dioxide. So Susan Solomon was using a spectrometer developed at the NOZAA Aeronomy Lab which used the Moon as a light source to measure levels of nitrogen oxides and chlorine dioxide. Catching moonbeams sounds romantic, but not in Antarctica where temperatures were on average $-40°C$. Yet she somehow overcame these handicaps and later recalled that 'I was colder in Chicago than I ever was in Antarctica because they give you such good parkas and boots'.

Other techniques employed during NOZE included the first attempts at measuring chlorine monoxide from the polar ice cap. Bob De Zafra and Philip Solomon (no relation) from the State University of New York had developed a microwave receiver which had been used at mid-latitudes. They tuned a microwave oscillator to listen to the frequency at which it was known that the chlorine monoxide radical rotated. As the hole opened, they expected to hear an increasingly larger signal.

This, then, was the group of personnel charged with the task of diagnosing what was happening to the ozone layer above their heads. They all knew that chlorine chemistry might not be totally to blame, and there were two other hypotheses which NOZE would have to test. The first was known as the dynamic hypothesis and at its heart was the idea that ozone-poor air from the tropics would rise upwards from the troposphere, fan out through the polar vortex and effectively reduce ozone levels as seen from the ground. This ozone-poor air would be richer in methane and nitrogen oxides, which could be used as convenient markers for the motion of air from the equator. There was some evidence that temperatures in the Antarctic stratosphere had been steadily decreasing, thereby increasing the likelihood of this upwelling of ozone-poor air.

The second was known as the odd nitrogen hypothesis and had as its cause the high levels of nitrogen oxides that result from increased solar activity. Every eleven years the Sun undergoes distinct cycles of activity: 1979 had been a particularly active maximum and it could have triggered catalytic cycles involving NOx to produce observed levels of ozone loss. Over subsequent years, these odd nitrogen radicals would gradually drift into the stratosphere and possibly deplete the ozone layer. As theories went it was a rank outsider and to some it had as much going for it as a headline which had graced a supermarket tabloid with a photograph of the Hollywood extra terrestrial ET below it: ALIENS DESTROY OUR OZONE LAYER. Nevertheless, the natural course of science involved testing such hypotheses, no matter how wild or zany they might seem.

No abnormally high levels of methane or nitrogen oxides were found in the stratosphere; in fact, measurements of nitrogen oxides generally revealed that their levels were among the lowest ever recorded and thus acted to enhance ozone depletion. Hofmann's balloon flight – thirty-three in all – showed that atmospheric particles seemed to be falling, not rising, which was hard to reconcile with the flow of ozone-poor air from elsewhere. Other observations showed that the total amount of chlorine bound up in reservoirs such as HCl or chlorine nitrate increased from September to October, possibly from the evaporation of condensed chlorine compounds in polar stratospheric clouds as temperatures warmed. The net result was that the Antarctic stratosphere was highly perturbed, chemically speaking, and the total chlorine in the gaseous reservoirs seemed to have increased as the vortex broke up.

The measurements of the two oxides of chlorine revealed there was a hundred times more active chlorine in the stratosphere above McMurdo than anywhere else in the world. It was difficult to see how such elevated levels of ClO could be formed without the breakdown from CFCs. Barney Farmer's observations showed that there seemed to be three times as much hydrofluoric acid, which results entirely from industrial emissions. In a paper published the following year in *Nature*, Farmer concluded: 'While this is not proof of a cause and effect relationship between halogenated source gases and the springtime ozone depletion, it is entirely consistent with such hypotheses.'

And the NOZE scientists said as much during a press conference, radioed from McMurdo Sound to the National Science Foundation in Washington in late October 1986, once they were happy with their initial analyses of data. They were, nevertheless, cautious and simply said that some sort of chemistry was responsible. The observations strongly discounted the odd-nitrogen hypothesis, and made it difficult to reconcile dynamics as the cause. Almost immediately the dynamicists and odd-nitrogen theorists grumbled. Most of their criticisms were along the lines of 'send chemists to the South Pole and of course chemistry will be to blame' but when fully analysed, the tentative NOZE conclusions were correct. 'When I look back at the job we did I can honestly say we weren't too far wrong,' Susan Solomon reflected.

For her, at least, the strangest memory from NOZE served as a comic counterpoint to the danger of the cold and the exhilaration of the work. On her first day at the McMurdo Sound base, as she changed in the sparse accommodation, she heard the sound of footsteps screeching in the snow outside. It reminded her of winters back in Chicago. Standing in thermal underwear and bra, she slowly realised that there was a peeping tom outside. 'The really amazing thing about it is how long he must have been out there,' she said. 'At forty below you don't get to stand out there just to peep!'

Scientifically speaking, the NOZE scientists were themselves nothing more than peeping toms, peering as they did through the bulk of the atmosphere into the rarefied heights of the stratosphere. It was clear that the next step forward would have to be nothing less than a direct assault.

Adrian Tuck was reflecting on his work and its increasing frustrations at home one day in 1986 when a transatlantic phone call came which changed his life for ever. For fourteen years, he had worked for the UK Meteorological Office, the organisation charged with handling weather forecasting in Britain. He was unusual in that he had started off as a chemist but had drifted into theoretical modelling and meteorology. After a doctorate at Cambridge, a Fulbright scholarship in San Diego followed (as did marriage). Returning home, he nearly didn't become a researcher at all. In the early 1970s he spent eighteen depressing months at University College London holding a 'supposedly prestigious research fellowship that wasn't index-linked to the tremendous inflation of the day. I was about to become a tax inspector as I had a wife and kid to support.'

Tuck had applied to work for the Met Office, though he knew full well that it recruited through the tortuously slow procedures of the civil service. But someone, somewhere, had seen his application, and Tuck was asked to join almost immediately. He loved it and worked through as full a gamut of research opportunities as the Met Office could offer: first as a gas-phase chemist, retraining as a meteorologist and then as an experiment builder. This experience would hold him in good stead in assessing the perpetual arguments between the dynamicists and the chemists in the stratospheric ozone debate. In 1985, he too had been called to cast his critical, peer-reviewing eye over the discovery of the ozone hole. 'Joe Farman sent me a preprint of his paper and I said: "I can't see anything wrong with this. You've got to publish – but I think your chemical interpretation is wrong."'

By this time, however, Tuck had begun to feel a marked sense of *déjà vu*. Under Margaret Thatcher's premiership, those who carried out scientific work in Great Britain had become ever more demoralised. A leader who prided herself on being a scientist (her greatest contribution to the sum of human knowledge was finding that ice cream tastes better if it contains more sugar) had created a sour legacy for her former colleagues. Science funding had haemorrhaged, leading to an undermining at the core of research. Tuck, of course, had dual nationality through his marriage, and the brain drain beckoned. When the NOAA Aeronomy Lab started to sound out the possibility of his emigrating, he jumped at the chance. He was sorry to leave. 'It wasn't the salary,' he explained. 'It was the demoralisation, especially of the younger scientists, and the certain lack of future resources in Britain.'

His departure was in motion that summer of 1986 when Bob Watson reached him at home in Wokingham, Berkshire. 'Because of the time difference I was sitting in a deck chair sipping away,' Tuck recalled. 'Basically he said: "Are you interested in being project scientist for aircraft missions to Antarctica?" I replied that I hadn't even arrived yet, but Bob wanted me for my aircraft experience and background.' In this way Adrian Tuck came to be the project scientist for the Antarctic Airborne Ozone Experiment, which NASA would embark upon the following year.

In his role as manager of NASA's upper atmosphere research programme, Watson oversaw a small fleet of research aircraft for the purposes of stratospheric research. They included a refitted DC-8 from Braniff Airways and three ER-2s, civilian derivatives of the illustrious U-2 high-altitude spy plane. AAOE would employ both in a fully fledged campaign which, unlike NOZE, would have the luxury of months of preparation. For the first time, aircraft would fly into the freezing cold of the Antarctic vortex. 'Nobody had ever attempted anything like it,' Watson said. 'But it was the logical next step.' Adrian Tuck's role would be to coordinate and plan the aircraft flights, something for which his Met Office experience had relevance, as Watson well knew.

'If we were going to do something as dumb as flying a wind-sensitive aircraft like the ER-2 from Cape Horn in winter,' Tuck said, 'then the forecasting was going to be crucial.' In this regard, freshly transported to Colorado, he arranged for two of his former Met Office colleagues to be on hand to do the forecasting. Peter

Salter and David Merrick became attached to the AAOE to make sure that the pilots and experimenters had the best chances of safety and scientific return. Salter had had experience in that part of the world: during the Falklands War he had been chief forecaster. This was the nearest experience anyone had had in understanding the complicated wind systems which act as a circumpolar collar around the southern landmass.

The choice of airport was crucial. A NASA team led by Bob Watson flew south to reconnoitre possible sites, shepherded through the antagonistic politics of South America by State Department officials. Argentinian Air Force bases were the most obvious choice, but to keep the Chileans happy they added Punta Arenas on the Straits of Magellan to the list. As the NASA aircraft would fly over Chilean airspace it was at least prudent to include it in the intinerary. For generations of mariners, the roaring forties had become a familiar hardship to endure in and around Cape Horn. Within a few weeks the NASA team would also be forced to take it into consideration for the Antarctic Airborne Ozone Experiment. As they inspected each new site in Argentina, the windiness of the bases seemed to increase. The ER-2 would need very calm conditions to take off and land as it was susceptible to cross winds (its wingspan is greater than its length). At one site which didn't seem too windy, an Argentinian colonel explained that some aircraft left outside had to have two-tonne forklift trucks strapped to them overnight or else they'd literally take off.

With an increasing sense of desperation, the NASA team reached the world's most southerly city, Punta Arenas, nestling in the Straits of Magellan across from Tierra del Fuego. It had the supposed disadvantage of being contained within the 'ferocious fifties', but it soon became clear that it was shielded by surrounding mountains and not nearly as bad as had been originally feared. 'It was ideal because it had adequate hangars and an adequate runway, both of which are dictated by the ER-2 operations,' Tuck remembered.

Both factors effectively clinched the choice of the city on which Tuck and his colleagues descended in August 1987. To their surprise, Punta Arenas turned out to be more like a pleasant Massachusetts seaport than a South American city. Outside the Cabo de Hornos hotel where they were staying, a statue of Magellan could be seen, which was said to bring luck to explorers. Luck was something AAOE would need, as Adrian Tuck well knew. He realised he was only as far south as his Norfolk birthplace was north – 53° – and yet they were the nearest to the ozone hole they possibly could be to sample it directly. Like the explorers of old they had little idea what they might find or of the dangers the pilots would have to face.

By the summer of 1987, the debate between the chemists and the dynamicists over the origins of the Antarctic ozone hole had reached a piercing crescendo. Though disagreements are part and parcel of scientific enquiry, ozone science had become so polarised that it took the form of an intellectual prizefight. And that was certainly how the media chose to portray it, often playing colleagues off against each other. One of the more vociferous dynamicists, NASA's Mark Schoerbel, led a Cassandra-like chorus against chemistry alone by telling *Atlantic* magazine: 'I'm amazed at how much people have lost their scientific objectivity because of political and funding pressures.'

For the AAOE scientists holed up at the Aeropuerto Presidente Carlos Ibañez, the freezing wind and cold were of greater immediate concern. Typically, the AAOE scientists would work a disorienting schedule dictated by trying to get the ER-2s into the air at around 9 a.m. The scientists and ground crew would have to arrive at 3 a.m. to prepare the instruments for flight. At 6.30 a.m., Tuck would brief the pilots, who would then have to suit up in readiness for the flights. At the tip of South America, there were precious few observations upon which to make accurate weather forecasts. It could lead to further frustrations, as Tuck would recall: 'There were days when there'd be a flat calm and we'd issue a forecast for 30 knots. It would actually come in at 57 knots. Day after day we'd be having to abort.'

At the start of the seven-week campaign, Adrian Tuck estimated that they would get six flights of the ER-2 (its chief pilot, Ron Williams, privately estimated three). In the end, they got a dozen flights, timed roughly every five or six days to coincide with the rotation of the vortex. Before the ozone hole opened up, polar stratospheric clouds would form in the freezing cold of the stratosphere and give rise to localised ozone depletion when the Sun fired the photochemistry. By timing the ER-2 flights with the rotation of the vortex, they would sample roughly the same air parcels to see what had happened to it since last they had passed through it. In this way, the AAOE scientists hoped to see what sort of large-scale chemical processing was developing as the Sun returned and fired the starting gun for wholesale stripping of the ozone over Antarctica.

In this they were aided by flights of the NASA DC-8 which looked up at the vortex from below, helping to define the background conditions against which ozone depletion would result. Susan Solomon and others returned to Antarctica for NOZE II to provide similar information. Whether rightly or wrongly, the ER-2s would get the most attention because they would provide the clincher – so far as CFCs were concerned. By directly flying into the vortex they would obtain unequivocal proof of the role of chlorine chemistry. To say the least, flying single-engine planes over Antarctica was an exercise fraught with danger. The three ER-2 pilots expressed the concern that their flights were more dangerous than those they had carried out during the Vietnam War. Engine failure or aircraft problems would condemn them to death in the freezing cold. But as the ER-2 was the only aircraft capable of reaching the place where all the 'chemical sex and violence' (in Adrian Tuck's vivid description of the lower stratosphere) occurred, they were willing to take the risk.

Once the ER-2 disappeared off into the skies beyond Tierra del Fuego, there was no radar coverage. The pilots were on their own. On more than a few occasions, as they nosed into the freezing mass of air of the Antarctic vortex, the ER-2 fuel lines started to freeze over. They had flown virtually every aircraft in all weathers, but the eerie experience of a misty haze above 12,000 metres was something new; they were the first pilots in the world to fly through polar stratospheric clouds. Despite the odds against success, they obtained exactly the data the scientists wanted. Five years later, AAOE veterans could not praise the pilots too highly. 'In some sense we were putting our scientific careers on the line,' said Adrian Tuck. 'They were putting their necks on the line.'

On 23 August 1987, the ER-2 found exactly what the Antarctic Airborne Ozone Experiment had set out to look for.

* * *

Jim Anderson had also been tracked down by the indefatigable Bob Watson the previous year. After the frustrations of the balloon failures, things were picking up and his Harvard team had been working flat out on developing a computer-controlled laser system to measure the presence of his first love, the hydroxyl radical. For the first time since he could remember, Anderson had finally managed a month-long vacation at his retreat in Idaho.

Whereas the expedition to the pole, NOZE, had been put together with what instruments had been readily available, the Airborne Antarctic Ozone Experiment had the luxury of time to develop instruments that could conclusively demonstrate chlorine chemistry as the culprit. As Watson fully realised, Jim Anderson's participation in the AAOE was crucial. He was one of the few scientists in the world who had experience of measuring that other *bête noire* in the fraternity of free radicals, chlorine monoxide. Correlating increased levels of the 'smoking gun' with distinct ozone loss would prove beyond all doubt that chlorine chemistry had punctured the ozone layer. Watson would set him an almost Herculean task: to design, build, test and successfully operate a new instrument within a year, even though he had never flown anything on an aircraft before. It was, he says today, 'a crazy proposition', but Anderson nevertheless agreed to participate because of its importance.

He had a number of aces up his sleeve. His former college roommate, Paul Soderman, was a dynamic modeller at NASA's Ames Research Center in California, from where the ER-2s operated. They had collaborated before on determining how air would flow around balloon packages dropped from the stratosphere. Now they would consider the altogether more difficult problem presented by the air flow around an ER-2. Because of the speed of the plane, there was a chance that the radicals would be destroyed even before they were sampled. After obtaining plans for the plane, Anderson informally consulted Soderman on where the inlet for the instrument should be. 'The only way to do those measurements is to slow down the flow,' Anderson remembered. 'It was a significant calculation to work out where the snout was. If the angle was wrong, you'd get turbulence and mix the air sample and destroy the free radicals.' What resulted was a 180-centimetre-long instrument which fitted under the ER-2's wing with a large primary inlet 12 centimetres in diameter. Contained within it was a concentric inlet, into which the pristine radicals would flow – or that was what theory predicted.

And, surprisingly, it worked. When Anderson arrived at Ames in May, the ClO instrument was a box of parts with no software written. Within three weeks, it was operating. 'We'd faced a lot of the problems in developing the balloon instruments,' Anderson said. On its first test flight, ClO was measured over California, and again on the first test flight out of Punta Arenas. Although there were nearly a dozen instruments aboard the ER-2, Anderson's would provide the most irrefutable proof of humanity's ultimate responsibility in creating the ozone hole.

On 17 August, the ER-2 took off into the unknown from Punta Arenas. When it returned, Anderson found to his horror the instrument had failed before the ER-2 had even reached the vortex. As possible culprits there were a couple of hundred connectors and thousands of electrical wires to choose from. But computer software diagnosed the problem as a 3-millimetre electrical connector whose central pin had failed in the freezing cold. Six days later, with the recalcitrant connector fixed, the next attempt to sample the ozone layer within the vortex was made.

Anderson's precious data were carefully unloaded and displayed on the computer screens within two hours of touchdown. It was late, freezing cold and, as Anderson later recalled, 'the hangar was rattling with winds up to fifty knots'. He felt his stomach tighten as the computer console displayed data in its undigested form – the number of ClO radicals present could be discerned by the number of 'hits' scored by fluorescing photons. The number of counts displayed would be a rough index to what was happening over the South Pole. The range of the numbers was shown: the minimum was one count recorded over Punta Arenas after take-off. Anderson 'reeled backwards' when he saw the maximum was 10,600 counts. And twenty seconds later, a trace showed the rise coincided with the arrival of the ER-2 in the vortex.

On subsequent flights, similar results were shown. As time passed on into spring, the ozone hole had opened and depletion within the vortex had intensified. This was corroborated from the DC-8, the NOZE II scientists at the South Pole and TOMS. But the ClO measurements clinched the role of CFCs. For Anderson, the defining moment of the AAOE occurred on 16 September, when the ER-2 entered the vortex and almost instantaneously returned measurements which revealed a steep wall of ClO compared to ozone, which had equally rapidly declined. 'We can talk about free radicals, spectroscopy, rate limited reactions till we're blue in the face,' said Anderson. 'Mention any one of those words in front of a Senate panel that doesn't include Al Gore and their eyes will glaze over. But show them the anti-correlation between ozone and chlorine monoxide and it has a huge impact.'

CFCs were the cause of ozone depletion, a result which underscored the timeliness of the signing of the Montreal Protocol on Substances That Deplete the Ozone Layer a few days earlier. Human beings had unequivocally opened up the ozone hole, and finally they were taking responsibility for it. The AAOE results would be used as the scientific basis for amendments to this legislation by bringing forward the timetable of phasing out the CFCs. Though it would take many months of analysis before they appeared in print, they implied something immediate and serious. Collectively, people have an almost infinite capacity to behave like ostriches, as the ozone story had shown. Antarctica had seemed very remote from the everyday occupations of most of the world's population. But the results from the AAOE spelled out that the ozone-depleting chemistry happening over the South Pole could quite easily occur over the north.

Would the same large-scale chemical destruction of ozone occur in and around the Arctic vortex? There was really only one way to find out.

There can hardly be a worse time to start work than 8 a.m. on New Year's Day, nor a less attractive place than Stavanger Airport on the Norwegian coast. But that was where the scientists of the Arctic Airborne Stratospheric Expedition found themselves that first morning of 1989. This latest mission, administered by Bob Watson and directed by Adrian Tuck, would once again involve impressive logistics: 300 scientists from a dozen research organisations, sixteen DC-8 flights and one less by the ER-2.

Whereas Chile had seemed difficult to reach, Norway presented much less of a psychological barrier. But there was no less pressure on the team, because of the more complicated atmospheric flows around the Arctic. There were more theorists in evidence whose wits would be pitted against the atmosphere, including those of Susan Solomon. To her, at least, trying to diagnose whether Arctic ozone variations were caused by chemistry or dynamics was like 'trying to take a drink of water from a gusher. So much is coming out that it's hard to take it all in.'

Their task was aided by a quirk of weather. Statistically, it was the warmest and windiest January ever in Stavanger, and minimum temperatures in the stratosphere were their coldest for thirty years. This was as the theorists expected, for if the surface is warm, the upper atmosphere cools to maintain a thermal equilibrium. What this meant was that polar stratospheric clouds formed before Christmas and during January and were present in numbers above the statistical average for the previous decade. In other words, the Arctic was primed for more intense chemical processing than the AASE scientists had originally thought.

By 1989, understanding of the exact mechanisms of ozone loss over the Antarctic had been matched by laboratory experiments with ice. There seemed to be two sorts of polar stratospheric clouds, whose occurrence was dictated by temperature. The first, known as PSC type 1s, form when nitric acid freezes out at $-77°C$ and ties up oxides of nitrogen which would otherwise be available to neutralise active chlorine. At $-85°C$, water ice freezes completely, causing the second type of PSCs, which are much larger and thus drop out of the stratosphere much more efficiently. When these icy particles fall Earthwards, a greater part of the atmosphere's natural immunity against ozone depletion is removed. In Antarctica, both processes – denitrification and dehydrification – occur at the same time. Because there is about a thousand times more water than nitrogen oxides available in the stratosphere, the water ice would form around nitric acid nuclei, after which ozone would be lost when sunlight became available to catalyse it.

'The biggest surprise out of Stavanger was denitrification on its own,' says Adrian Tuck. AASE measurements showed that nitric acid cloud particles could release active chlorine far more efficiently than had been thought. Later ER-2 flights showed that nitrogen in all its forms was reduced by up to 90 per cent at higher altitudes;

but at lower altitudes, as measured by the DC-8, they seemed to be elevated by three to four times. This suggested that cloud particles laden with nitric acid might have descended and released considerable amounts of active chlorine.

The groups of scientists in Stavanger had an informal competition to put their findings to some sort of musical accompaniment. By far the best effort used the theme to *Rawhide*:

> PSCs will surely titrate HCl and chlorine nitrate,
> Which with the Sun will make more ClO.
> We've established in Stavanger – Arctic ozone is in danger!
> And this year is likely not atypical.
> Our results do not lie! Do we all want to die?
> Get rid of CFCs!

('If W. S. Gilbert can rhyme "ozone" with "ohone",' says one AASE participant, 'then I don't see why we can't get away with rhyming "danger" with "Stavanger".')

Although the presence of PSCs was patchy, the chemistry of the Arctic stratosphere was perturbed throughout the region. Rapid fluctuations in localised temperature were caused by air flowing over Scandinavian mountain ranges, cooling as it did so: this resulted in PSCs forming locally and thereby creating mini-holes. Similar observations were made by the Franco-German CHEOPS experiment the next year. At the end of the mission, a summary issued by NASA concluded: 'The preliminary results of this mission have substantially increased confidence in the scientific understanding of the PSC-induced, chlorine catalysed, ozone depletion phenomenon. As a result, it is clear that enhancements of chemically active chlorine compounds do indeed occur in both the Arctic and Antarctic stratosphere.'

Once again, the teams of researchers would return to their home laboratories to try to understand the data which they had obtained. It was complex work and even today there is no complete understanding of why the denitrification alone occurred over the north polar regions. And, though nothing remotely approaching an ozone hole was found, something like 10 per cent of the ozone layer that winter was removed by chemical means. The observed denitrification implied that wider ozone loss across the northern hemisphere could result because it could take place at markedly lower temperatures.

Herein was another clear implication: writing in the 31 May 1991 issue of *Science*, a group of AASE researchers led by William H. Brune of Pennsylvania State University concluded that with expected chlorine loading of the atmosphere, ozone depletion could double by the year 2010 and cause a loss of total ozone of 10–20 per cent over the North Pole. And if the winters were cooler in the Arctic stratosphere there would be 'more frequent and widespread PSCs, which ultimately will lead to even greater ozone destruction. Thus an Arctic ozone hole, smaller and less intense than the Antarctic ozone hole, is possible in the near future.'

Within a few months, a follow-on mission would show just how close this was to coming true.

4

Eight Miles High

> Oh, I have slipped the surly bonds of earth
> And danced the skies on laughter-silvered wings;
> Sunward I've climbed, and joined the tumbling mirth
> Of sun-split clouds – and done a hundred things
> You have not dreamed of . . .
> And, while with the silent, lifting mind I've trod
> The high untrespassed sanctity of space,
> Put out my hand, and touched the face of God
>
> John Gillespie Magee, Jr, 'High Flight' (acknowledged as the best-known poem among test pilots the world over)

Out of sight of land, nestling the edge of the stratosphere in the dead of winter is no place to have an unexpected problem if you are the pilot of an ER-2 aircraft. But, as Jim Barrilleaux would reflect for months afterwards, he was lucky it didn't happen over the North Pole. As an experienced pilot of the strange, glider-like aircraft which are still among the highest-flying in the world today, he is constantly aware that there are only small margins for error. Malfunction is an omnipresent, unspoken fear on the missions which he and his colleagues have routinely flown to search out signs of ozone depletion high above the Earth's polar regions.

How Soon Is Now?

This was the frightening situation in which he found himself one morning in November 1991 on the second Airborne Arctic Stratosphere Experiment that was investigating the increasing likelihood of an Arctic ozone hole. Quite suddenly – and totally without warning or action on his part – the control flaps of the ER-2 extended. As mission rules dictated, Barrilleaux aborted the flight immediately and returned back to the airfield on the coast of Maine while he still had the chance. Fortunately, it was the only serious operational hiccup in the AASE II, once again choreographed by NASA. Like the European Arctic Stratosphere Ozone Experiment which was being carried out at the same time in Scandinavia, AASE II would also last for five months after nearly two years of planning. As the only aircraft capable of sampling the stratosphere directly, the ER-2 would discover something dramatic: levels of ClO *higher* than had ever been seen over the South Pole in springtime.

Three weeks earlier, Barrilleaux had inaugurated this second polar aircraft campaign by flying the ER-2 out of Fairbanks, Alaska, on an eight-hour flight to the North Pole and back. As November brought wind and snow to the airfield, so ER-2 operations moved to the rather more clement climes of Bangor, Maine, where they would be based for the rest of the campaign. It was Jim Barrilleaux's misfortune that the first flight out of Bangor resulted in problems with the wing flaps. With typical test-pilot reticence he depreciated its significance, suggesting that it was the sort of problem that could happen at any time and for which he had been trained to take evasive action. But Estelle Condon, NASA's manager for AASE II aircraft operations, was much less sanguine: 'Had he been at operating altitude, we would have lost the plane and probably Jim with it. We were all grateful he was able to return.'

It is a measure of the seriousness of the problem that once Barrilleaux had landed the ER-2 was effectively pulled apart, its electronics totally refurbished and all its hydraulic and mechanical systems carefully checked for any hint of the cause of the problems. To this day it has not been determined exactly why its flaps lowered, though none of the engineers believe it is ever likely to happen again.

One close call was enough.

* * *

Eight Miles High

The danger and glamour associated with the ER-2 often eclipses something more prosaic: it is the only research aircraft available anywhere in the world today which can sample the stratosphere directly. This fact, above all others, has ensured the ER-2's preeminence, for although there are many other aircraft involved in atmospheric research, they do so from well below the tropopause. As many of those same aircraft are identical to the ones upon which tourists find themselves herded on package holidays, it is hardly surprising that they are not so widely associated with the 'right stuff' of aviation heroism.

While European scientists have concentrated on balloon-borne experiments, most American research relies on aircraft missions. Though NASA still has a small balloon programme and other federal research agencies use them, its large-scale campaigns tend to utilise the DC-8 and the ER-2. Despite their lower cost, balloons have obvious limitations. The most interesting chemistry in polar ozone depletion, for example, occurs at the edges of the vortex where winds are at their highest. Balloons cannot easily investigate those regions for the wind conditions are simply too hazardous for them to fly through. EASOE scientists either had to look inside the vortex or outside it: balloons simply cannot cross the boundaries separating the two. Aircraft, on the other hand, will – within the limitations of range and ceiling – go more or less where they are wanted within or without the vortex.

The ER-2s are equipped with self-contained laboratories which sniff out virtually all the molecular fragments that are involved in ozone depletion. High in the stratosphere, they can take samples of the air which flows unhindered past their airframes and, either by high-precision gas chromatography or spectrometry, experimenters can determine the abundances of the families of radicals which are involved in ozone-destroying chain reactions. In addition, weather and temperature readings are simultaneously taken, along with routine samples of the particles and aerosols which are present in the air flows. Taken together, these multifarious readings enable the ER-2 scientists to quantify the chemical processing which has taken place in that slice of the atmosphere through which the aircraft has flown.

In particular, the ER-2 flights have refined the technique of measuring 'tracers' – relatively inert trace gases (such as the various oxides of nitrogen) which bear a well-established correlation with the ozone present. If there is a measured abundance of a tracer in a parcel of air, there should be an easily calculated abundance of ozone itself. When compared with actual ozone measurements, it is possible to estimate how much ozone depletion has occurred in that parcel of air.

In the literature of the results from campaigns like AASE II, it is customary to find that 'such-and-such instrument discovered that . . .' These statements often overlook the importance played by the pilots, who ensure that the instruments (some fourteen in all on the AASE-II) are switched on and operating as the ER-2 noses its way through the freezing cold of the Arctic stratosphere. After each flight, the instruments are removed and extensively tested and calibrated to limit the chances of failure before another flight.

To gain a wider overview of the conditions affecting the atmosphere in and around the vortex, the NASA DC-8 makes complementary observations from below the tropopause. It carries fourteen experiments which broadly fit into two major classifications. The first remotely senses the stratosphere by looking up through windows or portholes drilled in the roof of the aircraft. The second obtains *in situ* data with inlets for air samples which are later analysed by gas chromatography back in home laboratories. All these instruments can be operated by teams of scientists aboard so that they can continuously monitor conditions in and around the DC-8 in real time. They provide a useful background reading on the state of the atmosphere below those regions in which chemical processing is taking place.

The ER-2s and the DC-8 are not the only aircraft used today in atmospheric research. In Europe, there are various Fokkers, Transalls and Hercules transporters which are flown specifically for the task of investigating the troposphere. Where once it might have been possible to treat each part of the atmosphere separately, our polluting influence has inextricably linked the stratosphere and the troposphere in ways which are not well understood. The rise of ozone in the troposphere introduces the question of the atmosphere's ability to cleanse itself.

* * *

Eight Miles High

The highest-flying operational aircraft in the world have their genesis in a secret factory which was christened the Skunk Works in 1944. To the designers of the P-80 Shooting Star at the Lockheed Advanced Development Plant in Burbank, the smell from a nearby plastics factory reminded them of the famously noisome place in the L'il Abner cartoons. And so the name stuck, imbuing the inventors of one of America's greatest aeronautical contributions to the Second World War with a mystique that would serve them well in the subsequent years of the Cold War. Under the guidance of Clarence 'Kelly' Johnson – for whom 'aeronautical genius' was a regular accolade – the Lockheed team had their work cut out for them during the decade that followed.

The US military were severely limited in their attempts at assessing the perceived threat posed by the Soviet Union. Very little hard information was obtainable by traditional methods of espionage. As a result, the desire to spy upon the Soviet Union gave fresh impetus to the ingenious aeronautical innovators at the Burbank Skunk Works. The notion of a jet which would be capable of soaring out of range of surface-to-air missiles soon crystallised in Kelly Johnson's imagination. His basic idea was immediately agreed upon by the Pentagon and he was given ninety days to come up with a complete design. Something akin to a lightweight glider took shape on the drawing boards of the Skunk Works and was given the code name of Project Aquatone. Eventually it was renamed the U-2 and the task of turning the blueprints into engineering reality came under the direct aegis of the Central Intelligence Agency.

Then, as now, Lockheed's plant in Burbank was shrouded in utmost secrecy. To maintain the anonymity of this, its most secret project yet, a series of crates left the Skunk Works in the summer of 1955 with the label 'Arcticle 341' prominently displayed. Transported to a remote part of Nevada, they were assembled for a practice run. Within two years, Lockheed had a dedicated U-2 production line in operation and the first batch of thirty were dispatched to operational air fields around the borders of the Soviet Union in Turkey, Norway and Japan.

Any hopes the U-2's designers had for a life of muted anonymity were shattered on May Day 1960 when one was shot down and its pilot, Francis Gary Powers, captured. In vain did NASA claim ownership of the jet, alleging that it was equipped with weather-sampling equipment. That a plane flying above the steppes of Sverdlovsk was 'monitoring typhoons' was a fiction that nobody believed. The press soon found out that the true sponsor of the aircraft and its mission was an agency which was gleefully renamed 'Caught In the Act'.

With the cat out of the bag, no more flights would take place over the Soviet Union, though further U-2 flights were carried out by the US Air Force elsewhere in the world. U-2s thus provided invaluable photographic surveillance during the Cuban missile crisis in 1962, and they subsequently flew regular sorties during the extended conflict in Vietnam and the Arab–Israeli wars in 1967 and 1973. Many of the pilots who learned to fly the U-2s for the military at that time would eventually be employed by NASA, for the space agency realised just how useful they could be in high-altitude research. Three of the original production batch of U-2s were delivered to the space agency on 'permanent loan' from the USAF in 1971 with a trio of experienced pilots to fly them. According to legend, even Francis Gary Powers applied to join but the space agency politely (and probably wisely) declined to employ his services.

These first civilian U-2s were based at the NASA Ames Research Center at Moffett Field, south of San Francisco, close to Silicon Valley. They did exactly what NASA had claimed they were supposed to have been doing a decade before, testing prototypes of experiments which would eventually be flown on satellites, most conspicuously in the field of 'Earth resources' which anticipated the start of remote sensing from space. Most importantly, the U-2s inaugurated NASA's Stratospheric Research Program in November 1973 by allowing aircraft scientists to take air samples at heights of around 18,000 metres for the first time ever. Whereas balloons were limited by the winds which lifted them aloft, aircraft could, within safety margins, fly anywhere on Earth. The following summer the U-2 pilots carried out their first annual visits to Alaska, and entered the record books as the first people to photograph the whole of the state in one frame. Throughout the rest of the year, the U-2s would make routine profiles of the stratosphere across most of the continental United States. Most notably, they were employed to assess the environmental impact of vicious forest fires and the climatic effects of the eruption of Mount St Helens in 1980.

By this time, a second group of aircraft investigating the stratosphere was established at NASA Ames and formed its Medium Altitude Missions Branch. Originally it comprised two Convair 990s and NASA's experience with them dramatically demonstrated the inherent dangers associated with flying aircraft at the cutting edge of research. The first Convair was lost over the nearby Moffett Field golf course when air-traffic controllers directed two planes to land on the same runway in 1973. The surviving Convair would work well until the middle of the 1980s, when it caught fire on takeoff at Riverside Airport, down the Californian coast in Orange County. This time the crew luckily survived, although the plane had to be written off, leaving NASA without a medium-altitude aircraft just as the Antarctic ozone hole was announced.

The U-2s continued to perform flawlessly even though they were supposed to be more dangerous to fly since they had only one engine. Developments in avionics and electronics had led Lockheed's Skunk Works to build a new production run for the US Air Force in the early 1970s. They were known as the TR-2 series, for 'Tactical Reconnaissance', to distinguish them from the earlier versions. These improved aircraft had extended capabilities in terms of both range and additional payload-carrying capacity. NASA was interested in them for its high-altitude research programme, and so in 1981, three were delivered to Moffett Field with a fresh influx of USAF pilots (including Jim Barrilleaux). They were rechristened with the more politically correct label of 'Environmental Research' as the ER-2. NASA remains sensitive about the ER-2's image and its military heritage. Wherever possible it prefers to call ER-2 campaigns 'experiments' as opposed to 'missions', a word that has military connotations.

During the 1980s, NASA Ames played host to a permanent staff of six ER-2 pilots who would carry out over 200 missions each year. Of these, about two dozen investigated the stratosphere directly, while many more looked at interactions between the stratosphere and the troposphere. The old US Air Force hands loved the ER-2s because they were easier to fly and much more comfortable than the U-2s ever were. Operationally, the bulge-shaped 'superpods' over the wings could carry larger and more varied scientific instruments. Cameras or instruments that could sniff out trace molecules could be carried in the nose or below the pilot in the fuselage – or the Q-bay, as he would call it – of the aircraft. The ER-2s became indispensable tools in the routine monitoring of how the stratosphere was altering around the globe.

In January 1987 they landed in Darwin in northern Australia to take part in another stratosphere/troposphere exchange experiment. It was only seven months later that they were to form an integral part of what Bob Watson termed 'the single most important Earth science mission in a decade', the Airborne Antarctic Ozone Experiment out of Punta Arenas. They were joined by the replacement for the lost Convair 990, a DC-8 bought from Braniff Airways. ('Typical,' said one Ames engineer. 'NASA never buys new planes, only decrepit ones.') Ames technicians fitted it out to take teams of scientists and their complicated equipment in flight, installed a new air-conditioning system and new, higher-efficiency engines which would allow flights of up to twelve hours' duration.

Although the ER-2 received most of the attention from the media, the DC-8 was crucially important in helping to unravel the subtleties of chlorine chemistry over the South Pole. 'The pilots really pushed the envelopes of performance to accommodate us,' remembered Bob Watson, then overall manager of NASA's stratospheric research. 'They had to develop ways of stopping the fuel from freezing when we were flying in the vortex by literally sloshing it from one wing to another.'

The work of its support personnel was acknowledged a year later when the summary of results from the AAOE appeared in the *Journal of Geophysical Research:* 'The ground crews had the hardest, most unpleasant jobs of all, and the record 25 flights over Antarctica in the space of 6 weeks is a tribute to their skills.' For native Californians, working in aircraft hangars at the end of the world where the temperatures dipped below −10°C, was not an experience they would care to repeat.

Compared to the earlier NASA aircraft campaigns, the AASE II was relaxed and smooth in its execution. Though there were tensions from time to time, the 150 scientists who participated were old hands and worked together in reasonable harmony. Perhaps it helped that they were on US soil for the first time, which had an obvious advantage in lessening the psychological distance they had felt on previous campaigns. The first one, out of Punta Arenas, had been 'at the ends of the Earth with machine guns literally in our faces', in the phrase of one participant, with the additional, intense pressure to produce data to prove or disprove chlorine chemistry as the cause of the Antarctic ozone hole. The experience gained in Chile helped the subsequent Stavanger-based campaign, which was more relaxed for the operations people. There were, however, 'dynamic tensions' between the theorists whose job it was to interpret the scientific data. Bangor, by comparison, was 'a piece of cake' in the words of one participant, an assessment with which many others interviewed for this book agreed.

There was, however, a significant difference. The first Arctic aircraft campaign had been (like the Punta Arenas campaign) a six-week snapshot of conditions in and around the pole. Though it returned useful information, it had been carried out 'with smoke still coming out of the gun', in Jim Anderson's telling allusion to the chlorine monoxide radical. As mission scientist for the Bangor campaign, Anderson underscored its fundamental difference by extending his metaphor: 'With AASE II we wanted not only to see what happened when the trigger was pulled but also the state of mind of the criminal at that time.' By conducting flights over a five-month period, it would be possible to see how the Arctic vortex was established, how it evolved and what would happen when it broke up. For the first time it would be possible to assess the relative contributions from both atmospheric dynamics and chemical processing over a full winter season well into springtime.

Without the two types of aircraft it would not have been possible to assemble the data into a coherent whole. The flights of the DC-8 and the ER-2 essentially validate each other's observations while each performs something the other cannot do. Though the ER-2 has the obvious advantage of being able to fly through the lower stratosphere, it is limited to a narrow geographical slice because it can only fly into the vortex and out again over an eight-hour period. The DC-8 extends the range of measurements and carries them out for much longer, though from below the tropopause and indirectly.

So far as possible, flights of both aircraft were coordinated so that their data could be compared and contrasted directly.

Of the two, the ER-2 is uncommonly sensitive to the vagaries of the weather and is particularly susceptible to cross-winds on takeoff and landing. They set severely taxing limitations to its operations. Bangor International Airport, as it grandiosely calls itself by dint of flights arriving from Canada, was chosen as the base for ER-2 operations because it was comparatively free of snow and ice during the winter months. As a result, the ER-2 could fly more or less when the mission scientists wanted. Having clear runways was a necessity: if the ER-2 landed on snow-covered runways its precious landing system would be damaged because a side load would be imparted to its friction gears that would be too great for them to bear.

In Bangor, the daily routine was similar to that of previous campaigns, with planning meetings held at regular intervals. The pilots dealt solely with Jim Anderson rather than a gaggle of experimenters who might issue contradictory orders. They would plan their route and operations with him in some detail. There will always be some trade-off between the scientists' needs and the performance of the aircraft, but the pilots are always accommodating within the confines of safety. Though they lost the month of November after Jim Barrilleaux's ER-2 malfunctioned, the pilots made over twenty flights during the campaign. Not all headed towards the North Pole; there were also flights to determine background atmospheric conditions down towards the tropics. 'The numbers of takeoffs equalled the numbers of landings,' Jim Barrilleaux noted with the barest hint of a suitably laconic Texan drawl.

The DC-8 managed 187 hours of operations, flying as many times as it could and normally repeating a regular pattern of flights based out of the NASA Ames Research Center. Takeoff times were largely dictated by the need of absorption spectrometers to follow the Sun at an altitude of 12,000 metres. Having the Sun streaming into experimenters' eyes when they were expecting early-morning darkness caused havoc with even the most resilient of body clocks. The flight schedules were particularly gruelling. On the first leg of its two-week mission, the DC-8 would normally leave NASA Ames for Anchorage at 4 a.m. Because many of the onboard instruments required five hours to prepare for flight, most of their scientists skipped going to bed the night before. Though a direct trip to Alaska need only take four hours, the DC-8 would arc around the north Pacific so that it would be in the air for twelve hours en route to Anchorage.

When they arrived in Alaska, most of the scientists were exhausted, though resigned to the fact that they would be taking off again within a day for a transatlantic flight to Stavanger. By the time they arrived in Norway, they would be dead beat and allowed a few days to recuperate. The worst indignity for many participants came after they had landed at Bangor, when they were made to wait in line for customs officials to appear in the early hours of the morning. 'Everywhere else we were just waved through,' said one disgruntled DC-8 experimenter. 'I had my suitcases checked every time through Maine.'

The DC-8 appeared in Bangor once or twice in the first three months of 1992, so that the scientists aboard could compare their findings with those from the ER-2. Normally, the DC-8 scientists would work two weeks on the plane before returning to their home laboratories for the rest of the time to catch up on sleep and analyse their data in detail. Before returning home to Ames, the plane would make another 'local' flight north to the pole or else south to the Caribbean. As the DC-8 carried new, improved cruise engines it could fly for up to twelve hours. Federal Aviation Authority rules dictated that the aircrews had to be rotated after eight hours, 'but there's no similar rule for how long scientists can work,' said the NASA Ames aircraft missions manager Estelle Condon.

The only real contention during the DC-8 operations was over altitude changes. The experimenters who were taking air samples, for example, would often want to take profiles anywhere up to 12,500 metres and at all altitudes below to get a sense of the trace molecules circulating inside and outside the vortex. Because the DC-8 could not avoid commercial air lanes, it was often ordered to change altitude at short notice. Altitude changes are the preserve of air-traffic controllers many hundreds of kilometres away who have decidedly non-scientific concerns foremost in their minds. To ensure that there were no nasty surprises, however, all experimenters wore headphones to hear the aircrew announce possible changes in height.

How Soon Is Now?

Despite the concentrated effort which accompanied the DC-8 flights – compounded by the acute jet lag – stratospheric researchers in the United States were clamouring to get aboard. There were always more people wanting to fly experiments than space permitted. According to Estelle Condon, the only problem to bedevil DC-8 operations throughout the whole of the AASE II was as bizarre as the ER-2 problem was serious. After one transatlantic flight in February 1992, the aircrews had some difficulty in obtaining meteorological information to plan their route back from Stavanger. By prior arrangement, computers in the DC-8 support office at the airport could access the latest weather data from the UK Meteorological Office in Bracknell. Yet no data was forthcoming and three telephone companies, in Britain, Norway and the United States, were called in to investigate. They could not find anything wrong with their computer networks or satellite links, so software engineers in Stavanger searched for clues on the ground. Very quickly they found the root cause: a field mouse, presumably attracted to the warmth provided by their temporary offices, had nested inside a mass of electrical connectors and bitten through a wire.

'Whoever said that flying is 99 per cent boredom and 1 per cent terror got it about right,' said Jim Barrilleaux quite levelly. This was the nearest he would come during a two-hour interview to admitting the dangers he faces as an ER-2 pilot. His test-pilot sang-froid is based on consummate professionalism. Jim Barrilleaux simply would not enter the vortex unless a risk analysis had been performed. There is very little room for a gung ho attitude in the polar stratosphere.

'The Arctic and Antarctic missions are the most hazardous we fly because there's very little chance of survival should anything happen to the aircraft,' said John Arvesen, the high-altitude missions manager at NASA Ames. 'They're extremely reliable aircraft or we wouldn't do the missions. But they're single-engined so there's a risk.' As a result, the ER-2 pilots are unquestionably brave but they are certainly not foolhardy. Bob Watson of NASA Headquarters was quick to point out 'they don't want to be seen as *Top Gun* pilots' and, indeed, Jim Barrilleaux played down his role as something that has to be done and done well. To extend Watson's cinematic metaphor, this casts the ER-2 pilot closer to the sort of parts Humphrey Bogart used to play than to Tom Cruise's most famous performance.

In reality, flying an ER-2 is an extremely strenuous exercise. The ER-2 pilot has his work cut out for him for he also has to act as navigator, communicator and instrument operator during each flight. It is not just because of fuel limitations that each ER-2 mission is limited to a maximum of eight hours. Quick thinking and the ability to react to possible dangers will always be basic requirements, despite the safety net provided by the myriad technological marvels at the pilot's disposal in the cockpit. 'For some operations where you need positive actions fast,' Barrilleaux said, 'humans are much better.' This constant vigilance is matched by the eight ground-support crew who keep tabs on the aircraft after each flight and are constantly on the lookout for possible problems. The three ER-2 aircraft in NASA's service are virtually as good as new: after 1,000 hours of operations, each is stripped down and totally refitted. By comparison, commercial airliners are normally checked after 10,000 hours of flight.

Unlike Air Force 'jet jockeys', the NASA ER-2 pilots have no pet nicknames for their aeronautical charges. In the old days, the U-2s were known as the *Dragon Lady* and the *Deuce*. The ER-2s based at the NASA Ames Research Center are known rather prosaically as NASA 706, 708 and 709. Partly this is because there are no favourites among the pilots – and for good reason. 'If they aren't similar, the pilots want to know why,' said John Arvesen.

By now, polar aircraft operations are well established. The pilot begins his work in earnest three hours before takeoff, after consulting with the mission scientist on the trajectory which he is required to fly. He will draw up a detailed flight plan and check on the weather conditions which can be expected throughout. Should there be the possibility of winds stronger than 15 knots across the runway by the time of his return, the ER-2 mission would have to be unceremoniously cancelled. If the weather looks fine, the pilot will don his pressure suit an hour before the appointed takeoff time. It is exactly the same as the ones astronauts aboard the shuttle use, and each ER-2 pilot has his own bespoke suit. Should there be any problems with it before takeoff, a back-up pilot is ready and waiting to take over. Like their colleagues who fly in space, the ER-2 pilots have to undergo the tedious procedure of 'pre-breathing' before they head off into the skies. The aim is to breathe pure oxygen to purge the body of nitrogen, which will maximise the pilot's efficiency at altitude and ensure that he avoids the aeronautical equivalent of diver's bends.

Although the pressure suit would afford the pilot some protection in the event of his having to use the ejector seat, it isn't much use 'at ground level in Bangor in the middle of winter', Barrilleaux wryly noted. Once he arrives at the aircraft, he performs the traditional pilot's walk-around and 'kicks the tyres' before he is satisfied that he can take off. All the while he will be carrying a small Dewar flask of liquid oxygen for breathing. When he boards the ER-2, he will then be hooked into the plane's own supply. Another dozen tubes and connectors have to be plugged in before he is plumbed into the aircraft's own environmental control system. As a result, there is so little room for manoeuvre inside the cramped cockpit that the only refreshment he can get during a flight is in the form of small phials containing either juice or something similar to baby food. The pilot will insert a small straw into a valve in his helmet and suck on it to ingest.

Just getting the ER-2 into the air requires a delicately choreographed sequence of actions. Stretching some 31.5 metres across, its wings are wider than the fuselage is long to maximise the lift generated when it reaches the rarefied heights of the stratosphere. The relatively stubby body was designed to keep the total aircraft weight down, thereby increasing its range. Though both factors ensure the ER-2 will soar gracefully into the sky, the weight restrictions cause other problems. The relatively narrow gauge of its retractable landing gear, makes it very difficult to manoeuvre the ER-2 on the ground. For this reason, additional outrigger wheels are slotted into place under the wings. These 'pogos', as the pilots call them, cannot be carried aloft because of their weight and are kept in place only by the additional weight of the fuel in the wings at takeoff. Once the ER-2 heads upwards, they drop out and are retrieved by technicians for later use.

On most flights the aim is to get the aircraft to operating altitude as quickly as possible. Out of Bangor, the ER-2 was at the mercy of flight controllers dealing with commercial air traffic in one of the busiest sectors of North American flight lanes. But once they entered Canadian airspace on the edge of the stratosphere, there were no other aircraft to have to consider. As a matter of routine on the AASE II, the ER-2 pilots would have filed detailed flight plans with Canadian and Danish (Greenland) air-traffic controllers. Compared to other ER-2 missions, this was simplicity itself.

On the first AASE, the pilots had wisely avoided Soviet airspace but had fallen foul of Swedish flight controllers, who wanted notice of flight plans a few days in advance. As the ER-2 essentially tries to follow the shifting vortex, this was not possible. On the remote sensing missions which the ER-2s flew out of the USAF base at Alconbury in Norfolk in the summer of 1991, the ER-2 would routinely fly over French, German, Austrian, Italian, Swiss and Spanish territories in rapid succession. 'Even though there's no one else at our altitude, most still want flight plans to be filed,' said John Arvesen. 'In the case of Switzerland, we're often in and out before they've really noticed!'

It is normal practice for the pilot to actually fly the ER-2 up to 15,000 metres, after which point they switch to automatic pilot. The ER-2 has to fly within a narrow margin of air speed or else it would either stall or run out of fuel before the mission was completed. The fuel itself is a version of the Jet-A gasoline used by commercial airliners, although it has a necessarily much lower freezing point and vapour pressure so that it won't expand or explode in the wings. On the first flights out of Punta Arenas, the fuel actually started to freeze in the lines. At the nominal operating altitude of 19,000 metres, the ER-2 has to fly between 0.70 and 0.71 Mach in level flight and this is normally achieved by use of the autopilot. If it flew any slower, the ER-2 would stall: any faster and it would start to buffet and rock, so much so that the pilot might lose control.

The first U-2s required their pilots to navigate by compass or the stars, but the first ones operated by NASA, and subsequently the ER-2s, introduced an inertial navigation system. For the AASE II, the pilots had the benefit of one of the most enduring spin-offs from the US military's space efforts – a GPS receiver. A small network of Global Positioning Satellites orbit the Earth and constantly emit coded navigation signals which can be read by hand-held receivers (as was demonstrated by US combat soldiers during the Gulf War). A rather fancier receiver is mounted in the ER-2 cockpit, and by judicious reading of the GPS signals, a small computer can tell the pilot his height and position to within ten metres. This has proved an invaluable tool for navigation over the the poles, although the pilots do keep their eye on a 'whiskey compass' in case of a loss of signal.

The normal practice on an ER-2 flight is to profile the stratosphere in a predetermined up-and-down sequence from 19,000 metres down to 15,000 and back up again. During these sampling runs, the aircraft will also change direction and the details of these complicated manoeuvres are noted on a pad carried on the pilot's arm. It is little more than a stiff piece of card upon which details of the itinerary are scrawled. The pilot also has to operate some thirty-six control switches for the scientific instruments carried in the nose, wings or Q-bay. Some of them have to be switched on or off in flight or require mode changes, details of which are also noted on his pad. The pilot has no way of knowing if the instruments are working as they should, but if there were problems that might affect the flight, he can manually switch them off. 'The experiments have a remarkable record of success,' said Barrilleaux, 'given that they are designed to operate at over sixty thousand feet and in temperatures of anywhere between $-60°$ to $-90°C$.'

Once the required samples from within the vortex have been obtained, the pilot heads for home. The most demanding time for the ER-2 pilot is the landing, which represents by far the most precarious part of any mission. 'It's ironic that the most intense part of the flight coincides when you are at your most fatigued,' Jim Barrilleaux said. To land, the pilot effectively has to stall the aircraft a few feet above the runway or else it will take off again because of the immense lift which the wings continue to generate. He also has to ensure that he lands square onto the runway because the landing wheels allow so little room for manoeuvring. By the time he reaches the end of the runway, the control yoke will be protruding into his chest as the ailerons are used to steer the ER-2 on the ground. The ground crew will then fit the pogos back into their wing slots and the pilot will return the aircraft to the hangar for the next flight.

After eight hours of concentration, danger and the indignity of baby food for nourishment, the pilot will be totally exhausted. 'Certainly you don't have any trouble sleeping after one of those missions,' Jim Barrilleaux said with a smile.

If flying into the stratosphere provides the ER-2 pilot with excitement and satisfaction, the endless waiting and uncertainty over winds blowing across unseen runways is a counterpoint of mind-numbing tedium. Jim Barrilleaux, however reluctantly, has come to accept it as part of his job for the very good reason that there is nothing he can do about it. He invited me to watch his immaculate preparations for a flight in early June 1992 when he was due to fly one of the ER-2 aircraft out to the Azores as part of an investigation into the way in which the oceans interact with the lower atmosphere. On that first Friday of the month, he fully expected to leave Ames for another NASA centre across the Chesapeake Bay from Washington, DC, known as NASA Wallops. There he would refuel and head out across the Atlantic some time the next day.

At 5 a.m., I join him for a breakfast of the distinctly gut-busting variety at Denny's across the freeway from Ames. While I can just about manage a cup of coffee, he eats with obvious enjoyment: a cast-iron digestion would seem to be *de rigueur* for someone who will later have to endure baby food and orange juice. Despite his pleasant demeanour, he is not hopeful for a takeoff today. 'The weather isn't looking good,' he says, referring to a cold front on the other side of the continent. The sense of frustration is heightened by the fact the Sun rises into a perfectly cloudless Californian sky.

We head back to Ames, and Barrilleaux shows me around the aircraft hangar, where the oldest ER-2, NASA 706, is having a major disassembly. The hangar is strewn with aircraft parts. Some thirty-five technicians work for the High Altitude Missions Branch and it looks as if they have had a field day. All the paint has been removed from inside the ER-2's fuselage, and all its subsystems are being systematically checked for corrosion and cracks. All the electrical wiring has been stripped out, as have all the hydraulic lines. These major services will ensure that the aircraft is 'as good as new' when it takes off again.

Barrilleaux decides to stay at Ames and catch up on the paperwork which inevitably piles up while he is elsewhere. Whereas the other ER-2 pilots work for Lockheed and are contractors to NASA, Barrilleaux is a civil servant and has all the minutiae of bureaucracy to deal with. If the weather predictions are better tomorrow, he will come in early again and repeat the whole procedure, this time suiting up if it seems warranted. He knows that he may have to do this time and time again until it is possible to leave for the East Coast through some benign quirk of the weather. So much for the supposed glamour in the life of an ER-2 pilot: it seems a hell of way to make a living.

These frustrations will undoubtedly continue into the immediate future as there is little prospect that the ER-2s will be retired. At one time it had been thought that the SR71 Blackbird would replace them because they can fly higher and faster. But this latter capability alone would make it impossible for scientifically useful missions to be carried out. Travelling at three times the speed of sound, the Blackbirds heat the surrounding air and generate shock waves which render sensitive measurements of trace gases impossible. Instead, 'the future holds higher-altitude sampling with unmanned subsonic aircraft which should be able to reach 82,000 feet, or 26 kilometres,' Barrilleaux says.

In 1994, the ER-2s are once again flying into the Antarctic stratosphere for the purposes of ozone research. The Antarctic Southern Hemisphere Ozone Experiment, is, as its name suggests, investigating the way in which ozone depletion has taken hold over most of the austral hemisphere throughout its winter and spring. Seven years after the missions out of Punta Arenas, the ER-2s are once again sampling the freezing air which is chemically processed in the vortex and destroys virtually all the ozone within the lower stratosphere. This time they are based in Christchurch, New Zealand, and once again the flights will be carried out over an extended period of time.

Once again, pilots of the ER-2 will have to face the most inhospitable flying regimes possible. Jim Barrilleaux prefers to play down the dangers, and not, I suspect, just because it is the done thing among the fraternity of test pilots the world over. During our conversation, it strikes me that he is bored stiff by repeatedly having to elaborate answers to a question he must face with monotonous regularity. He also probably does not wish to alarm his relatives unduly: after telling reporters that their missions out of Punta Arenas were the most dangerous they had ever flown, more than one ER-2 pilot received anxious phone calls from their families. So the fear remains unspoken. The ER-2 pilots face it with characteristic self-depreciation. There is a cartoon attached to Jim Barrilleaux's noticeboard with the following words:

> Aviation in itself is not inherently dangerous. But to an even greater degree than the sea it's terribly unforgiving of any carelessness, incapacity or neglect.

> Below is a picture of a Sopwith Camel crashed into a tree.

The fiftieth anniversary of aircraft exploration of the upper atmosphere passed largely uncelebrated in 1992. The British Meteorological Research Flight has always steadfastly avoided publicity and, as a result, its work has hardly received the credit and attention it undoubtedly deserves. In a striking parallel to the creation of the U-2, the Met Research Flight was formed because of the direct needs of the military. In fact, its *raison d'être* – to sample the stratosphere directly – was outlined in a phone call to Gordon Dobson at the Clarendon Laboratory just after the Battle of Britain.

Dobson had started his career at Farnborough, and knew many of the Met Office high-ups very well. As a result of his work with Frederick Lindemann in using aircraft to predict weather patterns during the First World War, Dobson's ingenuity was highly regarded among the upper echelons of the British scientific establishment. No less exalted a person than Sir Nelson Johnson, the director general of the Meteorological Office, spelled out how Dobson could help solve a particularly difficult problem for the Royal Air Force. 'The long-distance photographers are troubled because they're making trails which give themselves away to enemy aircraft,' Dobson recalled Johnson saying in a memoir which was published posthumously by the Royal Society. 'We want to be able to forecast the conditions and heights when they will *not* make trails. Can you look into the whole matter?'

How Soon Is Now?

Thus entrusted with work of national importance, Dobson and his Clarendon colleague A. W. Brewer systematically set about the task. Realising that contrails formed so easily because the stratosphere had an unparalleled ability to hold water vapour, they designed a number of hygrometers which could be used to measure atmospheric humidity from aircraft which could be flown well within the lower stratosphere. Given his background work with photoelectric detectors, Dobson designed an instrument that would automatically record humidity readings onto a photodetector. Alan Brewer, however, produced an instrument that had to be manually operated, and the Met Office forecasters preferred this version as they had little time for newfangled equipment. Brewer might well have been less assiduous in the task had he realised that he would soon be dispatched to test his instrument from the unpressurised confines of a Boston bomber.

On this and subsequent flights, Brewer determined that the stratosphere was impossibly dry. When Brewer and Dobson reported their findings to the Met Office weather forecasters, very few were inclined to believe them. Dobson himself recalled that the tenor of criticisms was: 'It's bound to be saturated already – it couldn't possibly be as dry as that.' So, in late 1941, a committee was convened to look at how conditions in the stratosphere could best be examined in detail. From its deliberations came the creation of the High Altitude Research Flight, whose work began in earnest in August 1942 out of the RAF station at Boscombe Down in Wiltshire. Two Bostons and a Spitfire were employed at the beginning, and subsequently they were joined by a Flying Fortress and several Mosquitos (the Fortress was one of a half-dozen which President Roosevelt had personally 'given' to Winston Churchill). As an official history of the Flight records, there were more aircraft available than pilots at this time.

The scientists attached to the Flight had to suffer the effects of lack of pressurisation while making very exacting and difficult measurements. On repeated flights, Dobson and Brewer's earlier findings were nevertheless confirmed. Because of the temperatures and pressures regularly encountered in the lower stratosphere, emissions from aircraft engines would supersaturate it almost straight away. In effect, this was the equivalent of producing of a large target sign for German anti-aircraft gunners to follow. In time, regular flights in Mosquitos would discern the exact position of the tropopause, which would move up or down due to the weather systems beneath it. Meteorologists could thus predict whether aircraft would be flying through the relatively drier regions of the stratosphere and how likely it was that contrails would form. Indirectly, these high-altitude observations led to the first operational use of water vapour as a tracer for large-scale atmospheric motions.

The High Altitude Research Flight, renamed the Met Research Flight, moved to Farnborough in 1946 and employed the regular services of two Mosquitos and two Halifaxes with a permanent RAF air crew thrown in for good measure. The legends from this time have a *Boy's Own* ring to them. The most famous – variously attributed to a handful of different researchers – involves a Mosquito pilot turning to his scientific sidekick and saying: 'I say, old boy, you don't happen to know the flaps setting for this crate?' Before he could answer, the plane hit the runway with teeth-jarring force. Philip Goldsmith, who joined the MRF at around this time (and recently retired as head of Earth Observation at the European Space Agency) believes that it is an exaggerated story. 'There were a couple of regular pilots for four aircraft,' he said. 'And when replacement pilots did spells of duty they often didn't know the engine settings.'

The Mosquitos were favoured by the crews because the pilot and the scientist sat side by side. 'We were so close that we knew what each other had to do,' Goldsmith said. 'I'm sure the pilots could have done my job and I felt certain I could have flown the aircraft!' By the end of the 1940s, the Mosquitos were partially pressurised, which provided their occupants with greater comfort when they reached the stratosphere. On a flight investigating the properties of cloud physics, Goldsmith recalled, the Perspex window at his side blew out at 10,000 metres with a sharp explosion. As the air rushed outwards it expanded and cooled thereby filling the cockpit with a curious haze as the water vapour reached saturation point. 'So I can honestly say that I'm one of the few meteorologists to have actually flown *inside* a cloud chamber,' he said.

Repeated profiles of the lower stratosphere on MRF flights from

RAF bases in Europe and Africa during the first postwar years led Alan Brewer and Gordon Dobson to postulate a mechanism by which the stratosphere would be dried as it was continually replenished by ozone production at the equator. The tropopause at the equator is markedly higher than at other latitudes because of the effect of tropical storms which push it up higher. (The tropopause is marked by the flat tops to 'anvil' cumulonimbus clouds where water vapour cannot permeate any higher.) Because the tropical tropopause was higher it was also colder, so as air naturally circulated upwards and polewards, it would effectively be 'freeze-dried'. Yet Brewer and Dobson both realised that some sort of unknown photochemistry could be drying out the stratosphere at higher altitude. As Mosquitos could go no higher than about 12,000 metres before their engines stalled in the rarefied air, it simply was not possible to test what was already known as the Brewer–Dobson mechanism.

When a Canberra bomber arrived in Farnborough in 1952, the Met Research Flight scientists could hardly wait to fly up to its ceiling altitude of 17,000 metres. The indefatigable Philip Goldsmith was one of the scientists aboard and later recalled watching the MRF pilot progressively open the jets' throttle with increasing altitude, as was normal practice aboard the Mosquito. But suddenly at 11,000 metres, the compressors in both engines overheated and the Canberra's jets cut out. 'We stayed calm and at 5,500 metres the pilot was able to re-ignite the engines,' Goldsmith recalled. 'So instead of making the measurements we became the first people in the world to glide in a Canberra.'

Subsequent flights in the Canberra were rather more successful and MRF scientists regularly reached 17,000 metres to determine that there were no unsuspected drying mechanisms in the stratosphere. Alan Brewer and Gordon Dobson were essentially correct in their proposed 'freeze-drying' at the equatorial tropopause. And throughout the rest of the 1950s, the work of the Canberra was expanded to pioneer investigations into the mysterious physics of clouds and how heat was transported through the upper atmosphere. Work by the MRF ushered in the now universal method of using trace gases to infer large-scale flows of air masses in atmospheric circulation generally. Many MRF veterans refer to the 1950s as the 'halcyon days' of aircraft research.

For the RAF air crews assigned to the MRF for extended tours of duty, its relative informality and the curious, often bizarre, behaviour of its scientists took some getting used to after the strict regimentation of most airfields. But they were more than compensated for by the challenges presented by the unusual flights which they were asked to carry out in the name of science. Chasing weather systems normally avoided at all costs – such as thunderheads to search out lightning – was virtually the next best thing to being a test pilot. At times it was no less dangerous. In the early 1960s, the Met Research Flight suffered its first accident when the Canberra was coming in to land at RAF Leuchars just north of St Andrews on the Scottish coast. Suddenly an engine burst into flames as it was on its final approaches over the North Sea. The Canberra lost all power and crashed into the sea, though fortunately the crew were rescued. A replacement Canberra ensured that the Met Research Flight was still in business throughout the 1960s and 1970s.

In the late summer of 1992, the skies above Farnborough often held a remarkable sight for even the most casual of observers. Two aircraft made a series of flights to commemorate the fiftieth anniversary of the MRF over sizeable portions of southern England. One was instantly recognisable as the country's saviour in its most celebrated air battle: the Spitfire is distinguished by its curiously curved, elliptical wings. Next to it was a larger, four-propellered aircraft which has come to symbolise the tragedy of the former Yugoslavia. Yet this particular Hercules transporter was clearly not intended for UN-related mercy missions, as a long, red-striped nose probe could be seen ahead of the aircraft. This garish proboscis protruding from an otherwise normal airframe was an indication of the unusual nature of its work.

The Hercules is listed by the Royal Air Force as C-130 XV208, though it is more often than not called *Snoopy* (because of its nose probe) by the MRF scientists who fly aboard it. It has been a mainstay of the work of the Met Research Flight in the twenty years it has been on loan from the RAF and is now the only aircraft in use after the replacement Canberra was retired in 1981. As the Hercules cannot reach the stratosphere, the high-altitude capability of the MRF has been, however reluctantly, lost. Direct sampling of the ozone layer has thus been deferred to NASA's ER-2s. Nevertheless, *Snoopy*'s ability to sample air masses at any altitude from 15 to 11,000 metres still means it has many useful contributions to make to current research. In the mid-1990s, *Snoopy* is being used to investigate the rise of ozone in the troposphere which has important ramifications for the 'health' of the atmosphere generally. Tropospheric ozone is yet another complication in the ozone issue and, as we will see, has largely disquieting implications.

Today the Met Research Flight's main function is to provide detailed information for many areas of atmospheric research where it is lacking. To achieve this, *Snoopy* and the sixty personnel of the MRF are not exclusively based in and around Farnborough all the year round. They are often sent on extensive detachments overseas as British contributions to long-planned international experiment campaigns. In recent years, *Snoopy* has flown out of Ascension Island, Bahrain, Norway and Kuwait – the last-named at short notice in March 1991 when it investigated the oil fires left behind by retreating Iraqi forces. Its fifty hours of flying through the smoke-filled air highlights the ability of the Hercules transporter to fly through just about all weathers and conditions.

Snoopy itself has been transformed into one of the most versatile and sophisticated 'flying laboratories' anywhere in the world. It can fly continuously for twelve hours and has a range of 5,000 kilometres. Instruments for each mission are tailored to the individual needs of scientists, and those who investigate atmospheric chemistry predominantly employ gas chromatographs which return data in real time. For more complex chemical species – such as oxidants like ozone which are produced by the action of sunlight – air samples are returned in stainless-steel bottles for more thorough analyses on the ground. During each flight, one scientist sits on a jump seat just behind the pilots to liaise with them directly. A handful of experimenters will be in close proximity to their equipment around the fuselage while a 'flight leader' will sit in a soundproof cabin to direct the overall scientific scope of the flight.

During 1993–94, *Snoopy* was taking part in an extended

campaign known as OCTA – the Oxidising Capacity of the Tropospheric Atmosphere – to investigate the way in which ozone in the lower atmosphere is on the increase. Though this rise is much less pronounced than the losses in the stratosphere, it gives an important insight into how human beings are changing the state of our atmosphere. Tropospheric ozone production is partly caused by the action of sunlight on increasing artificial pollutants such as the oxides of nitrogen and hydrocarbons. However, the way in which ozone production results from these precursors is not well defined. There is also some evidence that ozone from the stratosphere has increasingly been impinging downwards owing to changes in circulation across Europe. Measurements from the ground have been geographically limited to a few sites and those within the troposphere have come from intermittent balloon profiles. To fill in these considerable gaps, *Snoopy* would regularly fly out over the Bay of Biscay and back across central Europe to see how the background atmosphere changes over the different seasons. Most photochemical ozone production occurs during the summer, but information is needed on the 'natural' tropospheric conditions which precede and follow the summer months.

'Undoubtedly there is transport of ozone from the stratosphere down into the troposphere,' says the OCTA project scientist, Dr Danny McKenna of the Met Research Flight. 'Tongues of ozone-rich air are brought down to ground level behind frontal systems.' But the background state of the atmosphere and its ozone content as it flows across Europe are not well known. Repeated flights by *Snoopy* will help determine exactly where tropospheric ozone is coming from: if it is relatively dry, then it will have descended from the drier stratosphere above; if it is richer in organic compounds, then it is more likely to have been chemically produced from the surface. Extensive flights by *Snoopy*, choreographed with correlative measurements from balloons and from the ground, will help determine the relative contributions from these sources. The exact details of the photochemistry which produces ozone by the action of sunlight on pollution remains elusive. 'It is very easy to identify processes,' said Danny McKenna of the myriad chemical reactions possible. 'It's very difficult to put numbers on them. OCTA will help us to do that.'

How Soon Is Now?

Snoopy is also involved in a long-standing investigation of interactions between the sea surface and the atmosphere in the Pacific (known as TOGA-CORE). At the start of 1993, *Snoopy* was totally refitted before it flew down to Australia for the first flights in that campaign. Other projects are in the offing and *Snoopy*, like the NASA aircraft, will always be oversubscribed. Many experimental campaigns require its unparalleled ability to observe small-scale processes which affect larger atmospheric motions. A recent MRF report also notes that 'the need for satellite instrument development, and subsequent validation of space-borne measurements, is likely to expand as a greater proportion of global data comes from satellites' and so far as possible, in these financially recessed and post–Cold War times, *Snoopy*'s future is assured for many years to come.

Anyone presenting scientific data from Wank Mountain in the English-speaking world is liable to be greeted by raised eyebrows or giggles. But, for German scientists who sometimes show their results at conferences, it presents no self-conscious embarrassment. Evidence from this mythical-sounding place in the Bavarian Alps has proved unequivocally the role that human pollution has played in the increase in tropospheric ozone. That ozone can be produced at the ground because of the action of sunlight on smog has been well established since Haagen-Smit's work at Caltech in the 1950s. (See p. 35.) Where photochemical smogs in cities routinely involve tropospheric ozone levels reaching the 300 parts per billion (ppb) range, the background level of naturally occurring ozone is of the order of 10 ppb. The fact that tropospheric ozone is increasing because of photochemical pollution is a sign of the degradation in the capacity of the atmosphere to cleanse itself. Ozone is itself one of a handful of oxidants which serve in this capacity: as they all work to remove other trace gases, they are an increasingly important consideration in the science involved in global change.

Ozone measurements have been made in and around the Austrian Zugspitzer since the 1930s and is a legacy from the time when 'taking in the ozone' was generally perceived as being good for human health. Observations from around Wank Mountain, in particular, show a clear trend that ground and mid-level ozone has doubled over the course of the century. At higher altitudes, the trend is not so clear, so it is fairly obvious that the rise below is underpinned by industrial pollution. Similar observations have emerged from balloon flights over Switzerland, Germany and Belgium. Many models of the troposphere predict that the rise of ozone will continue to increase by anything from 0.3 to 1 per cent per annum over the next fifty years.

However, many of these models are severely lacking because of uncertainties in the veritable mêlée of chemical reactions which accompany the abstruse photochemistry involved. If the chemistry of the stratosphere could be called complex, then that in the troposphere is downright Byzantine. Partly it is due to the fact that the trace gases involved are much less abundant. And although the rise in tropospheric ozone is now well established, the same cannot be said for other oxidants. In particular, atmospheric chemists are interested in the OH radical (sometimes known as the hydroxyl radical) which is a key factor in much of this underlying photochemistry. Where tropospheric ozone is present in the parts per billion range, the hydroxyl radical is there in parts per *quadrillion*. This is why it has been christened the 'Howard Hughes radical' and its detection has been the cause of much professional heartache for atmospheric chemists on both sides of the Atlantic.

The hydroxyl radical is produced when ultraviolet light decomposes ozone into molecular oxygen and energetically excited oxygen atoms (almost the reverse of ozone production in the stratosphere). A small fraction of the latter will react with the rather more plentiful water vapour in the lower atmosphere, and will form two hydroxyl radicals as a result. There the certainty ends, for OH is so highly reactive that it often cannibalises itself and is thus extremely short-lived. William H. Brune of Pennsylvania State University outlined the problems it causes for the unsuspecting atmospheric chemist in a review in *Science* magazine in May 1992: 'The hydroxyl radical is destroyed by reactions with a large number of other trace gases and by collisions with the ground, plants and other surfaces. These reactants are so numerous that the hydroxyl radical disappears within a few seconds in the shadow of a thick cloud passing overhead.'

Atmospheric chemists have developed half a dozen techniques to determine the abundance of naturally occurring hydroxyl radicals in the atmosphere. They have generally not been successful because the measurements themselves have interfered with the presence of the radical. Laser-induced fluorescence, which has helped unravel the chemistry of the stratosphere, largely failed until the 1990s because the reactions they generated tended to swamp the naturally occurring hydroxyl signal. Two new techniques, however, show promise in the task of stalking the elusive OH radical. One method involves sending a laser beam across a wide distance – a valley between mountains, for example – and by tuning a spectrometer to frequencies characteristic of absorption by the OH its presence can be inferred.

Another method is to introduce reagents where the chemical pathways involving OH can be predicted accurately and, as a result, its abundance can be calculated from measurements of these other reactants. This local, or point, measurement method involves the selective ionisation of reactions when hydroxyls are transformed into sulphates and their presence is determined by a mass spectrometer. Informal comparisons between both these methods in the Colorado Rockies has shown an emerging consensus on the abundances of OH. As they are based on different physical principles, atmospheric chemists are increasingly certain they are homing in on the correct answer. 'The future is really bright for measurements of the OH radical,' Brune said in 1993. 'Within a couple of years, I think we'll be in a position to measure it with 90 per cent accuracy.'

In a sense, the oxidising capacity of the troposphere could be thought of as a complicated chemical maze. The current work is unravelling the maze piece by piece, or, chemically speaking, process by process. Photochemical models of the troposphere have continually overpredicted the presence of OH and its diurnal variations: it is obvious there are unknown chemical reactions somewhere in the maze. In the summer of 1993, a further campaign in the Rockies led by William Brune was helping to find out more about these considerable unknowns in tropospheric chemistry. It is not just an isolated problem. Another nickname for the OH radical is the 'tropospheric vacuum cleaner' because of its unparalleled ability to remove many of the trace gases which play a part in stratospheric ozone depletion and global warming. 'Tropospheric ozone and the hydroxyl radical are keys to the understanding of many other problems we face,' William Brune said with obvious concern.

The observed increases in tropospheric ozone are not the same the world over, nor is industrial pollution the only culprit. There is a distinct bulge in the production of lower-level ozone over the south Atlantic during the late summer and early autumn. Its predominant cause is the annual biomass burning in Brazil and Africa, and to investigate it more closely the NASA DC-8 was totally refitted after the AASE-II campaign. In late September 1992, it left Ames and headed south to embark upon a new mission known by the acronym TRACE-A (TRAnsport and Chemistry near the Equator in the Atlantic). Though some of the AASE-II instruments remained on board, new ones were incorporated to sniff out the obviously lower abundances of trace gases in the troposphere.

The DC-8's task was to determine the exact extent of this observed Atlantic ozone bulge, for it is not exactly pronounced and amounts to only some 30 Dobson units in total. Indeed, some atmospheric scientists had wondered how real the bulge actually was, for hitherto, it had only ever been observed from satellite measurements, which are much less detailed by comparison. One satellite instrument measured the total ozone column, while another on a different satellite estimated the amount present in the stratosphere only. When the two values were subtracted, there seemed to be a bulge – but it was perilously close to the margins of error of measurements in both instruments. 'This technique gives you the tropospheric ozone signal but it's not terribly accurate,' said Robert J. 'Joe' McNeal, the manager of tropospheric chemistry research at NASA Headquarters. 'You're subtracting one large number from another.'

So a more detailed *in situ* investigation by aircraft was obviously warranted and TRACE-A has conclusively shown that tropospheric ozone levels are indeed elevated over the south Atlantic at that time. In early October 1992, the DC-8 travelled along the Brazilian coast to monitor ozone and other chemicals that result from the burning of the rainforest, which is then at its greatest. On board the DC-8 was the NASA experimenter Jack Fishman, who was astounded by the measurements which were being made. On landing, he told Marlise

Simons of the *New York Times* that just north of Brasilia, 'every instrument went off the scale. It was unbelieveable. People went wild. We were approaching a Stage 2 smog alert in Los Angeles and that at 11,000 feet.'

Subsequent flights of the DC-8 showed that tropical storms over Brazil raised the pollution up towards the tropopause. Here it mixed in with the normal atmospheric circulation patterns, heading first towards southern Brazil and from there across the south Atlantic. Near the African coast, levels of tropospheric ozone increased again to three times their normal values. Flying out of Windhoek, the DC-8 directly sampled emissions from forest and vegetation burnings over Angola, Zambia, Zaire and South Africa towards the end of October.

To determine the relative contributions from either side of the Atlantic, the DC-8 would fly in patterns destined to make the normal aircraft passenger feel more than a little giddy. Off the coast of Africa, the aircraft would start sampling at around 900 metres and then rise up to 12,000 metres to catch the westerlies off Brazil in a matter of minutes. This same sampling routine would be repeated many times in the space of a day across the coast along parts of southern Africa.

In the spring of 1994, the TRACE-A results were still being analysed, but there is definitely a tropospheric ozone bulge with biomass burning as its cause. According to Anne Thompson, another NASA researcher, 'Eventually we will be able to unravel the pieces, flight by flight, to determine the extent and relative contributions from Africa and Brazil to the observed bulge.'

In the future, the NASA DC-8 will also continue to investigate the inexorable rise of tropospheric ozone. The research will be concentrated on the islands of the central Pacific, because they are among the least polluted areas on the face of the planet, as has been known from a NOAA-operated 'clean air' monitoring station since the 1960s on Samoa. However, this may not last for very much longer, for the economies around the Pacific rim are among the fastest growing in the world. Industrial pollution will assuredly follow, and so, over the decade of the 1990s, repeated flights of the NASA DC-8 will observe to what extent the tropospheric atmosphere is degrading as that pollution begins to take hold. A series of dedicated DC-8 flights generically known as the Pacific

Exploratory Missions will quantify just how the chemistry of the troposphere is changing.

Methane is another trace gas whose importance has been realised in recent years. It is perhaps best known for its starring role in the great global warming debate, because, molecule for molecule, it has twenty-five times the heat-trapping ability of carbon dioxide. Yet it plays another, rather more subtle part in ozone depletion, because once it reaches the stratosphere it is broken down into its constituent atoms of hydrogen and carbon by photodissociation. Both are rapidly oxidised, and the hydrogen then forms H_2O, or common or garden water vapour. The net result is that the dry stratosphere is starting to fill with water vapour and, although this is taking place in the parts per million range, it is more than enough to encourage the formation of ice clouds similar to polar stratospheric clouds. In other words, more surfaces are becoming generally available upon which ozone-destroying chlorine activation can take place.

This is an area of study to which Sherry Rowland and his University of California at Irvine colleague Donald Blake have addressed themselves. 'Some people argue that the presence of PSCs is a recent phenomenon because of the increase in methane,' said Rowland. 'Certainly we have shown beyond all doubt that atmospheric methane has been on the increase.' Since 1978, they have meticulously collected air samples every three months at more than sixty remote locations throughout the Pacific. As we have seen, because this is a largely unpolluted region of the planet, the samples taken represent a 'background' reading of the atmosphere's cleanliness. In 1990, Blake and Rowland reported their findings: methane has been increasing steadily at about 1 per cent per annum. Global concentrations had reached 1.71 parts per million, some 13 per cent higher than when the study began in 1978. Blake believes that as much as half the water vapour in the stratosphere may have resulted from this rise in methane. He also estimates that the stratosphere is 30 per cent wetter than it was forty years ago when the first humidity measurements were made by Brewer and Dobson. Air samples taken as part of the AASE-II and TRACE-A campaigns will provide interesting comparisons for the UC Irvine researchers.

How Soon Is Now?

Trying to figure out exactly where this methane is coming from is yet another puzzle for tropospheric chemists. Methane is produced by rice paddies, forest fires, natural gas emissions, termites and the digestive tracts of animals (particularly cows). Though there are many field experiments investigating the contributions from each of these sources (and their individual trends), it is almost impossible to follow the course of individual methane molecules as they inexorably rise into the stratosphere. There are a myriad chemical reactions in between which are subtly altering – and, indeed, being altered by – the oxidising capacity of the troposphere.

There is some evidence that methane may reach the stratosphere more efficiently than had been originally thought. The hydroxyl radical serves to keep a balance between carbon monoxide and methane production in the increasingly polluted atmosphere. That balance has changed mainly due to the build-up of carbon monoxide from car exhausts, which may destroy up to 80 per cent of the hydroxyl radicals which would otherwise be available to scavenge other pollutants across the globe. This means methane molecules may effectively last longer because their path into the stratosphere is increasingly unimpeded. Work by A. Ravishankara and G. Vaghjiani at the NOAA Aeronomy Lab in Boulder, Colorado, has shown that naturally occurring levels of hydroxyl radicals react with methane a quarter less efficiently than had been thought. This suggests that the average lifetime of methane molecules in the atmosphere will increase from ten to fifteen years.

These reactions however have been observed in the relative comfort and marked simplicity of the laboratory. The NOAA scientists have also used models of the troposphere in which there are many unknowns. The atmosphere is unendingly more complicated, and work by Sasha Madronich and Claire Granier at the National Center for Atmospheric Research, also in Boulder, has determined that the lifetime of methane may actually be shortening. Increased levels of ultraviolet radiation resulting from stratospheric ozone depletion have also permeated down through the troposphere to destroy naturally occurring tropospheric ozone. Paradoxically, this will have the curious result of making available more excited atomic oxygen to create greater numbers of our elusive friend the hydroxyl radical when it comes into contact with water vapour. This increase in the availability of OH would then serve to shorten the lifetime of methane by reacting with it more efficiently and removing it from the troposphere altogether. Madronich and Granier have calculated that this would reduce the annual 1 per cent increase in methane reaching the stratosphere down to 0.6 per cent.

Much more work on the complex chemistry which takes place in the troposphere is needed before any conclusions can be drawn. 'There is no question that the rates of methane entering the atmosphere have decreased in the early 1990s,' Sherry Rowland said. 'It is an important area of study for which we need more information.' And although the rate at which methane is entering the stratosphere has slowed down in recent years, it is still increasing in total abundance. This also means it will have a direct impact on stratospheric ozone depletion, although – surprise, surprise – there are further complications. Though methane reacts with ozone to form water-ice crystals, there is some evidence that it will also react with the CFCs which are also photodissociated in the stratosphere. This will serve to remove the active chlorine which would otherwise be available to catalyse ozone destruction itself. As yet there is very little available data on which effect of methane is the more efficient, but as chlorine loading passes through its maximum at the turn of the century, it is clear that thereafter water-ice formation will increasingly have the edge.

As if to reinforce the perplexing role of atmospheric methane, there was a piquant – some might say pungent – twist to its importance raised at Christmas 1992. Perhaps the most curious aspect of its rise in the atmosphere involves the euphemistically correct 'gastric emissions' of animals – cattle in particular. There is very little by way of hard data on this source and thus, its eventual contribution to the formation of polar stratospheric clouds. This very point prompted the environment writer Fred Pearce to remark in *New Scientist*: 'Farting cattle are unlikely to be the main cause of the ozone hole – but they may well contribute.'

An Australian doctor raised the bizarre possibility that human flatulence, inflamed, as it were, by yuletide scoffing of turkey, plum pudding and other excesses of Christmas indulgence, may also play a part. Terry Bolin, an otherwise entirely respectable gastroenterologist from the Prince of Wales Hospital in Sydney, suggested that increased 'emissions' from humans would be an additional, important source of methane over the Christmas period. However, Tom Wigley, a well-respected climate researcher from the University of East Anglia, blew a raspberry in response. Wigley calculated that if all the turkey-eating people in Christendom produced an extra two litres of methane as a result it would only amount to an increase of one fifty-thousandth the total annual methane released annually into the atmosphere (some 500 teragrams in total). 'From a climate point of view, don't worry about farting,' Wigley memorably informed the *Guardian* newspaper. And for good measure he added: 'For the ozone layer, fart as much as you possibly can.'

Farting aside, the inexorable rise of methane continues, despite the complications from the oxidising chemistry below. The fact that methane will tend to fill the stratosphere with water vapour portends a worrying development for both the ozone layer and global warming. For as the surface of the Earth warms – perhaps by as much as 2°C over the next thirty years – the stratosphere will cool to maintain thermal equilibrium. The likelihood is that ice-cloud formation will increase as a result and lead to increased ozone depletion throughout the world. The details of how the greenhouse effect and ozone depletion will antagonise each other will be examined further in Chapter Six.

Out across the Mohave Desert in southern California, the sight and – usually more memorably – the sound of experimental aircraft is nothing new. The dried salt lakes which make up a startlingly bleached backdrop for the Edwards Air Force Base and its environs have played host to test flights for many generations of aircraft, the SR-71 Blackbirds and the Space Shuttle among them. But at the nearby El Mirage Dry Lake throughout 1992 and 1993 there have been intermittent sightings of one of the most peculiar aircraft ever seen. It looks like a large vacuum cleaner to which upward-curving wings have been attached along with an outsize propeller at the rear which makes the sorts of noises more commonly associated with domestic lawn mowers. This curious vehicle is self-evidently uncrewed, and the keenest-eyed would have discerned the name 'Perseus' written on its large tail fin. Strange as it may seem, this lightweight, high-performance 'drone' promises to usher in the next phase of aircraft studies of the ozone layer.

Although the ER-2 has revealed many of the mysteries of ozone depletion, its observations are constrained by altitude. Its measurements are, to quote Jim Anderson, 'nibbling at the bottom of the stratosphere'. Given the 16–19 kilometre ceiling of ER-2 operations, atmospheric scientists, have been able to observe the chemistry directly in only 10–15 per cent of the total ozone column. Anderson used the analogy of an upturned iceberg: we have been able to scratch away at the tip, but the bulk of what is going on remains resolutely hidden from view. 'The rest of the ozone column is a mystery,' he said. 'We have little information on the exact details of the chemistry which occurs throughout. We need to get higher, sooner in the spring over the poles to see the initial conditions of the material which is descending.'

Bemoaning these frustrations to colleagues from the Massachusetts Institute of Technology, Anderson was offered a solution in 1989. John Langford is an aeronautical engineer who had been an integral part of MIT's Daedalus Project, which had built a lightweight human-powered glider that was literally pedalled between Crete and Santorini on Thira, the nearest Greek island to the north, the summer before. At Anderson's prompting, Langford realised that the technology which had made Daedalus a reality could quite easily be adapted for high-altitude scientific research. For the first time, there was a possibility that aircraft measurements could be made at an altitude of 25 kilometres, 'where the action is' in Langford's phrase. He subsequently left MIT to set up Aurora Flight Sciences in Virginia to develop the project. Funding was not as forthcoming as he hoped, but when NASA became interested in uncrewed drones in 1991, the future of his company was placed on a surer footing.

The basic design which resulted from Anderson's discussions with Langford is an aircraft whose capabilities have been directly dictated by the needs of atmospheric scientists. Perseus is a linear descendant of the human-powered Daedalus aircraft but employs robotic technology and, most importantly, 'a removable nose instead of a cockpit', to quote John Langford. Within it is a self-contained instrument package which can easily be removed, or replaced, in a matter of minutes. Modular payloads specifically tailored to the sorts of investigations required in stratospheric research will be a key element in Perseus flight operations. The drone will effectively fly à la carte research missions into specific mysteries of stratospheric research.

Perhaps the most remarkable aspect of the drone is its engine: it is rated at only 80 horsepower and uses a closed-loop cycle of a design commonly used in torpedoes and submarines. The exhaust products are fed back into the combustion chamber and excess heat is radiated away while the engine as a whole is cooled by liquid oxygen. 'At the highest altitudes we wish to fly,' said John Langford, 'we have to carry our own air supply because the atmosphere is so rarefied. We make use of that liquid oxygen.' As a result, Perseus will have a 20 per cent greater performance than any high-altitude vehicles currently in existence. This means that Perseus will be able to fly subsonically in the stratosphere, and as it noses through the rarefied air at an altitude of 24,000 metres, the trace molecules and radicals that it is sampling will be preserved in their naturally occurring abundances. To ensure that the air flow is not disrupted, its propulsion system is obviously located at the rear.

A typical ozone mission would involve a winched launch because the 440-centimetre variable-pitch propeller needed for high-altitude flight cannot clear the aircraft when it is on the ground. Once in the air, the propeller will unfurl and cut through even the most rarefied air with sufficient thrust to keep the drone in level flight at speeds of around 320 kilometres per hour. To ensure that the wings can generate sufficient lift, they are some 9 metres across each but weigh in at only 118 kilos in total (the aircraft as a whole weighs only 600 kilos). Total flight time will be around six hours, with an hour for air sampling at 24,000 metres. Unlike a balloon, the Perseus is not restricted to air currents and can actively search out regions of polar stratospheric clouds and observe the most interesting chemistry in level flight. Though Perseus automatically flies by a sophisticated onboard computer (which senses where it is by use of global positioning satellites), it can be controlled from the ground by teleoperator flight systems. Project scientists will literally be able to chase polar stratospheric clouds or regions of unusual chemistry through which they have flown. 'The scientists can sit next to the pilot,' Langford said. 'They like that. It puts them in the cockpit without ever having to leave the ground.'

By 1994, Aurora Flight Sciences had built five Perseus aircraft. The first was a 'proof of concept' vehicle, which successfully made three test flights over the Mohave Desert. Two more are owned by NASA and were earmarked for its southern hemisphere campaigns in the spring of 1994. Two aircraft are owned by Aurora directly and will be leased to scientists who require them for specific purposes. A key element in the development of Perseus was the desire to keep it affordable for the scientists themselves. At approximately $1.5 million for each aircraft, it is not so prohibitively expensive that only larger agencies can afford to buy or fly them. Indeed, ozone science is one of the many research possibilities offered by uncrewed drones. Aurora is developing a variant of the high-altitude Perseus (known as Perseus B) which could fly for 72 hours in continuous flight at 16,000 metres without recourse to a liquid-oxygen tank. Aurora believes it would have many uses in global change research, such as obtaining data to predict the onset of hurricanes, determining how sea surface temperatures are changing and monitoring emissions from future fleets of supersonic aircraft. 'There is a great deal of interest in these sorts of areas,' said John Langford.

Towards the turn of the century, Aurora Flight Sciences is hoping to build even larger drones for the purposes of atmospheric research. What could be described as Perseus' big brother is a larger, twin-turbocharge-engined craft that could reach 30,000 metres and stay there for approximately a month. This Theseus drone would literally fly around the polar vortex during the time in the spring when most chemical processing takes place. Further into the future is the intriguing possibility of 'the mother of all drones' – a nuclear-powered behemoth known as Odysseus which could stay in the stratosphere quite happily for up to a year. 'For the moment we're quite content to get Perseus off the ground,' said John Langford, anticipating the remarkable advances it will bring about in the understanding of ozone depletion.

Despite the arrival of uncrewed drones like Perseus, the ER-2s will continue to play their part in atmospheric research for many years to come. In fact, their working life is expected to stretch well into the twenty-first century. John Arvesen at NASA Ames pointed out that the first U-2s were built in 1955 and not retired until 1989. 'The only reason our U-2s were retired was because something better came along to replace them,' he said, adding that they could well have flown for another thirty years. As for their replacements, the trio of ER-2s, he suggested that 'if nothing better comes along to replace them, they could operate for the next fifty years'. Their unique capabilities will keep them busy in monitoring the stratosphere throughout the world as chlorine loading reaches its peak in the next ten years or so.

In intelligence circles, it has been estimated that the U-2s which flew during the 1950s and 1960s saved the United States many billions of dollars from not having to build strategic aircraft to face down what they found to be a nonexistent Soviet threat. Perhaps a hundred years from now, when levels of stratospheric chlorine have returned to their normal, pre-industrialised levels and Antarctica will no longer be threatened by seasonal ozone depletion, people will recognise that its descendant, the ER-2, played as crucial a role as the U-2s.

5

Global Perspectives

To see the Earth as it truly is,
Small and blue and beautiful in that eternal
silence where it floats,
is to see ourselves as riders on the Earth together,
brothers on that bright loveliness in the eternal cold –
brothers who know now they are truly brothers.

'A Reflection'
Archibald MacLeish

SECURITY is palpable as you enter the windowless building just north of the Beltway where it splits off towards Baltimore. Outside is the marrow-numbing cold of Maryland in January; inside is the reassuring comfort of an air-conditioned sanctuary. The name plate outside announces it simply as Building 2, although it is usually known by the acronym POCC. Both serve to guard its anonymity and only the large NASA logo at the front gate testifies that there may be more to it than at first meets the eye. Unlike the facilities at Cape Canaveral or in Houston, NASA's Goddard Space Flight Center remains for the most part unknown to the general public. The satellites which its scientists and engineers build and operate are unmanned and, at face value, not as exciting as manned spacecraft. Yet today, Goddard is intimately involved with the latest Space-Shuttle mission, which will deploy a telecommunications satellite without which crewed spaceflight would simply not be possible.

The Payload Operations Control Center represents the holiest of holies. It acts as a veritable nerve centre for most of the United States' worldwide network of satellites, threaded with a mass of electronics which extend mission control out beyond the upper atmosphere. At any one time, Goddard controllers will be monitoring the health of about half a dozen spacecraft which silently orbit the Earth. Working in shifts throughout the day, they are constantly vigilant for the slightest electronic hiccup which may herald serious problems. Contact with the outside world is restricted to closed-circuit television became some of those satellites are controlled by the Department of Defense. Occasionally, the television cameras pull back to reveal a bank of clocks above the main tier of control consoles which indicate local time around the globe dotted against an impressive Mercator projection of the world. A perplexing spider's web of orbits crisscross and spoil its otherwise geometric precision.

Today, security is tighter than normal because the POCC is taking part in the tracking of the fifty-third shuttle mission. Entry within requires a series of elaborate security checks and once inside, security guards are liberally sprinkled around many smaller offices whose functions cannot easily be discerned from the many acronyms which grace the hallways and doorways. In a small room along a basement corridor a handful of people are intimately involved with 'UARS Operations'. Their undivided attention is focused on the Upper Atmosphere Research Satellite, orbiting some 600 kilometres above their heads. It is the most powerful tool yet devised to investigate the stratosphere with a truly global perspective.

After launch from the Space Shuttle in September 1991, UARS has lived up to the expectations raised by NASA's description of it as 'the most complex satellite devoted to atmospheric research ever sent into space'. Weighing nearly 7 tonnes, it is certainly one of the largest. Its unique vantage point from low Earth orbit means that it can observe slices through the upper atmosphere across whole continents and scrutinise its chemistry, dynamics and wind patterns with unprecedented sensitivity and precision. Despite the hype, UARS has had its fair share of problems, some of which have been potentially debilitating for the $700-million mission. Baffling mechanical failures have been a feature of even the most successful space projects.

Today the dozen or so engineers associated with UARS operations are puzzling over the protracted failure of one of its ten scientific instruments. The Improved Stratospheric and Mesospheric Sounder (ISAMS), built by physicists at Oxford University, is one of four instruments designed specifically to look at the chemistry of ozone depletion. It has repeatedly suffered problems with a propeller-like 'chopper' used to filter infrared radiation entering the instrument. After repeated failures, the chopper is working again – but it is spinning far too fast and overheating its motor in the process. Now the Goddard controllers will be in contact with UARS for just a few minutes, dictated by other demands on NASA's worldwide telecommunications network. It is almost permanently oversubscribed by other users in the POCC, including UARS' predecessor, Nimbus 7, the satellite that confirmed the existence of the Antarctic ozone hole.

The UARS controllers will attempt to load a set of commands to the ISAMS microprocessor to see if the chopper wheel can be slowed down. The way in which they are executed should help diagnose which part of the chopper electronics has failed. Tension mounts as UARS comes into range and suddenly engineering 'housekeeping' data from the satellite streams onto the controllers' computer screens. They calmly enter the commands and watch as the UARS output slowly changes in response: it is akin to playing chess blindfold and having the pieces manipulated in another room by someone who doesn't know the rules.

The screens are awash with data which scroll past very quickly. Only these cognoscenti in the POCC know what the perplexing numbers actually mean and seem mildly encouraged by the response to their commands. They have partly succeeded in trying to power down the chopper motor. By the time of the next 'downlink' – when UARS data will be dumped back here at Goddard – they should know how successful they have been. It is routine work for mission controllers. 'If the voltage drops tonight, let me know,' says an engineer whose shift is about to finish. He can instantly tell that the innumerable symbols portray an otherwise perfectly healthy satellite. As UARS passes out of range, he has reason to hope that ISAMS will be working again.

April Fool's Day is hardly the most auspicious time to begin any new enterprise, but one excited meteorologist was later to exclaim, 'we went from rags to riches overnight' after 1 April 1960. The occasion was the launch of an otherwise obscure satellite known by the acronym TIROS, which today would hardly merit any attention at all. But in those days the first Television and Infra-Red Observation Satellite was nothing short of revolutionary for it ushered in the age of the weather satellite. It also brought the prospect of a truly global perspective on the workings of our atmosphere nearer to reality.

In time, TIROS would be superseded by new generations of satellites, many necklacing the equator in geostationary orbit. They would play a central role in an international research effort known as the World Weather Watch. To the populations below, progress could be discerned by the replacement of those first, flickering black and white images with the computer-enhanced swirls which now routinely grace television weather forecasts the world over. Very soon the idea of the Earth as a planet became ingrained in the public consciousness. By far its greatest achievement was the precious Kodachrome booty returned by the Apollo astronauts, showing their home planet, in the phrase of one, 'as a round and bounded globe that is utterly alone in the wilderness'. It could be argued that today's preoccupations with all matters environmental were born in that moment when astronauts could cover the planet from whence they came with an outstretched hand. As the American biologist Norman Cousins famously remarked, 'on the way to the Moon, mankind discovered the Earth'.

With each new, ever more complicated satellite, weather forecasts improved accordingly. In all, ten experimental TIROS satellites were launched, followed by a further six by the end of the 1960s. By then information returned by weather satellites was an integral part of the US National Weather Bureau's daily archive and was overseen by what was then called the Environmental Science Services Administration. By 1970, it had changed its name to the National Oceanic and Atmospheric Administration, whose acronym was adopted as the name of the next generation of US polar-orbiting satellites. Weather satellites, NOAA was keen to point out, were no longer experimental but operational, a distinction which was entirely fitting with NOAA's low status in the hierachy of the federal government. Ostensibly part of the Department of Commerce, NOAA was so anonymous that it was rumoured that most Secretaries of Commerce didn't realise what it did until after they had left office.

The rather more visible National Aeronautics and Space Administration was charged with prompting new developments in meteorology from space. Scientists at NASA Goddard proposed the largest, most complex satellites yet built and named them after the Latin word for raincloud. Under the guidance of Bill Norburg – who, ironically, was to die of skin cancer – the Nimbus satellites became a flying test bed for new scientific instruments which would then be flown as a matter of routine on the next generation of operational spacecraft. All were built by General Electric in Philadelphia and from a distance bore a resemblance to a giant butterfly as their solar panels extended out from the main body of the spacecraft. All were launched from the Vandenberg Air Force base in northern California so they could be inserted into polar orbits without fear of launch mishaps: the next land they would cross would be Alaska for a northerly launch and Antarctica for a southerly one. As they orbited the poles – like the NOAA weather satellites – they would in effect watch the whole of the Earth as it spun beneath them.

Nimbus 1 scored an immediate hit after launch on 28 August 1964, when its new television cameras spied Hurricane Cleo which was then ravaging Florida. It was to record some 27,000 TV and infrared images of Earth at higher resolution than had ever been obtained before. Nimbus 2, launched in 1966, carried an infrared radiometer which measured light in five different bands, pioneering the detection of water vapour throughout the upper atmosphere, a technique which is now commonly employed on weather satellites. Nimbus 3, however, failed to reach orbit in May 1968, when its launcher rocket malfunctioned and deposited the satellite unceremoniously into the northern Pacific. It was a painful reminder, if it were needed, of the pioneering nature of space enterprises. This did not diminish the excitement among

Nimbus scientists of being at the forefront of research. As one pioneer put it: 'If there were mishaps, we didn't really care. We knew there'd be another bus coming along fairly shortly.'

When Nimbus 4 was launched, in April 1970, the press was mainly interested in a radio tracking device which was used to monitor a bear in a cave in Montana. More importantly, the satellite also included a full range of scientific instruments enabling ozone to be measured for the first time from space. Though it was hardly as exciting as tracking bears, Nimbus 4 and those which followed pioneered techniques for detecting trace molecules in the atmosphere. The fact that they are routinely used today shows just how remarkable a breakthrough they truly were.

Essentially, there are four ways to look for trace molecules from orbit. The most obvious was a variation on Dobson spectrophotometry: by comparing the ultraviolet output from the Sun with that reflected back to space, it would be possible to determine how much UV had been absorbed by ozone in the stratosphere. Another method involved looking at the Sun or stars and seeing how trace gases selectively absorbed radiation across parts of the infrared spectrum. Each molecule had a characteristic emission spectrum which could be searched out at different altitudes as the Sun or stars set or rose. Two other techniques essentially tuned into the molecular signatures of ozone: thermal emission radiometry in the infrared, and microwave spectroscopy.

John Houghton, who had studied under Gordon Dobson at the Clarendon Laboratory in Oxford and later headed the department, worked on Nimbus 4. It was the first time that non-US technology had been flown on an American weather satellite. Along with Desmond Smith from Reading University, Houghton had developed an infrared radiometer which gave scientists the chance to measure temperatures throughout the upper atmosphere. On 9 June 1966, they had test-flown a prototype of the instrument on a balloon over Wiltshire. 'I can remember the date quite clearly,' Houghton said nearly thirty years later. 'It was the deadline of a proposal to NASA to fly a version in space. Talk about brinkmanship!'

Once their instrument had been accepted by the kindly space agency across the Atlantic, they prepared a version to be flown in space. Its development was dogged by technological headaches right up to the time of launch; the instrument did not behave as it should have done and repeatedly produced 'strange kinks' in test runs of its radiometer. 'We devised all sorts of programmes for the test data,' Houghton recalled, 'but we never did find out what the cause was.' Their fears turned out to be groundless and the instrument behaved far less recalcitrantly in orbit than it had ever done on the ground. Proof came on the night after the launch when the first data were literally read out over a transatlantic phone line.

Houghton's experience was typical in those heady days before the American space programme lost its innocence. Technological hiccups and niggling frustrations were part of the price to be paid for making revolutionary measurements. As the first arguments in the supersonic aircraft debate were raised in the early 1970s, Nimbus experimenters soon found themselves elevated to new heights of importance, for they had at their disposal the wherewithal to monitor ozone across the whole of the globe as routinely as weather satellites tracked clouds. By the time Nimbus 7 came to be launched, in October 1978, ozone depletion by CFCs was still a controversial theory for which little confirming evidence had been obtained. None of the scientists working on this, the last in the Nimbus series, realised just how crucial its data would be in confirming that evidence, nor that it would still be working fifteen years later.

Every few weeks, keen-eyed Los Angelinos will see streaks of light burning up into the night over the horizon north of the city. They are the launches of satellites from the Vandenberg Air Force base, some of which have involved top-secret espionage. On the pleasantly cool evening of 24 October 1978, Nimbus 7 successfully lifted off, witnessed by a handful of Goddard experimenters. Given the crucial role he would play in the Nimbus 7 mission, it is an odd fact that Arlin Krueger was not there to see it head into orbit, for he had overseen the development of two ozone-mapping instruments aboard. He was in Colorado on a sabbatical from NASA Goddard and busily writing up a PhD on space-based observations of the stratosphere.

By the time of the Nimbus 7 launch, Krueger was one of the most experienced old hands in the stratospheric ozone business. In the late 1950s he had graduated from the University of Minnesota and joined its balloon group, which was pre-eminent in making measurements in the stratosphere. After the International Geophysical Year, Krueger joined the Office of Naval Research and concentrated on making ozone sonde measurements and then using sounding rockets, some of which were launched from ships. 'Academic studies of ozone didn't exactly go far in the navy,' he remembered, 'and they wanted me to do other things like building bombs or designing warheads instead.' So in the late 1960s he moved to NASA Goddard, where he felt his talents would be put to better use in executing its burgeoning Nimbus programme.

In those days, NASA still launched sounding rockets from its facility at Wallops Island in Virginia, and Krueger happily continued to build ozone-measuring instruments which would shoot up to 70 kilometres altitude and return data as they parachuted back to Earth. Shortly, he was to adopt his skills to develop an instrument which served as a prototype for the routine monitoring of ozone from space. The Backscatter Ultraviolet (BUV) instrument was developed by Krueger and his colleagues and worked – like the ground-based Dobson instruments – by determining how much ultraviolet light is selectively absorbed by ozone in the stratosphere. Above the Earth's atmosphere, a satellite can measure the amount of ultraviolet light radiated from the Sun directly: at certain wavelengths, ozone will absorb that radiation either weakly or strongly. At each wavelength, the ratio of the solar radiation to that scattered back (or, as scientists would have it, backscattered) gives an indication of how much ozone is present in the stratosphere.

With increasing wavelengths, ultraviolet radiation will penetrate further towards the Earth's surface as seen from space. So by measuring across the ultraviolet spectrum, the backscattered radiation at each wavelength reveals how much ozone varies with altitude. As a result, BUV enabled scientists to determine the total amount of ozone present in the stratosphere and how it varied with altitude. To do this required a fiendishly complex process of analysis but it was well worth the effort for it yielded unprecedented details on the structure of the ozone layer. 'We tried out profiling and total ozone retrievals and they worked pretty well,' Krueger said. 'You must remember, though, all this was pretty new.'

BUV's moment of glory came in August 1972 when there was a euphemistic 'solar proton event'. The Sun is not quite the benign, quiescent object that it seems to the naked eye: every eleven years it undergoes distinct cycles of activity, measured by the proliferation of sunspots on its surface which result from intense magnetic fields. This activity can also generate giant surges of solar energy which burst into space and eventually slam into the Earth's magnetic field, cascading down towards the polar regions. Theory predicted that the most intense activity would destroy ozone quite markedly at higher latitudes as a result. In August 1972, BUV actually saw it happen and observed 20 per cent of the ozone being lost in and around the North Pole over a period of a few days.

With such dramatic confirmation of the validity of the backscatter technique, Krueger was keen to develop an advanced version of BUV. This meant missing the next two Nimbus 'buses' for there was a five-year turnaround from initial proposal to the time it appeared as hardware – if it was accepted at all. But by the early 1970s, there had been a sea change in NASA's interest in the otherwise esoteric chemistry of the stratosphere. 'Up to the time of Nimbus 4, this was pure science,' Krueger said. 'It was very hard to sell these proposals to NASA. After chlorine was brought in, it became more political.' In 1976, NASA was mandated by Congress to lead the work of monitoring and studying the stratosphere, which gave further impetus for increasingly complex satellite observations.

Although supremely useful, the BUV data were limited because the instrument looked only at a spot in the stratosphere directly below the satellite. Nimbus 4 was orbiting the Earth thirteen times a day, so Krueger's team would get thirteen linear traces of the presence of ozone across the Earth running from north to south. 'We could see there was structure in the ozone,' he said, 'but we couldn't resolve it.' Krueger and his colleagues proposed an instrument that would map the ozone by scanning sideways. A small mirror would sweep out a 50-kilometre swathe of backscatter measurements at right angles to the satellite's track across the surface of the Earth. 'We realised that NASA wouldn't fly two separate UV instruments,' Krueger added, 'so we combined the two into one instrument.' The Total Ozone Mapping Spectrometer shared electronics and optical components with an improved BUV, known as the Solar Backscatter

Ultraviolet instrument. Where SBUV could profile ozone throughout the stratosphere, TOMS used wavelengths that would theoretically reach the surface. When the amount of backscatter was determined, it revealed the total amount of ozone present throughout the whole of the atmosphere (termed a 'column abundance').

Within days of the launch of Nimbus 7 in October 1978, TOMS returned information of staggering clarity and quality. In stratospheric research it was akin to the moment where *The Wizard of Oz* changed from black and white to colour. 'It was incredible! There was amazing structure in the ozone in the atmosphere,' Krueger said. 'TOMS gave us a road map of where ozone was formed.' In time, the TOMS observations showed on the scale of continents what had been suspected earlier: that ozone varies with the weather. Ridges of high pressure pushed the tropopause higher, effectively thinning out the flow of ozone above it. Troughs of low pressure would bring down tongues of ozone into the lower atmosphere in their wake. In the late 1970s, as the TOMS data were being calibrated against ground-based Dobson instruments, the Goddard teams worked in conjunction with airlines who regularly flew in and around the tropopause. According to Krueger, when the chief meteorologist of Northwestern Airlines saw the TOMS data he was ecstatic. 'It was the first satellite data that made any sense to him,' Krueger recalled. 'He could see what was going on at the altitude where airplanes actually flew.'

With 196,000 observations a day, the TOMS data were archived at Goddard, and occasionally dipped into to check up on the state of the stratosphere. In April 1982 both TOMS and SBUV observed something strange over Mexico: it appeared to be a rapid loss of ozone in the stratosphere. TOMS in particular watched as a 'cloud' of low ozone spread in the prevailing circulation and eventually girdled the whole of the equator. At first, it looked as though something had gone wrong with the instrument, causing Krueger and his colleagues no end of puzzlement. Ground-based Dobson instruments could neither find nor confirm this curious occurrence.

Eventually the culprit was found to be a volcano in Mexico called El Chichón which had spewed 7 million tonnes of material into the stratosphere, mostly in the form of sulphur dioxide aerosols. They served to backscatter solar radiation less efficiently and thereby give the totally false impression that a great deal of ozone had been lost. 'When TOMS was designed, it was believed that sulphur dioxide would be a trivial absorber of ultraviolet,' Krueger recalled. 'It was a shock to learn that volcanoes produce so much of it.'

In time, Krueger developed a technique to discriminate between sulphur dioxide plumes and the presence of ozone. 'We've seen at least fifty different volcanic explosions since then,' Krueger said. 'El Chichón was definitely our "learning volcano".' He has spent a large fraction of his time looking at how these different explosions have resulted in quite different eruption plumes and how the stratosphere has responded to them. El Chichón was in many ways a curtain raiser for a more powerful volcanic eruption a decade later. By then, measurements from space would show just how markedly volcanoes could affect ozone in the stratosphere by intensifying a phenomenon whose frightening existence Nimbus 7 had confirmed – the Antarctic ozone hole.

Had Nimbus 7 failed to reach orbit or malfunctioned prematurely, ozone studies during the 1980s would have been severely deficient because its replacement took so long to be launched. Work on what eventually came to be known as the Upper Atmosphere Research Satellite had started in earnest in 1976 after designs for the Nimbus 7 experiments had been 'frozen'. It was already clear that a larger, more complicated satellite would be required to learn more about the subtleties of the stratosphere. At this time, it was envisaged that UARS would be ready for launch in 1984 but, in the event, it was not destined to reach orbit until 1991. This hiatus was mainly due to NASA's blind faith and utter dependence on the Space Shuttle, whose 'assured access to space' took a severe blow when *Challenger* exploded in January 1986.

When the UARS project was being defined in the late 1970s, two spacecraft were envisaged to assume the role of a long-term monitoring tool for the upper atmosphere. Cost cutbacks meant only one spacecraft was actually built by General Electric. It was still quite a behemoth compared to earlier satellites: whereas Nimbus 7 stood 3.5 metres tall, UARS dwarfed it at 9.8 metres. UARS weighs six times as much as Nimbus 7, and in this case, bigger means more versatile for it carries a greater diversity of instruments. UARS was designed to give as wide a range of observations as possible on the perplexing variety of chemical reactions in the stratosphere. 'Nimbus 7 was designed before we realised how important chlorine chemistry would be,' said Robert J. 'Joe' McNeal, the UARS Program Manager at NASA Headquarters. 'If we had flown more Nimbuses, we would never have got the overall picture at the same time and the same points in space. UARS was sized for the job it had to do.'

UARS scientists often refer to it as 'the observatory' because it uses ten complementary instruments which can be used to look at the same part of the atmosphere. Four look at solar output – an important consideration in monitoring the underlying photochemistry of ozone depletion – and one monitors the charged particles trapped by the Earth's magnetic field. Two other instruments look at upper-atmosphere winds and monitor how changing dynamical conditions affect the chemistry of the stratosphere. The remaining instruments – almost inevitably known as the 'Gang of Four' – are specifically concerned with those chemical species which act as tracers for the complex coupling between atmospheric dynamics and ultraviolet radiation from the Sun. The various chemical reactions release heat which alters the temperature and affects the dynamics of the upper atmosphere. The complementarity of the instruments enables the UARS investigators to try to untangle this mess of underlying factors.

Whereas TOMS and SBUV look down, the UARS instruments look sideways at the crescent, or 'limb', of the Earth as the spacecraft moves forward in orbit. The Gang of Four cover virtually the whole of the infrared spectrum all the way into the microwave region. As a result, they can keep better tabs on the chemical merry-go-round of ozone depletion. Both long-lived species – which act as 'tags' for atmospheric circulation – and fast chemical reactions are monitored simultaneously. UARS experimenters are able to take as wide a sample as possible of the processes that can affect ozone in the stratosphere. In the near infrared, at wavelengths which are too long to be seen by our eyes, most of the radiation is in the form of scattered sunlight. Thereafter, thermal radiation due to the vibration of molecules becomes significant, as eventually does radiation caused by the actual rotation of the molecules. Molecular rotation accounts for most of the radiation in the longer wavelengths of the far infrared and through into the microwave region of the spectrum.

The signal from the atmosphere that the Gang of Four measure is thus a hotchpotch of these various processes. The fact that useful information can be extracted from them is a remarkable intellectual triumph. 'It still astounds me to think how we use satellites like UARS,' reflected Professor John Harries, a veteran atmospheric researcher who is head of space projects at the Rutherford Appleton Laboratory in Oxfordshire. 'From nearly 500 kilometres away we're trying to look at the radiation absorbed or emitted by one molecule in a billion where the atmospheric pressure is a fraction of a percentage of that at the surface.'

When *Challenger* exploded, UARS had been scheduled to be launched in late 1987. At that time, the individual instruments were being rigorously tested at General Electric before they were fitted aboard the body of the spacecraft. Before a satellite is launched, engineers have to make sure that none of the instruments consumes more power than the others, and that no thermal hot spots will occur which might then interfere with their observations. The instruments also have to be tested to ensure that they don't electronically interfere with each other and that there are no mechanical imbalances that would alter the spacecraft's centre of gravity and its ability to point towards the atmosphere. It is mind-bogglingly tedious to the outside observer and can take many months of exacting work.

The *Challenger* accident meant that the whole process would have to be repeated over again when the launch was rescheduled. In the interim, each instrument had to be returned to its home laboratory and kept under wraps until that unspecified time in the future. This enforced hibernation was not a trivial task for the sensitive electronics had to be shielded and kept clean, often in nitrogen-filled, hermetically sealed vaults. The frustration of having to wait for the rescheduled shuttle launch was nothing compared to the rigour of having to keep the instruments up to scratch and the last-minute headaches in getting everything ready again. In 1991, hardware that had been stored in safekeeping was finally brought across country and integrated together, first at General Electric and then at the Kennedy Space Center. UARS was scheduled to be launched on the forty-third shuttle mission in the autumn of 1991.

Nearly all its principal investigators were present at Cape Canaveral when the shuttle *Discovery* lifted off, their precious cargo contained inside the shuttle's cavernous belly. For the UARS project scientist Carl Reber, a veteran of many space missions conceived at Goddard, getting UARS off the ground involved a no less gruelling schedule than the other satellites with which he had previously been involved. This time, however, it was much more intense in its emotional impact because of the human element. 'Watching the launch of the shuttle is something else,' he said, 'as we'd got to know the astronauts.' The day before launch, his team had spent four or five hours inside the shuttle payload bay in a special gantry during last-minute tests.

Lift-off occurred at dusk, Florida time, on Thursday 12 September 1991. After the hype and embarrassing problems with the Hubble Space Telescope, NASA tried not to be hyperbolic in its publicity statements. The nearest anyone came was when NASA's chief scientist Lennard Fisk told gathered reporters that UARS would 'inaugurate the environmental era of the space programme'. As during all shuttle launches, there were last-minute technical hiccups which added to the tension: a leak in one of *Discovery*'s thrusters, followed by minor problems with its communications.

At 7.11 p.m., the 'stack' lifted into the fading Florida light and with it went the hopes and fears of more than a hundred UARS scientists and their families. They had gathered to view the spectacle from an observation stand a few kilometres away. A minute into the flight there was a heart-sinking moment when flames appeared to engulf *Discovery* similar to those which presaged the *Challenger* accident, but this was an optical illusion caused by high-altitude clouds. More than one onlooker sighed in relief when *Discovery* reached orbit within eight minutes. It was a sentiment altogether shared by the astronaut crew.

'You don't really have much time to enjoy the ride,' said the commander, John Creighton, a veteran of two previous flights. 'When you get on orbit is the first opportunity you have to really look out of the window.' Even he was amazed by the sight that greeted him – 'The first words out of my mouth were amazement at how high we were.' It was because the shuttle was in one of the highest (540 kilometres) and most highly inclined orbits to which it could be launched. Both were dictated for the safe dispatch of the Upper Atmosphere Research Satellite into orbit.

Choice of orbit is crucial for space missions. For UARS the relatively high inclination (57°, against the more usual 28° for shuttle flights) meant that its limb-viewing instruments could sample virtually the whole of the atmosphere during the course of a day. It could also observe the polar regions of both hemispheres in detail for the first time from space. Nimbus 7 was in a sun-synchronous orbit, always flying over the same part of the Earth at the same local time: this meant the Sun would always be at the same angle in the sky and under the same aspect of solar illumination. But UARS scientists wanted to see how the photochemistry of the upper atmosphere would change over time. They specified that each time the satellite was monitoring the same part of the atmosphere, the Sun would be in a different position relative to the spacecraft.

This meant UARS would have to fly what is known as a precessing orbit, which presented a fearsome technical challenge to the spacecraft designers. Whereas the Sun stayed in the same point relative to Nimbus 7, it would appear to move through the sky as seen by UARS and so its solar panels would have to track the Sun. Because the Sun moved, it would eventually encroach upon measurements made by the Gang of Four. Trying to isolate the signal from the stratosphere would be severely affected by the radiation from the Sun. Not only could this swamp the measurements being undertaken, it would permanently damage the sensitive electronics. The simplest way of getting round this was to turn the spacecraft around in orbit. Simply to avoid the Sun, this had to happen every thirty-six days or so.

If that sounds complex, it was simplicity itself compared to actually deploying the UARS from the cavernous payload bay of the shuttle. Like extraterrestrial removal men, the shuttle astronauts had to be certain that nothing was wrong with the satellite before it left their hands. UARS wasn't scheduled to be pulled out of the payload bay until the third day of their mission, when they were awoken by the oleaginous, yet strangely appropriate, strains of Engelbert Humperdinck and 'Please Release Me'. Up to this point, the satellite was plugged in to the shuttle's internal power supply; to release it involved charging up the satellite's internal batteries and then racing to get the solar panels open before the power dropped. The harsh environment of space, where temperatures range from 200°C in sunlight to -200°C in darkness, is no respecter of power failures.

The satellite was duly lifted out of the payload bay by the orbiter's manipulator arm, controlled by astronauts from inside *Discovery*. As the shuttle came into daylight on its thirty-fourth orbit, the six-panel solar array unfurled and started to track the Sun to power up the spacecraft. For once, events proceeded smoothly, so much so that the satellite was up and running an hour ahead of schedule. As the manipulator arm let it go, astronaut Creighton told an expectant Mission Control: 'It's on its way.' After fifteen years' planning, and an unscheduled wait of five years, UARS was finally in orbit.

The astronaut crew could relax a little. For them the next day involved changing *Discovery*'s orbit to avoid hitting a spent rocket stage, and on the fifth day they participated in a radio talk show. Two days later, they returned to Earth, their mission accomplished. For the mission engineers at Goddard, work was only just starting: there would be weeks of exhaustive checking that the satellite was working as well as it should and the equally laborious process of ironing out any teething troubles. For the scientists with instruments aboard, the successful deployment of UARS was a last chance in catching up on sleep before the first streams of data started to come through in earnest. As they would find out soon enough, the only cloud on the horizon was literally in the stratosphere.

Until the summer of 1991 – just three months before UARS was launched – Mount Pinatubo was known only to a handful of volcanologists and perhaps the most devoted players of Trivial Pursuit. Few, if any, of the people working on the Upper Atmosphere Research Satellite had heard of the largest volcano in the Philippines, which had lain dormant for 600 years. And while many of their number were doubtless aware of the dramatic impact that volcanoes could have on climate, few suspected it was something that would affect them directly. In early June 1991, as each individual UARS instrument was delivered to eager technicians at the Kennedy Space Center, all that changed.

Had they not been so preoccupied with integrating their instruments aboard the satellite, UARS scientists would have noted that newspapers were carrying reports of tremors along the so-called 'Ring of Fire', the series of volcanoes straddling the Pacific rim. An eruption at Mount Unzen in Japan on 8 June portended something dramatic along the tectonically sensitive ring. Geologists were able to alert civil authorities before the explosion of Mount Unzen by judicious sampling of seismometer read-outs, and in consequence it claimed only thirty-four lives compared to the 10,000 that were lost when it erupted without warning in 1792. Geologists watched as earthquakes and aftershocks from Unzen reverberated around the ring, eventually culminating the next day with spectacular activity in the Philippines.

Mount Pinatubo erupted for over fifteen hours and then a further three times during the next five days, pouring out a horrendous deluge of mud which destroyed nearby villages. Newspapers and television bulletins recorded the frightening spectacle of a huge cloud of grey dust hanging menacingly in the sky above it. Within hours, some of it had rained back to the ground and covered the Philippines in an unpleasant layer of dust several centimetres thick which rendered them a passable imitation of the surface of the Moon. The timely evacuation of 300,000 Filipinos minimised the death toll to 350, and 1,500 US servicemen were also evacuated from the Clark Air Force Base before flights in and out of the islands were suspended. This was a wise precaution in view of the fact that fifteen airliners reported engine blockages after flying through the dust on their final approaches to Manila Airport just minutes after the explosion.

Weather satellites recorded the dust plume erupting from Pinatubo in the infrared. Ash and dust were observed moving upwards and outwards, eventually being injected into the stratosphere, reaching altitudes as high as 32 kilometres above the Earth. Instruments aboard Nimbus 7 monitored the plume of the volcanic ash as it girdled the globe at speeds approaching 120 kilometres per hour. By 8 July it was 5,000 kilometres in extent and TOMS was 'saturated' by its sheer volume. Temperature measurements by other satellites confirmed that the dust was effective in shading the surface of the Earth from the Sun's rays. A perceptible lowering of the average global temperature resulted in the months which followed, giving rise to an earnest debate about Pinatubo's long-term effect on climate. As will be shown in the next chapter, it also led to an appreciation of how its volcanic aerosols might directly affect ozone depletion.

In the weeks after the eruption, Australian newspapers carried reports of sunset glow lasting for more than two hours. Beautifully coloured sunsets were also observed throughout the Pacific (some of them were alleged to have put hens off laying) as a result of the dust preferentially scattering blue light in the stratosphere. As the cloud spread, so the atmosphere perceptibly changed colour. Throughout the northern winter of 1991, the more observant of people would have noted a purplish afterglow during twilight at higher latitudes.

Three months after the Pinatubo eruption, UARS was launched. Its scientists soon realised that the volcanic dust which continued to encircle the stratosphere would significantly complicate and hinder their work. Whether they liked it or not, they had been drawn into what one of them wryly termed 'the volcano culture'. The satellite's chemistry instruments – the Gang of Four – would each reveal subtle and unexpected details of how the ozone layer around the world would be affected in the months that followed.

The Jet Propulsion Laboratory is unique among the many American space-research establishments in that it has flown spacecraft to all but one of the planets in the Solar System. But even Pluto will one day come in for scrutiny; JPL engineers are now defining a mission to fly there early in the twenty-first century. In some ways, the Earth was left behind in this heady rush of exploration, a situation which has now been rectified by the many JPL scientists who spend their days within the motley collection of buildings which make up the laboratory. Today JPL boasts a strong interest in Earth sciences, and, in particular, different aspects of stratospheric research.

Despite its pre-eminent reputation, JPL appears like a university which ran out of money while it was being built. It is visible in its piecemeal entirety from Highway 210 which straddles the San Gabriel foothills just north of Pasadena. For many years, staff overspilled into temporary trailers, but now most have been relocated in more recent and more permanent buildings. Building 183 is the scientific powerhouse of JPL, where astronomers, physicists, geologists and research chemists go about their business. On the seventh floor one comes across a group of atmospheric scientists who have developed a technique often described as 'radio astronomy of the Earth', which promises a remarkable breakthrough in stratospheric research.

Its latest, most advanced expression is a UARS instrument known as the Microwave Limb Sounder, or MLS, nursed through its development by a patient, equanimous scientist named Joseph W. Waters. Originally hailing from Tennessee, Joe Waters describes himself as 'a farm boy', but unlike most, he had sufficient aptitude and intelligence to study radio astronomy at one of the country's foremost academic institutions, the Massachusetts Institute of Technology. After moving to JPL he assumed leadership of a group who are the leading practitioners of monitoring the Earth at microwave frequencies. 'We are, to some extent, the new kids on the block,' Waters said, for although JPL had contributed a number of microwave instruments aboard earlier Nimbus spacecraft, this was the first time a large, limb-sounding version had flown in space for the specific purposes of studying stratospheric chemistry. Previous MLS prototypes had been tested on aircraft and balloons, but never flown from the perspective of low Earth orbit.

The microwave portion of the electromagnetic spectrum is found between infrared and the longer wavelengths of radio waves. The Microwave Limb Sounder essentially monitors the natural microwave emission from the Earth's stratosphere. The strength of the signal from what might be termed 'radio stratosphere' reveals how many molecules of trace gases are present. The MLS tunes into their emission by use of a dish antenna on one side of the UARS spacecraft. The problem is that the signal from the stratosphere is very weak and can effectively be swamped by the molecular emission from the rest of the atmosphere. To help get a signal from the background noise, MLS uses something known as the heterodyne technique to convert the incoming signal to a lower, more easily readable frequency.

A simple analogy is to consider what happens when someone talks to you when you are outside on a summer's day. The background noise from neighbours mowing their lawns and children playing may swamp what your friend is saying. By giving your friend a loudhailer, his or her voice can be amplified so you can hear better. Electronically, a similar amplification happens to the signal received by the MLS. The emission from the atmosphere is mixed in with a stronger 'carrying' signal generated within the instrument itself. Because its characteristics are well known, any differences detected between the mixed-in signal and the amplified signal alone will result from the stratospheric molecules themselves. The emission from the atmosphere is mixed in at three specific wavelength ranges dictated by the pronounced emission from the free radicals implicated in ozone depletion. There are three separate 'radio receivers' which enable the MLS scientists to detect the presence of ozone, water, sulphur dioxide and chlorine monoxide. The receiver which tunes into water-vapour emission was built in Britain and involved the participation of scientists from the Universities of Edinburgh and Heriot-Watt as well as the Rutherford Appleton Laboratory.

The Microwave Limb Sounder is the only UARS instrument that can measure chlorine monoxide, the 'smoking gun' in ozone depletion. As a result it can see when and where the trigger is pulled and obtain a truly global perspective on the 'gunshot wound' of ozone depletion which assuredly follows. Waters described the major function of his instrument as 'to watch when chlorine becomes active and how it destroys ozone' and within two days of the deployment of UARS in September 1991, MLS was 'off and running', according to Joe Waters. 'Our first surprise was the amount of sulphur dioxide from Mount Pinatubo,' he said. 'We weren't aware of how much volcanoes pump into the stratosphere.'

Absorption by aerosol particles is much less at longer wavelengths as it is indirectly proportional to the fourth power of the wavelength. In English this means that if you increase the wavelength by a factor of ten, the absorption is 10,000 times weaker. As the other three members of the Gang of Four observe at shorter wavelengths in the infrared, they were more affected by the Pinatubo aerosols than MLS in the microwave region. Indeed, these other instruments had severe difficulties in distinguishing exactly what was happening in the stratosphere. MLS, however, could see right through the Pinatubo aerosols as they slowly dispersed during the rest of 1991.

By the start of 1992, analyses of the MLS data had revealed distinct tongues of low ozone over the tropics extending northwards. At the same time, new data coming in were pointing to chlorine monoxide increasing around the Arctic to levels seen by MLS over Antarctica during the most recent austral spring after the satellite had been launched. As we will see in Chapter Seven, these observations at the start of 1992 raised the worrisome spectre of an ozone hole forming over the North Pole.

The Clarendon Laboratory at Oxford University has a unique place in the history of atmospheric ozone. It was here that Gordon Dobson and his colleagues faithfully assimilated daily ozone measurements from around the world, which also led Sydney Chapman to determine the mechanism by which the stratosphere is continually replenished with ozone. Since Dobson's day, the physics department has mushroomed into a host of new buildings as its activities have diversified. Just behind the Clarendon Lab is the Atmospheric, Oceanic and Planetary Physics building, whose existence stems from the pioneering work of John Houghton and the early Nimbuses. Today that work continues with an instrument aboard UARS known as the Improved Stratospheric and Mesospheric Sounder, 'the great-grandson of the first ones we built in the sixties', in Houghton's phrase.

Fredric Taylor is now head of the department and principal investigator for the ISAMS instrument. A Northumbrian by birth, he studied at Liverpool University for no better reason than 'it was 1963 and Beatlemania had broken out. There was also the northerner's fear of going any further south that sent me there.' He overcame that fear to move to Oxford as a postgraduate in the late 1960s, where he developed a new technique known as pressure modulated radiometry. This is one of the most ingenious methods of filtering out unwanted spectral signatures from the stratosphere and has reached its apotheosis on ISAMS.

'Other radiometers use filters to look at the trace gases they're interested in,' Taylor explained. 'We use a technique of selectively labelling the gases we're interested in without losing any of the signal.' ISAMS makes use of cells of gases which it is attempting to measure in the stratosphere, including ozone, the oxides of nitrogen, methane, water and carbon monoxide. In the case of, say, nitrous oxide, the radiation from the atmosphere is passed through a cell of nitrous oxide before its spectrum can be measured. This spectral signature from the stratosphere is swamped by the gas in the cell. But the pressure of the gas in the cell is repeatedly decreased and increased with an elaborate piston mechanism. As the gas pressure inside the cell decreases, the spectral signature from the atmosphere will become more pronounced. It will fade once more as the pressure is increased. By repeatedly modulating the pressure in this way, it is possible for the ISAMS detectors to home in on the spectral lines in which Taylor's team are interested. They will repeatedly flash on and off in time with the modulation.

Fred Taylor has good reason to recall the first test flight of a pressure modulated radiometer because the completion of his doctorate depended on it. With a sense of great trepidation, his Clarendon colleagues mounted it aboard a balloon early one morning in 1969 above Salisbury Plain. Two earlier attempts had failed, but that freezing morning high above Stonehenge, the PMR performed flawlessly and its faithfully recorded data were as he later reflected, 'more beautiful than the *Mona Lisa* to the perspiring supplicant for the D Phil on the ground beneath'. After receiving his doctorate, Taylor joined the brain drain and spent a greater part of the 1970s at JPL, working in the same building as Joe Waters. Whereas the MLS people had their eyes on the stratosphere, Taylor and his group were also concerned with the other planets. The PMR technique would eventually lend itself to the investigation of the atmospheres of Venus and Mars.

By the time NASA issued proposals for instruments to fly aboard UARS in 1981, Fred Taylor had returned to the Clarendon Lab and proposed the most advanced version of the PMR technique. A Stratospheric and Mesospheric Sounder had flown on Nimbus 7, but newer and more sensitive hardware could increase its resolving power by a factor of five and its sensitivity to radiation a hundredfold. A suitably impressed review panel accepted it for inclusion aboard UARS and not just because it could be flown 'free' so far as NASA was concerned (its development costs had been shared by Oxford University and the Rutherford Appleton Laboratory).

ISAMS also brought about another interesting innovation. To observe what is essentially a heat signal from the atmosphere, the radiometer itself has to be cooled to minimise its own thermal interference. Normally this is achieved by use of cryogenic liquids at supercold temperatures contained in slightly fancier versions of Dewar flasks in which the instrument detectors are located. However, cryogenic liquids will eventually boil away and thereby set a limit on the instrument's lifetime. ISAMS effectively cools itself, in Fred Taylor's phrase, 'by using pressure modulation to actually cool the instrument'. The pistons which modulate the pressure inside the gas cells alternately radiate heat away and cool the instrument directly when the gas expands. The temperature of the whole instrument can be reduced to $-195°C$. 'The benefits of mechanical coolers are enormous,' Taylor said. 'They are compact and effectively last for ever.'

The two final members of the Gang of the Four enhance the range of observations that can be made from space. One of the most serious limits on any infrared instrument is the 'background noise' generated by the instrument itself. There is no way of telling whether the infrared radiation it is trying to measure originated from the atmosphere or within the instrument itself. The most obvious answer is to cool the instrument so that it will have as little thermal impact as possible. Whereas ISAMS uses pressure modulation to narrow down the range of wavelengths measured, the Cryogenic Limb Etalon Spectrometer (CLAES) essentially filters the incoming signal by cooling the whole of the instrument. It was built by the Lockheed Palo Alto Research Laboratory, and its principal investigator is an Irish-born scientist named Aidan Roche.

The cryogen of its title is a block of solid neon which surrounds the instrument as a whole. As the neon melts and vents to space, it removes heat from the instrument. It is so efficient at this process that CLAES can be kept to within 15 degrees of absolute zero, $-258°C$. Originally, solid hydrogen was going to be used but the

Challenger accident enforced stringent safety considerations. As a result, one of the most dangerously flammable materials known to man was replaced by one of the most inert. The neon is surrounded by solid carbon dioxide – 'insulating the insulator', in Aidan Roche's phrase – to shield the neon from increased heating as the satellite comes into sunlight during each orbit.

Because it is 60°C cooler than ISAMS, CLAES can detect greater numbers of species. But unlike ISAMS, it cannot operate indefinitely because it has a finite supply of solid neon and carbon dioxide. CLAES can determine the presence of fifteen different chemical species which Roche classes as source molecules for chlorine (including CFC-11 and -12), the key radicals and reservoirs which tie up active chlorine. 'We can do a particularly good job on diurnal constituents like the oxides of nitrogen,' Roche said. CLAES is the most sensitive UARS instrument because it is that much cooler, but it is also the one that died first. 'CLAES will only get one look at the Antarctic ozone hole,' Roche told me in June 1992. 'By the next southern spring, we'll be gone.'

The final chemistry instrument aboard UARS differs from all the others. It is an absorption instrument and looks out in the opposite direction to the emission instruments on the other side of the spacecraft. The HAlogen Occultation Experiment (HALOE) observes how the spectrum of the Sun is attenuated owing to absorption by atmospheric molecules themselves. Its quarry is those ozone-destroying compounds containing bromine and fluorine which are particularly adept at doing so at warmer temperatures than most CFCs. Where the other instruments are shrouded in darkness, HALOE constantly seeks the Sun. 'We look north when the other instruments look south,' said its principal investigator, James Russell III of NASA's Langley Research Center in Hampton, Virginia.

In principle, HALOE should be capable of greater sensitivity because the Sun emits many more photons than the atmosphere. From its position on UARS, HALOE sees a sunrise and sunset on each of the fourteen orbits the satellite completes per day. As the Sun descends or ascends through the atmosphere, HALOE obtains a profile of molecules which are absorbing the sunlight in the infrared spectrum. It uses four broad-band filters to restrict observations to

NO_2, water, carbon dioxide and ozone itself. Another group of filters are used to measure halogenated species in the stratosphere. To make these spectral lines more noticeable, HALOE splits the light into two, and passes one half through a cell of the gas in which the scientists are interested. The other half is passed through a vacuum. Electronically, the gas-cell signal is subtracted from the vacuum signal, so that the spectral lines will emerge.

This technique does not require pressure modulation, but the optics have to be precisely aligned. Curiously enough, the gas cells that are used to look at halogenated species like hydrogen chloride and hydrogen fluoride are made of gold because they are so corrosive. 'That's the purest gold any of us will probably ever see,' Russell wistfully added. But over the following Antarctic winters, observations by HALOE would reveal subtle details about the ozone hole and its interactions across the whole of the southern hemisphere.

Once a satellite is up and running, engineers cannot blithely assume that the spacecraft and its instruments will behave themselves perfectly. It is for this reason that NASA maintains a constant vigil by use of its intricate network of telecommunications around the world. For the scientists who will attempt to coax meaningful results out of the raw deluge of data, the months following the launch of any satellite will be devoted to the time-consuming and – it must be said – tedious process of validating and calibrating the instruments. So, after UARS was launched, there followed a process of 'correlative measurements' which involved the meticulous choreography of balloon and aircraft flights to coincide with satellite observations. The former would fly through parts of the atmosphere that were being scrutinised remotely by UARS from orbit.

It would take the better part of a year after launch before the principal investigators would feel confident that their data were sufficiently in agreement with each other to be released to the outside world. The first raw UARS data were not made available to other researchers until the second anniversary of its launch in September 1993. Partly this was due to intellectual property rights in space missions, but also the validation process was hindered and prolonged by the intervention of the dust from Mount Pinatubo.

'The Pinatubo aerosol acts as an absorber of the ozone signal,' said Aidan Roche, whose instrument was notably affected by the volcanic plume. 'It is also an emitter in its own right.' By the spring of 1992, however, the Pinatubo cloud had dispersed to such an extent that all the Gang of Four had started to return data from which it was possible to compare their respective observations of the stratosphere. A workshop was convened at the Clarendon Laboratory in March 1992 where intercomparisons of the data were discussed over the course of nearly a week.

All the various idiosyncrasies, hiccups and inconsistencies of the instruments were compared and analysed in detail. The main conclusion of the workshop was, to quote from the report which was produced shortly thereafter, 'that the data are still in a very preliminary stage, and much more work remains before they will be ready for scientific application'. The next workshop would take place in October 1992 in Boulder, Colorado, by which time, unfortunately, one of the Gang of Four was in the throes of a strangely protracted death.

One topic of conversation at the March meeting in Oxford was the failure of the Improved Stratospheric and Mesospheric Sounder a few weeks before. As an example of the strange and unforeseeable pitfalls of using space technology, it probably cannot be bettered. The problems with ISAMS concerned the 'chopper wheel' at the front end of the instrument. To the untrained eye it appears similar to a small windmill-like propeller; its function is to convert the incoming signal into a more digestible form for the instrument's electronics. It modulates the signal entering ISAMS at higher frequencies than the pressure modulation (around 1 kilohertz) so that it can directly be converted into an AC current and more easily read off by the ISAMS radiometer.

On 18 January 1992, the daily return of ISAMS data revealed that something was seriously amiss. Though a signal was intermittently being recorded, it was obvious that the chopper was out of synch. By sending commands up to the instrument via the POCC at NASA Goddard and subsequent monitoring of the way in which the power seemed to dissipate, the Clarendon team were able to diagnose the cause of the problem. Over the following weeks, constant monitoring of ISAMS's vital functions seemed to suggest that the wheel was sticking. Most of the Oxford team were astounded that their problems would stem from something so simple as a chopper wheel. 'These sorts of choppers are bread and butter in remote sensing of the atmosphere,' Fred Taylor said. 'I cannot begin to tell you how we sweated blood over the pressure modulators and the closed-cycle coolers.' And, needless to say, these inordinately more complex components have worked perfectly since ISAMS was switched on.

So Fred Taylor and his colleagues repeatedly switched the chopper motor on and off in rapid succession to try to 'hammer' the wheel loose. It worked. On 27 March 1992, the chopper started up again and ISAMS operated successfully until the end of July, when the wheel failed completely. This time the engineering data seemed to show that there was an inherent fault in its motor's electronics. 'If we had the circuit in front of us we could diagnose the problem,' Taylor said, but the Oxford team were hampered by having to do it from 600 kilometres away, and the fault was in a part of the instrument which was difficult to analyse remotely.

They tried to get the wheel working again throughout 1992 but to little avail. An attempt in January 1993 resulted in starting the motor, but the chopper began to spin so fast that no signal at all was received by the instrument. Despite Goddard's optimism that something could be salvaged, the chopper would not slow down. Taylor and his colleagues decided to halt all attempts at resuscitating their recalcitrant instrument at the end of January 1993. In 1994 they were still trying to analyse the cause of the problem, with little hope of being able to get it working again.

The ongoing saga of the chopper wheel is one of the most peculiar Fred Taylor has ever had to face in two decades of research. 'Of course, we're sorry that ISAMS didn't run smoothly,' he told me in March 1993. 'If it hadn't got such a lot of data first, then we'd have been pretty upset.' And proof that his instrument was not an ignominious failure could be seen in the form of neatly stacked print-outs of 150 days' worth of data. Clarendon researchers were busily digesting and analysing their instrument's bountiful return so that it could be published and made available to the rest of the scientific community.

ISAMS has clearly returned a fascinating insight into the workings of the upper atmosphere. At the end of December 1991, as the Microwave Limb Sounder was monitoring a steady build-up of chlorine monoxide to levels greater than in the Antarctic springtime, ISAMS detected a very cold region and its associated chemistry just outside the vortex, which was then wobbling over the North Sea. As measured by ISAMS, the vortex was as cold as −77°C, the temperature required for nitric acid to freeze out in the stratosphere. 'This is the first spectroscopic study of a polar stratospheric cloud from space,' Taylor said with considerable pride. 'As far as we know, the region of chlorine monoxide which MLS saw resulted from the chemical processing that we saw.'

It is not, he was the first to admit, news of the Earth-shattering variety, but it is an important piece in the stratospheric jigsaw puzzle of ozone depletion. UARS was explicitly designed so that its different instruments would produce data that could be consolidated into a coherent whole. Suitably cross-referenced and meshed together, the data from all the UARS instruments will enable scientists to draw the most extensive portrait of ozone depletion. The ISAMS observations will enable researchers to diagnose the exact details of ozone loss within the Arctic. Detailed analysis will allow a limitless 'action replay' of the conditions necessary to destroy it on the periphery of the vortex.

The problems with ISAMS are not alone, for the spacecraft as a whole suffered a serious mishap which has limited the numbers of remaining instruments that can be operated at any one time. As a result of its precessing orbit, the Sun continuously moves across the sky relative to the satellite. To provide as much power as possible for its onboard systems, the satellite's solar panels (usually referred to as 'the array') have to be square on to the Sun to receive the maximum amount of sunlight. To maintain this geometry, they continuously have to track the Sun and employ two motors expressly for this purpose (one as backup). When the satellite has performed its turnaround, which it does every thirty-six days, the Sun will seem to be moving in the opposite direction so the motor is reversed.

On 1 June 1992, UARS was scheduled to perform a turnaround manoeuvre. The spacecraft as a whole (and its instruments) were powered down to 'hibernation mode' as is normal when the spacecraft is going to be without solar power for a time. The spacecraft batteries would be fully charged and, automatically, its computers would anticipate the time when the panels could once again track the Sun. The batteries would then be recharged and its scientific instruments could start to perform observations again. Normally this procedure would take a few hours at most, and all seemed to be well, until controllers at Goddard commanded the solar array to commence tracking the Sun again.

The clutch on the primary motor failed in such a way that it dragged on the second motor. Rather than risking damage to either motor, all attempts to let the panels track the Sun were abandoned at the end of June 1992. After intense analysis at Goddard, it was decided to continue using the first motor, despite the fact that its clutch could not be put into neutral during each turnaround manoeuvre. Subsequently, UARS engineers have learned to stall the motor carefully before starting it up in the opposite direction. In this way, they can largely avoid the extraterrestrial equivalent of the grinding noises which bedevil domestic motor cars with faulty clutches. To date, this approach has worked very well, though the effective loss of a backup motor is still a cause for concern for the long-term prospects of the UARS mission.

By the time the solar arrays began tracking the Sun again, on 17 July, some seven weeks had elapsed since the problem had first emerged. During that time it had been planned that UARS should have been looking south and scrutinising the way in which the Antarctic vortex prepares itself for the dramatic springtime loss of ozone. Rather than power the satellite down completely during this crucial time period, the solar arrays had been parked at the 'high-noon position' so that for a few days, at least, enough power could be generated for the instruments to remain in operation. In time, as the Sun drifted away, many of them had to be switched off, although it was possible to keep both MLS and CLAES on throughout the austral winter and spring of 1992.

Apart from separate, persistent problems with the recharging of the satellite's batteries, UARS had not suffered any major failures by 1994 and looked set to reach its third anniversary in orbit as planned. Though power margins were still worrying controllers at Goddard, the Upper Atmosphere Research Satellite has returned some remarkable insights into the mysterious workings of the stratosphere. Project scientits have been shown just how little was previously known about the conditions that give rise to the ozone hole while Antarctica is shrouded in midwinter darkness.

At the start of the austral winter of 1992, knowledge of how the Antarctic stratosphere primed itself for the massive ozone loss which occurred in the spring was, in Joe Waters's phrase, 'similar to knowledge of America at the time of Columbus'. Observations from stations on the ice are limited by their geographical distribution. TOMS and SBUV can only work with direct sunlight, and so cannot view the Antarctic stratosphere through the winter season as the continent below remains in darkness. In any case, the geometry of Nimbus 7's orbit is such that the Sun illuminates the poles very badly throughout the rest of the year. So scientists could hardly wait for the first ever data on 'polar night precursors' from the Upper Atmosphere Research Satellite, which is not hampered by such restrictions. The 'precursors' in question are those chemical species which play an important role in creating the ozone hole when the sunlight returns in the springtime. Researchers involved with the Gang of Four were not to be disappointed.

The first surprise was the build-up of chlorine monoxide so early in the season. Temperature measurements revealed that stratospheric temperatures had dropped to the threshold at which polar stratospheric clouds form as early as 10 May. Throughout the summer, whenever MLS was looking south, it observed active chlorine south of 65°S until the break-up of the vortex towards November. But the biggest surprise for Joe Waters was that at the start of the spring season 'there was enough chlorine for the ozone to go in a month'. In the absence of sunlight, however, there was no resultant ozone loss, although, towards the edge of the vortex where air was intermittently catching sunlight, swirls of ClO were observed to be processing the air there. MLS can measure both ozone and ClO at the same time and saw something very peculiar: *both* were on the increase during the first half of the winter. Previously, intermittent measurements of OClO from McMurdo Sound or even ER-2 profiles of ClO had only given limited geographical coverage across the South Pole. Now, thanks to UARS, the full extent of the mysteries of the Antarctic vortex was revealed.

The answer to this riddle turned out to be that ozone was coming in from above and descending down through the vortex in the manner of water disappearing down a plughole. After the vortex had broken up, Waters and his colleagues were able to analyse the season's MLS observations to perform the equivalent of an extended action replay of the chemical processing which had taken place. 'There is a race between the ozone-rich air and the active chlorine,' Waters explained. 'For the first half of the winter, the dynamics win. Then, when the Sun returns, the chemistry takes over.' And how – for, as we will see, the ozone hole which formed in 1992 did so far quicker than before and grew to cover the largest area on record. Other UARS measurements showed that there was a strong depletion within the vortex from September until the middle of October, and thereafter low ozone values persisted even after it broke up in November.

At the end of September, a tongue of low ozone was seen extending towards the north by MLS. It appeared to have been thrown off from the vortex rather like water from a sprinkler. An obvious question presented itself: how far did the chemical processing seen by the MLS extend beyond the vortex? Here CLAES and HALOE observations came in, because, unlike MLS, they can measure tracers – gases that bear a known ratio to ozone before it is destroyed within the vortex. They act as chemical 'fingerprints' for the motion of air parcels which have been chemically processed. By looking at methane and hydrogen fluoride respectively, CLAES and HALOE indicated that air with similar composition to that seen inside the vortex extended as far as 30°S well into the spring season before it had broken up. But it was not cut and dried, as there were other unexpected observations. Hydrogen chloride, for example, was observed to be very low at high southern latitudes in the last week of September, consistent with elevated levels of ClO. But then it suddenly increased during the first week of October while ClO levels remained the same.

This observation cannot be explained by current understanding of chemistry alone so dynamic processes may be the cause. The answer may be that strong vertical descent is occurring throughout the vortex, a conclusion supported by HALOE observations of tracers in the early spring (CLAES was looking north at that time).

These results have far greater implications for ozone depletion throughout the southern hemisphere: they suggest that the vortex isn't as contained as it was once thought. 'We have clearly seen that the chemical processing in the Antarctic springtime is not restricted to within the vortex,' said Jim Russell, the HALOE principal investigator. 'It implies that dynamics play a stronger part than we had previously suspected.'

MLS has implicitly confirmed this – and something rather more subtle besides. Ozone is clearly descending into the vortex as a result of the natural, dynamic transportation of ozone from higher latitudes and altitudes. This fresh influx of ozone has masked the extent of the chemical processing which is taking place in the Antarctic stratosphere. 'There is more ozone loss than we actually see,' Joe Waters said by way of explanation. If the vortex is indeed 'leakier' than had originally been thought, there is the possibility that ozone depletion may be more widespread across whole portions of the southern hemisphere.

In October 1993, Nimbus 7 reached its fifteenth birthday in orbit. It had long since exhibited signs of wear and tear, but now, results from its most infamous instrument were noticeable by their absence on television screens the world over. The abhorrently fascinating, false-coloured spectacle of the Antarctic ozone hole was not seen during the austral spring of 1993, for the Total Ozone Mapping Spectrometer had finally stopped working. Earlier that May, TOMS had succumbed to an electronic failure which had rendered it useless. But though it is gone, it is hardly forgotten, for TOMS had also alerted the world to a phenomenon whose importance, it may be argued, has already eclipsed the springtime losses over Antarctica. In nearly fifteen years of operations, TOMS bore continuous witness to a steady stripping away of the ozone layer throughout the rest of the globe. Though it only represents a small fraction of the total ozone in the atmosphere, it portends something very worrying, for chlorine loading of the stratosphere continues unabated and promises to reach a maximum after the turn of the century.

Global Perspectives

The resilience of the Nimbus 7 TOMS instrument has transformed it into a workhorse for the routine monitoring of ozone from space. New and improved versions of the basic TOMS design have already been, and will continue to be, launched to keep track of this most dangerous time of ozone depletion. Measuring trends, however, has not been a matter of merely accumulating observations and crunching out meaningful numbers shortly thereafter. 'When we proposed TOMS in 1973, we never thought we'd have the need to measure trends,' said Arlin Krueger. 'It was only designed to last for two years.' Long before it failed, the electronics and optical sensors used by TOMS had started to degrade significantly, causing innumerable complications in trying to determine what was happening to ozone losses worldwide. Circumventing these problems required considerable ingenuity by Arlin Krueger and his colleagues at Goddard.

Because the TOMS instrument compares the ratio of ultraviolet radiation from the Sun to that backscattered by the atmosphere, it has to take a reading of the Sun's ultraviolet output on a regular basis. The best time to do this is when the Sun is at right angles to the motion of the spacecraft. This orbital geometry occurred when Nimbus 7 was flying over the North Pole. Although the surface below would still be plunged in darkness during the winter season, sunlight is still visible from 1,000 kilometres on high. 'If we could have directly pointed the instrument towards the Sun, life would have been a lot easier for us,' Arlin Krueger said. As they could not, a 'diffuser plate' was mechanically deployed to reflect sunlight into the instrument's field of view. And though this technique worked perfectly well after launch, time and the vagaries of low Earth orbit took their toll of the plate's ability to reflect sunlight and, thus, the instrument's accuracy in determining long-term trends in ozone measurements.

It comes as a surprise to learn that the Earth's atmosphere even in these tenuous upper reaches, can cause severe problems for satellites. Lone atoms of oxygen, ionised by the interaction of the solar wind with the Earth's magnetic field, are particularly good at corroding the most sensitive satellite components. Most dramatically this takes the form of eating at electrical connections between the individual cells upon which the capability of solar panels to generate power depends. This ghostly remnant of our atmosphere can also oxidise the exposed surfaces of diffuser plates, which would otherwise remain in pristine condition. It was a point brought home in the early 1970s when the diffuser plate aboard Nimbus 4 lost 80 per cent of its reflectivity within two years through constant exposure to space.

So for Nimbus 7, Krueger wanted the diffuser plate to be deployed only once per orbit. However, a secondary aim of the Nimbus 7 mission was continually to monitor the ultraviolet output from the Sun. A compromise was reached where the diffuser was to be deployed only once per orbit, and, for the first few years after launch, all seemed to be well. But by 1983, TOMS was recording progressively greater losses of ozone globally than was being seen by the ground-based Dobson network. Unlike the venerable spectrophotometers, satellite instruments can be calibrated only once: they are at the mercy of degrading influences which are difficult to quantify exactly. By the mid-1980s, it was clear that diffuser-plate degradation was seriously affecting TOMS measurements.

After the ozone hole was announced and TOMS subsequently confirmed its existence, its measurements were drifting more markedly from the Dobsons. 'But underneath this instrument drift it was clear that there was a signal finally emerging,' said Krueger's colleague Richard Stolarski of the elusive evidence for trends in worldwide losses in ozone. It was clear that data from both the ground and space would have to be combined in some way to determine what the trends really were. Metaphorically speaking, it involved sewing new patches onto an old blanket. The ground-based Dobsons could only give snapshots of ozone loss overhead and it was not easy to extrapolate observations at all points in between the stations. Satellites, on the other hand, had the distinct advantage of a truly global perspective. Comparing the data from each source and investigating the precise details of the differences between them would thus resolve the problem.

This task fell to Stolarski within NASA Goddard's Atmospheric

Chemistry and Dynamics Branch. He was acutely aware that the extent of the global decline in ozone levels was no longer just an academic question. Their correct determination would have important political ramifications for the timing of the phase-out of CFCs. The re-analysis of the Arosa data by Sherry Rowland and Neil Harris in 1986 had proved beyond all doubt the usefulness and validity of the long-term record of the Dobson network. To quantify the satellite drift, Stolarski's Goddard team meticulously compared TOMS and Dobson observations whenever Nimbus 7 passed over selected stations in the northern hemisphere. The difference between their measurements was quantified, and overall, it was possible to gauge exactly how far TOMS had drifted. Although on average there was a statistically significant 6 per cent drift, when this was taken into account, it was clear that the northern hemisphere mid-altitudes had been losing ozone at an average rate of 3 per cent by the start of 1988.

There was, however, still a fundamental weakness in the correction process. 'If the Dobsons are wrong, we're wrong,' said Rich Stolarski. 'We wanted to find out what the drift was independently of the Dobsons.' So a technique termed 'wavelength pair justification' was developed at Goddard to iron out conclusively any errors in trend measurements. It uses observations from the SBUV instrument across the ultraviolet spectrum: in the same way that the ratio of incident and backscattered wavelengths is used to determine total ozone, a handful of wavelength pairs across the ultraviolet spectrum are compared.

In Stolarski's phrase, the 'brown goo' which builds up on the diffuser plate should affect all wavelengths to roughly the same extent. So the ratio of these carefully selected pairs should be the same across the spectrum if the plate's degradation is independent of wavelength. By ascertaining exactly how these ratios altered at specific points across the spectrum, it was possible to model how the diffuser plate would affect TOMS measurements. It is hardly the easiest of calculations and Stolarski admitted that 'the math do get a bit involved'. But he added: 'The effort is worth it as we have a method of internally calibrating the data.'

How Soon Is Now?

By the turn of the 1990s, however, the task of measuring trends over the course of the next decade would be considerably eased by the launch of new TOMS instruments into space. In the early 1970s, as a graduate student at the University of Michigan, Stolarski had shown that the Space Shuttle posed little threat to the ozone layer. Two decades later, he would have the benefit of observations from an SBUV carried aboard the shuttle to show that and, more importantly, help confirm Nimbus 7's observations that ozone levels were changing globally. And the first replacement for TOMS itself would bring about one of the strangest episodes in Arlin Krueger's thirty years of adventures in the stratospheric ozone business.

The degradation of space-based instruments is an inevitability to which their users have had to inure themselves. It had been anticipated by Ernest Hilsenrath, a Goddard colleague of Krueger and Stolarski, who is the project scientist for the Nimbus 7 SBUV instrument. He was a key player in a concerted effort to ensure that backscatter instruments would be in place to continue stratospheric monitoring as and when Nimbus 7 started to fail. Consequently, a number of SBUV instruments have been flown on NOAA polar-orbiting weather satellites since the mid-1980s.

The first 'operational' SBUV was flown on NOAA 9, launched in April 1984, and worked successfully for two years. It was followed in 1988 by NOAA 11, and a further instrument followed on board NOAA 13 towards the end of 1993. But these replacements were not the panacea which they might at first seem.

Taken together, their observations cannot simply be dovetailed with those from the Nimbus 7 SBUV. Further steps have to be taken in a convoluted analysis procedure. Operational weather satellites have different requirements from experimental ones like Nimbus 7, particularly in the way their data are archived and made available to researchers. Though NOAA weather satellites are also in Sun synchronous polar orbits, they cross the equator at different times from Nimbus 7. This means that their SBUV data record is displaced in time. As experience has shown, space hardware rarely behaves perfectly, and each and every instrument has its own characteristic kinks and funny little moments. True to form, both the NOAA SBUV instruments thus far have exhibited inherent errors in their observations which have further hampered trends analysis. Quite simply, assessing the NOAA and Nimbus 7 data records is a case of not comparing like with like.

'Our biggest headache concerns small changes in instrument characteristics against small ozone changes,' Hilsenrath said. 'They can be of the same magnitude, so they could easily get overlooked.' To get around this problem, Hilsenrath also proposed in the late 1970s that the Space Shuttle be used to fly an improved SBUV (known as the Shuttle SBUV or SSBUV) to keep tabs on the operational instruments for a few days at a time. The idea was to repeatedly fly the same instrument whose characteristics were well known, so that its observational record could be used as an independent check on the long-term efficiency of the NOAA and Nimbus 7 SBUV instruments. For NASA, it would be a reasonably inexpensive way of cross-checking the calculated drift in instruments which had been in space for many years.

Remotely operating a small instrument does not sound a particularly exciting task for Shuttle astronauts. In the past, Shuttle crews have oftened privately railed against what they call 'Larry Lightbulb' experiments (pointless PR tableaux designed to make the evening news usually involving high-school kids and tomato seeds).

However, the Shuttle crews have enjoyed using the SSBUV instrument because of its obvious importance in monitoring the ozone layer. And Hilsenrath was paid the ultimate accolade by the astronauts who used the first SSBUV on the thirty-fourth flight of the Space Shuttle in October 1989. The crew of *Atlantis* scrawled the following tongue-in-cheek message on their mission portrait photo which hangs on Hilsenrath's office wall:

> To Ernie
> We all know that SSBUV was the *real* science on STS-34.
> Thanks,
> The crew.

STS-34 had made banner headlines around the world because it was carrying the nuclear-powered *Galileo* spacecraft into orbit en route to the planet Jupiter. Protesters were out in force at Cape Canaveral to draw attention to the hazards of launching plutonium into space. 'They probably weren't aware there was an environmental instrument on board!' Hilsenrath said.

There were subsequent flights of SSBUV in October 1990, August 1991, March 1992 and April 1993. Hilsenrath's main task is to cross-reference the SSBUV results with the operational ones. There are differences, he said, but 'on the whole they are in good agreement'. As other operational SBUVs are planned for further NOAA weather satellites, the SSBUV flights will ensure that there is a continuous record of ozone trends towards the turn of the century when ozone depletion will reach its maximum. Twenty years after it was first tested aboard Nimbus 4, the backscatter technique is assured of its place in the constant vigil on ozone depletion.

Mikhail Gorbachev's downfall was incubating in the febrile imaginations of a handful of plotters when Arlin Krueger and his NASA colleagues arrived in what was still the Soviet Union in early August 1991. Just a few short years before, flying an American instrument aboard a Soviet satellite would have been unthinkable. But now, *glasnost* and swiftly executed opportunism would enable a spare Total Ozone Mapping Spectrometer to fly into space as the first replacement for that carried aboard Nimbus 7. By an unfortunate set of circumstances, however, its launch and Gorbachev's temporary downfall would almost coincide. Being foreign nationals in and around what had formerly been a sworn enemy's missile launch site was tense enough for NASA scientists without a coup to contend with.

It was Gorbachev's moves towards democratisation that had led to the idea of flying TOMS aboard a Soviet weather satellite in the first place. It had come about because of the historic Intermediate Nuclear Forces Treaty signed by Gorbachev and Reagan in Washington at the end of 1987. An accord had also been signed for further cooperation in space, which, for a NASA still in the psychological doldrums after the *Challenger* accident, presented too great an opportunity to be missed. By this time, it was clear that Nimbus 7 would not last for ever and Goddard scientists had started to ponder how to keep a continuous record going. With little chance of getting a replacement TOMS into space on a homegrown launcher, the offer of a Soviet launch was something of a godsend.

Global Perspectives

However, the immutable laws of orbital mechanics, aided by a set of strange circumstances left over from the Cold War, would conspire to make operations of the replacement TOMS that little bit more complex. Since the 1960s, the Soviets had launched handfuls of a polar-orbiting weather satellite called Meteor. In time a small constellation of them have been launched, so that there are weather measurements being taken over the same part of the Earth's surface every three hours. The Meteors have been launched from what was formerly one of the Soviet Union's most secret launch centres, Plesetsk, some 170 kilometres south of Archangel. As it was a prime launch site for Inter-Continental Ballistic Missiles it set special limitations on the orbits of the satellites that it also launched.

The Earth rotates from west to east (and as a result, the Sun moves in the opposite direction through the sky). Most satellites are launched towards the east to use the spin of the Earth to provide extra momentum for launch. But if a satellite is launched east out of Plesetsk, it will head into orbit over Alaska and within half an hour be heading south over sizeable portions of the continental United States. As American early-warning radars could mistake such launches for missile attacks (particularly if their boosters failed), the Soviets wisely decided to play safe and launch their satellites in the opposite direction in what are known as prograde orbits. Their space-based weather monitoring network is thus based on satellites which do not fly Sun-synchronous orbits. They fly over the same part of the globe at progressively earlier local times, but there are so many Meteors that there is always one satellite over that same spot at the same local time each day.

TOMS was specifically designed for a Sun-synchronous orbit. But it was clear to Krueger that it could still return useful information with a little judicious 'juggling' of the orblt of the Meteor satellite which would carry it into space. Although the instrument would not be flying over a constantly illuminated globe, its observations could be tied in with those of Nimbus 7 at the most important times of the year. The local time when the Meteor crossed the equator would progress through a 24-hour cycle in 220 days. 'Coming up to 6 a.m. or 6 p.m. we effectively "lose" a hemisphere to darkness,' Krueger said. NASA wanted to ensure that the Meteor would be in sunlight for the whole of the Arctic and Antarctic springtimes when ozone depletion was at its greatest. NASA Goddard duly calculated that the optimum time for launch would be in the late summer of 1991.

'We had to set the launch time to optimise coverage of the polar regions,' Krueger explained. This could be achieved if the Meteor was launched at 12.15 p.m. on Thursday 15 August 1991, although the track of the orbit flown would cause problems for the Russians' domestic weather measurements. Whereas many people who have dealt with Soviet space officialdom tell bloodcurdling stories of bureaucratic inertia and incompetence, Arlin Krueger and his Goddard colleagues cannot praise them too highly. 'They bent over backwards to help us,' he said, and in the end the Meteor satellite took off within 0.4 seconds of a launch date set three years before, and whose exact timing had been set only six months previously.

The TOMS team had arrived two weeks earlier to ready their instrument for lift-off and were later joined by various NASA high-ups and VIPs on the launch day itself. To the Americans, long accustomed to rigorous safety strictures at Cape Canaveral, Plesetsk was a revelation. Not only could they stand next to the rocket as it was being rolled out to the launch pad, they were even allowed to wander all over the pad as it was being erected. At the Cape security fences meant that once a satellite and its payload were integrated on its launch vehicle, project engineers effectively had to say goodbye to the fruits of their labour. Not so in Plesetsk: an hour before launch, Krueger and his compatriots were evacuated to a viewing stand one or two kilometres away. As a safety precaution, they were issued with gauze masks while the upper stages of the Meteor launch rocket were fuelled with hydrazine, one of the most poisonous of propellants used in modern rocketry.

That Thursday was 'a perfect day for launch', according to Arlin Krueger, although some of the onlookers were disappointed that the rocket passed behind clouds just a few seconds after launch. But its premature disappearance from view was more than compensated for by the fact that TOMS was powered up and showing signs of rude good health even before the Meteor satellite had completed its first orbit. Before its protective coverings and internal electronics could actually be switched on, however, the Goddard team would allow time for TOMS to 'outgas' any residual atmosphere which had accompanied it into space. In the near vacuum of Earth orbit, the high voltages commonly used in satellite instruments might cause powerful electric arcs to spark through any oxygen molecules which had lingered in close proximity. If they didn't short out the instrument as a whole, they certainly would damage its sensitive electronics. Krueger and his team planned to switch on TOMS on the following Thursday and would return home shortly thereafter.

Events at ground level would overtake them. That evening, they flew to Moscow for press conferences on the Friday and Saturday, and on the Sunday, they went to an air show. They had noticed large concentrations of troops en route, as well as having to pass through endless military checkpoints in and around the capital, but presumed that this was normal for the Soviet Union. Krueger thought nothing more of it and looked forward to another day's sightseeing in Moscow before he would spend anxious hours at the Kaliningrad mission-control center waiting to switch TOMS on. When Monday morning dawned, no tourist buses arrived and there was a palpable sense of disquiet. 'Our translator told us that there had been a coup,' Krueger recalled. 'Our dilemma was: how would we switch our instrument on?'

Krueger was keen to stay, come what may, since the Russians had guaranteed the safety of the NASA team. In the event, they realised this promise could count for nothing if there was widespread civil insurrection and, wisely perhaps, the director of Goddard ordered his TOMS team home as soon as they could. Krueger and his colleagues spent a calm day in their hotel, where, bizarrely, they were able to follow CNN coverage of the coup. More significantly, they were able to continue to communicate with the outside world via electronic mail, which convinced Krueger, at least, that the coup hadn't been as successful or draconian as Western commentators had feared. Around lunchtime, a rumour went around that at 4 p.m. the Soviet equivalent of the Beltway around Moscow would be closed. With its closure, any chances the NASA team had of leaving the country would have come to an end. In the event, it wasn't and early the next day the Goddard scientists reached the airport unhindered. They were able to leave for the Unites States via Finland without any problems. Strange as it seems, they did not see any tanks or visible signs of protest. 'One flare – that's all we saw!' Krueger said.

By the following Thursday, Krueger and his team were back in the Payload Operations Control Center at Goddard and switched the TOMS instrument on exactly as planned, independently of Kaliningrad. But in a sense, they did it blind as they could not see the full complement of the Meteor satellite's engineering data. The only way they would know if their command sequences had worked would be by the receipt of a sudden stream of data displayed on their computer consoles in the POCC. It worked. 'Data came in beautifully and there wasn't even a glitch,' Krueger recalled with obvious relief. 'And I'm pleased to tell you that it's been like that ever since!'

After the Herculean struggles to obtain meaningful trend observations from the various TOMS and SBUV instruments, scientists within Goddard's Atmospheric Chemistry and Dynamics Branch are by no means out of the woods. For the idiosyncrasies of technology are matched by the quirks of the upper atmosphere and, in particular, its interactions with the Sun. Over an eleven-year cycle of activity, ozone levels will drop by around 2–3 per cent owing to an increase in oxides of nitrogen which result from maximum solar activity. This drop has to be accounted for in pulling out meaningful trends, as does the natural variation in ozone transportation due to dynamics. Springtime values for ozone over the northern hemisphere are normally around 20 per cent greater than during the autumn, for example. These and other factors have to be eliminated before a picture of the underlying change in ozone becomes evident from within its annual variability.

In the summer of 1993 for example, observations by a totally unconnected NASA project led to a rash of strange headlines around the world. The Solar Anomalous and Magnetospheric Explorer, whose remit was to look at interactions of the solar wind with the Earth, showed that bursts of solar radiation were destroying ozone at 80 kilometres' height above the Arctic. Pictures of a psychedelic doughnut around the North Pole in the same false colours as those from TOMS duly appeared in newspapers. It was not an ozone hole, nor was it as significant a discovery as it seemed at first:

after all, Nimbus 4 had shown something similar two decades before.

In April 1992, Stolarski and colleagues had reported that 'statistically significant decreases are now being observed in all seasons in both the northern and southern hemispheres at middle and high latitudes' in a paper which appeared in *Science*. Most markedly, the mid-latitude loss in the northern hemisphere over the 1980s was 6–8 per cent. But the downward trend over the following year increased apace: ozone levels over the northern hemisphere were lower than had ever been seen in 1992–93. In the second half of 1992, ozone levels were 2–3 per cent lower than any previous year and 4 per cent lower than would be normally expected. The 1992 low values were particularly apparent over the mid-latitudes of the northern hemisphere and they persisted well into the new year. In December 1992, for example, they were some 9 per cent below their expected values; in January 1993 they were 13–14 per cent below normal and preliminary analyses of the March values revealed they were 11–12 per cent below normal.

When these low values were returned, some of the Goddard scientists wondered if the Nimbus 7 TOMS drift had suddenly accelerated. But they were confirmed by independent observations from both the Meteor TOMS and the NOAA SBUVs for the first time ever. This marked thinning of the ozone layer was not only independently corroborated, it was much greater than had been predicted. It was obvious that gas-phase chemistry alone at mid-latitudes was not solely responsible. The most likely culprit for enhanced ozone destruction was aerosols from Mount Pinatubo (as will be discussed in greater detail in Chapter Nine). But there was also the possibility that chemical processing over the North Pole had encroached upon lower latitudes, reducing the total ozone observed. Similarly, blobs of ozone-poor air could also have been thrown off and mixed in with air at lower latitudes, not just at the time when the Arctic vortex broke up. Although TOMS could not make observations with sufficient detail to clarify the situation, UARS could and – as we will see – did.

Towards the end of 1994, a B-52 bomber will take off from Wallops Island, Virginia, and herald the next era in satellite observations of the ozone layer. Appropriately enough, Arlin Krueger's NASA career will finish where it started off in the days when he fired sounding rockets into the stratosphere. In those days, B-52s brought war but today they promise peace and an interesting innovation in performing ozone observations from space. The newest version of TOMS will be flown on its own dedicated mini-satellite known as an Earthprobe. To reach orbit, it will use a booster known as Pegasus, looking like a slightly larger version of a cruise missile, which will be carried underneath a wing of the B-52. The Pegasus will be released at around 10,500 metres over the Atlantic and will fire its rocket motor to head over the coast of Panama and, hence, into low Earth orbit. Earthprobe TOMS will be injected into a Sun-synchronous orbit that will, so far as possible, mimic that of Nimbus 7 to continue the record of its observations.

'This is the first TOMS instrument designed specifically for trends,' said Arlin Krueger. As well as employing improved electronics, the Earthprobe-Tow TOMS will carry an onboard calibration light source which will continuously check on the reflectivity of the diffuser plate. TOMS will thus be able to measure how the plate is degrading directly. In any event, NASA hopes that this degradation will be minimised as the instrument will carry three diffuser plates, each to be deployed in turn on successive passes of the North Pole. Krueger believes that this development may reduce instrument drift to a third of that suffered by Nimbus 7 over the decade of the 1980s. 'Small is beautiful' is a design philosophy which has been eagerly embraced by many sections of the space community: as the Earthprobe carries just one TOMS instrument, there is very little that can go wrong – and, more to the point, if the mini-satellite as a whole fails, it can be replaced fairly easily. The same could hardly be said of UARS, and the tendency for space projects to become too unwieldy will directly affect the future of ozone-monitoring from space.

Once the Earthprobe TOMS has been launched, complementing the Meteor TOMS, NASA will have the means to monitor global ozone levels throughout the 1990s. Continuity, as Arlin Krueger well knows, is just as important a contribution from space-based instruments as their lofty perspective. 'You can't really argue with nearly fifteen years of data,' is his eulogy for the Total Ozone

Mapping Spectrometer aboard Nimbus 7, the instrument with which his own career has been inextricably linked.

The future of ozone monitoring from space in the twenty-first century is less rosy. NASA has decided to stake its future on a series of complex platforms, larger and carrying more instruments than even the Upper Atmosphere Research Satellite. The Earth Observing System is a prime example of 'big science', which most scientists see as a worrying development in modern research. Big science involves gargantuan projects whose costs and administrative unwieldiness invariably snowball and use up money that would otherwise be available for smaller projects. Two of the largest big-science projects in the United States have been distinctly ill-starred: in 1993, the Superconducting Supercollider was cancelled by Congress because of escalating costs, and NASA's space station barely survived the first cutbacks ushered in by the Clinton administration. More than one federal official declared that it was open season on big science in Washington.

In any case, NASA has come under closer scrutiny and not just because of costs. 'We no longer just do "satellites",' said one researcher involved with space instruments with understated bitterness. 'We do "programmes" that take forever.' In many ways, UARS has come to symbolise what the space agency's critics say is all too wrong about its approach to science. It took fifteen years from the time it was first proposed to its eventual launch into space. Where once there were regular Nimbus launches, NASA effectively put all its eggs in one basket with one satellite. And while there are valid reasons for launching bigger satellites, it has led to a fundamental change in the way space research is conducted.

The professional life of today's space scientist can often involve planning meetings for projects which will reach fruition a decade hence. Worse still, as satellites become bigger and more complex, up-and-coming students miss out on the experience of building and operating new instruments. They also have to wait for extended periods of time to get their hands on fresh data, upon which, ultimately, their doctorates and subsequent contributions to the subject depend.

Such problems threaten to become more acute with the Earth

Observing System. This project was touted to Congress as 'Mission to Planet Earth' and its rationale was to think big and launch what have memorably been called 'Battlestar Galacticas' to monitor the environment from space. When originally proposed in 1989, NASA envisaged that EOS would comprise large, 15-tonne platforms to remotely sense the Earth from the perspective of polar orbit. They would, it was claimed with no small fanfare, scrutinise the face of the Earth as never before. Each platform was envisaged to carry a multitude of sensors to observe the atmosphere, land or oceans in unprecedented detail. Many of these instruments would be complementary and make cross-referenced observations rather like the Gang of Four on the Upper Atmosphere Research Satellite.

Originally, EOS would have involved the launch of two identical sets of three platforms over a fifteen-year period starting in 1996. One set of platforms would have focused on the land and its climatic interactions with the lower atmosphere. The other series would have exclusively observed the upper atmosphere and oceans. But in 1990, as NASA reeled from the embarrassment of having launched the $2-billion Hubble Space Telescope with a faulty mirror, Congress demanded that the space agency cut back its plans for the Earth Observing System. Accordingly, NASA was forced to prune its total EOS expenditure from $16 billion to $11 billion until the end of the century. It was also ordered to subject its EOS plans to an extensive design review by scientists from outside the space agency led by Ed Frieman of the Scripps Institution for Oceanography in La Jolla, California.

The Frieman panel suggested NASA should employ a new philosophy for the future and recommended that EOS had to be drastically 'descoped' in size and complexity. The panel also argued that the space agency's future research into global change should be expanded to include 'science-driven process studies using small and intermediate-sized space systems, remotely piloted aircraft, *in situ* and ground-based programs'. To some extent, NASA has complied with these recommendations and, rather than staking all its future on the largest platforms, it will now use smaller, less complex satellites in between. One scientist interviewed for this book said: 'NASA has finally realized that it isn't going to square up to global change with big science.' Congress has also ordered the space agency to rethink its research priorities to look at global warming.

A fundamental part of the sea change in NASA's attitudes has come from its administrator, Daniel Goldin, who was appointed in March 1992. His desire to streamline the agency is encapsulated in the mantra-like slogan 'Better, Faster, Cheaper'. The former aerospace executive has become an enthusiastic supporter of uncrewed drones, for example, and went so far to tell *Science* that he was 'overwhelmed by the possibilities' they offered. Nevertheless, there are sophisticated and unprecedented observations to be made by platforms like EOS. The EOS budget has been reduced to some $8 billion, but if NASA now diverts money to smaller programmes, this will further delay the EOS programme. NASA managers would prefer to put EOS first, which will affect the timing and extent of future experimental programmes, including aircraft campaigns like the AASE II and the launch of small satellites such as the Earthprobes by the end of the century.

As currently envisaged, the Earth Observing System overall will involve the launch of larger platforms interspersed with smaller satellites. The larger platforms will come in two essential designs, the first crossing the equator in the morning and the second in the afternoon. As a result, they are known as EOS-AM and EOS-PM respectively. Any differences that have occurred at the surface or in the atmosphere in between the passage of these platforms over the same part of the Earth will instantly be discerned. The first EOS-AM platform is expected to be launched in June 1998, six months earlier than originally planned.

Two smaller missions looking at ocean colour and atmospheric aerosols will follow, before the launch of EOS-PM in 2001. This will be followed by an ocean circulation and ice mapping mission in 2002, after which the first dedicated stratospheric chemistry satellite, CHEM-1, will be launched. In a presentation to Congress, NASA outlined the scientific rationale for this satellite as: '[Ozone] depletion in the lower stratosphere can affect the heating or cooling of the troposphere. Thus the CHEM−1 spacecraft, which launches essentially ten years after UARS, will also carry instruments to check the accuracy of models of stratospheric chemistry.'

How Soon Is Now?

By the time CHEM-1 is operating, chlorine loading of the atmosphere will have peaked. Although ozone depletion will continue well into the next century, the space agency has decided to de-emphasise stratospheric research against global warming studies. This decision is not universally popular for, as one space scientist said: 'I am personally concerned that there's a feeling the ozone problem has been solved. There's very little by way of stratospheric monitoring when chlorine reaches a peak. I think we've got to be watching.'

In another building at NASA's Goddard Space Flight Center is the World Data Center for Rockets and Satellites A. The earnestness of its title suggests the sort of place where you might expect to find Buck Rogers and, indeed, it does date from a much earlier time when the very idea of space flight was a novelty more suited to B-movies. The Center is perhaps the most enduring contribution of the International Geophysical Year to the canon of modern science, for it was built with the express purpose of archiving every piece of data returned by US scientific satellites since the dawn of the space age. The Data Center is thus a veritable treasure trove of priceless information on the stratosphere as seen from space.

Within its meticulously arranged, thermally controlled archives are all the raw satellite observations of the ozone layer. They include the first read-outs obtained by the early Nimbuses right through to the precious observations returned by ISAMS during its glorious but short career. Whether one is trying to assess trends from TOMS or search out the finer points of 'smoking gun' chemistry from MLS, the barest, irreducible electronic form of those observations may be found here. For a minimal copying fee, the kindly archivists will dispatch it anywhere in the world. In space research, at least, the utopian ideal of freedom of information has been a reality for many years.

No matter how the ozone layer will respond to future chlorine loading, the population of the Earth can breathe slightly more easily for knowing that satellites will be keeping a constant vigil. 'The phasing-out of CFCs is a tremendous victory for the inhabitants of the Earth,' said Joe Waters of the Jet

Propulsion Laboratory. 'Most of the industrialised chlorine in the stratosphere will be there throughout the next century. We feel we owe it to our fellow inhabitants to see what, if any, danger threatens.'

6

Pieces in a Puzzle

The bright Sun was extinguish'd . . .
. . . and the icy Earth
Swung blind and blackening in the moonless air;
Morn came and went – and came, and brought no day,
   And men forgot their passions in the dread
   Of this their desolation; and all hearts
Were chill'd into a selfish prayer for light.
   . . . No love was left;
All earth was but one thought – and that was death,
   Immediate and inglorious; and the pang
   Of famine fed upon the entrails.

Lord Byron, 'Darkness'

THE Sun is somehow brighter in Boulder. At 1,640 metres above sea level, the city does receive appreciably higher levels of ultraviolet radiation, as attested by the routine measurements made by the many atmospheric scientists who live and work there. Well before the discovery of the Antarctic ozone hole and the subsequent worldwide worries about skin cancers, Boulder's citizens took care to limit their time in the Sun. Today most visitors need no prompting to avoid the Sun, although they are most likely to encounter a more immediate health problem in the form of mild altitude sickness until they become acclimatised to Boulder's relatively rarefied air. The city's 95,000 residents, at least, are used to it and many will smile when they tell you theirs truly is a mile-high city.

Boulder nestles the edge of the Rockies where they fan out onto the plains of Colorado below. Though not as famous as Aspen and the skiing resorts further south, it too boasts breathtaking vistas. Three large slabs of rock, known as the Flat Irons, form a remarkable backdrop for the city below. Boulder has the unmistakable feel of a pleasant college town thanks to the thriving student population that attends the University of Colorado on its sprawling campus. Despite the state's recent veering to right-wing politics, there is an unmistakable whiff of political correctness in the air and a homespun sort of ecological philosophy unique to Boulder. Only in the fictional home of *Mork and Mindy* could one expect to find an environmental support group called 'Hugging A Mountain' or see people chided for leaving their car engines running. The city council worries that it is becoming too expensive for people to afford properties there and has earmarked land for new houses that will be made available to its poorer citizens.

Boulder is also an intellectual hub unparalleled in ozone research and boasts some of the most important atmospheric research centres in the world. One can be found at the southerly end of Broadway, the main thoroughfare which bisects the city's boundaries, where a collection of buildings announce themselves as the US Department of Commerce Environmental Research Laboratories. Behind the main building, and slightly out of sight from the main road, is the NOAA Aeronomy Lab where a hundred or so research chemists work. It was here in 1971 that Hal Johnston first showed that NOx would be important in the ozone debate and fifteen years later that the first National Ozone Expedition took shape after a suggestion by Bob Watson. Downtown, close by the Crossroads Shopping Mall, is another NOAA building, which includes the physicists of the Climate Monitoring and Diagnostics Laboratory who operate the worldwide network of 'clean air' stations which have recorded the inexorable rise of CFCs in the atmosphere.

By far the most illustrious institution, however, is silhoutted against the Flat Irons sitting atop a small promontory known as

How Soon Is Now?

Table Mesa. To get there requires a journey up a winding road where the recently arrived visitor will find his or her ears popping because of the increase in altitude. Only near the top will the National Center for Atmospheric Research reveal itself as an appropriately futuristic edifice. Hewn out of local rock, it appears like a space age version of Stonehenge, which is exactly the effect its designer, the noted Chinese-American architect I. M. Pei, strove for. Whereas Stonehenge was assembled from uneven blocks, NCAR is a triumph of Pei's geometrically precise style. It has been described as the first post-modernist building, but Pei himself called it 'monastic, ascetic but hospitable'. He was directly inspired by the village of Mesa Verde in southwestern Colorado, built eight centuries before by the local Anasazi Indians out of a rockface so that it would be sheltered by stony overhangs. I. M. Pei drew up a precise concatenation of vertical rectangles and curves which even today, some thirty years after it took shape on his drawing board, still looks futuristic: the discerning cineaste may also recognise it as the building off which Woody Allen abseiled in his comedy *Sleeper*.

The future is indeed something with which the researchers at NCAR, NOAA and the University laboratories are intimately concerned, especially the way in which the atmosphere will evolve in the next century as human pollution takes its toll. The significant contributions of scientists in Boulder may be partly due to the close-knit nature of the scientific community: there is a good deal of cross-fertilisation of ideas. But Boulder scientists' pre-eminence in peering far into the future stems mainly from the fact they have at their disposal one of the most spectacular expressions of humanity's technological prowess: a Cray supercomputer.

It is hard to reconcile what appears to be a haphazard collection of unconnected, unfinished hexagonal wardrobes with the most powerful tool available to climatologists in their careful unravelling of the complexities of the atmosphere. Yet in the neon-lit, air-conditioned inner vestibule which is the NCAR Computing Center, that is exactly how a Cray Y-MP appears. To many visitors, it provides a bitterly disappointing experience, for the concept of supercomputers lends itself to flashing lights, an uneasy electronic malevolence and attendant white-coated scientists of a distinctly

Mephistophelian aspect. The Computing Center is reminiscent of every other clean and modern office you've ever come across: only the quiet hum of activity and racks of large computer disks nearby give the barest hint of what is achieved here.

In the field of atmospheric research, Cray supercomputers are almost *de rigueur*. They are employed to describe the general circulation of the atmosphere as an elaborate mathematical abstraction, a kind of three-dimensional board game where only the laws of physics are strictly adhered to. The Cray's unparalleled power of analysis allows unprecedented accuracy in understanding the complex web of feedbacks between disparate phenomena – such as the way in which clouds redistribute heat, how oceans change wind patterns and how ice and snow affect the global heat balance, to name by far the most perplexing. Without the power of a supercomputer, these interactions would be completely impossible to estimate.

Supercomputers enable atmospheric scientists to diagnose the symptoms of global change by effectively allowing them to assume the powers of a deity. They can spin the world faster, slow it down, turn the Sun off, make it warmer, cool the equator or warm the poles. They can alter each of these factors individually to take a measure of their relative importance, or else change them collectively to assess their full climatic impact. The mathematical equations which describe atmospheric conditions are abstractions with enough basis in reality to show how the Earth's environment might be altered on a global scale. In the Cray's mind's eye, the Earth's surface is criss-crossed by a geometrically precise lattice of reference points, or grid points. For climate modelling, the grid is coarse: at the surface they are separated by about 300 kilometres on average. The points extend vertically upwards into space and slice the atmosphere into anything between ten and fifteen layers. At each of these points, the humidity, pressure and temperature are calculated. The Cray's number-crunching ability is used to grind through millions of physical equations to determine how each of those factors will change at monotonously regular intervals.

NCAR is equipped with one of the most powerful machines which Cray Research can offer: a Y-MP. A cursory glance at its abilities will amaze even the most hardened technology fetishist. It can perform 2.5 billion calculations every second, store 60 billion bytes of information on disk and instantly recall 64 million words from memory. By use of a technique known as parallel processing it can perform eight aspects of a complex job at any given moment. It is equally well suited to studying the way in which violent storms are generated in mountainous regions, predicting when activities on the surface of the Sun will affect our atmosphere or running through the complex chemical cycles which are continuously taking place throughout the atmosphere as a whole. And for the first time ever, the NCAR Cray has enabled climate researchers to link the future evolution of the oceans and atmosphere as a true symbiosis.

Supercomputers have also made possible a significant breakthrough in studies of ozone depletion. By taking all the hard-earned data from balloons, aircraft and satellites, they can generate a coherent and accurate portrait of the stratosphere around the globe. Analysis of the experimental data by supercomputers helps check if our understanding of the upper atmosphere today is essentially correct. More importantly, perhaps, it also allows climatologists to play out scenarios of ozone depletion far into the future. The possible future atmospheres that are generated present a sobering warning for policy-makers and legislators in their decisions on which path to choose.

Boulder is not the only atmospheric research centre equipped with these remarkable technological crystal balls. Others may be found across the United States in centres of learning like Berkeley and Princeton, as well as half a hemisphere away in otherwise unremarkable towns strewn across southern England's commuter belt. The visions of the once and future atmospheres that they generate are as different as the cities where their supercomputers are found, and their predictions are as distinct as the accents of the scientists who use them.

The following quote from the Florentine political theorist Machiavelli appears on a noticeboard in the Climate Modelling Section at NCAR:

> There is nothing more difficult to plan, more doubtful of success, nor more dangerous to manage than the creation

of a new system. For the initiator has the enmity of all who would profit by the preservation of the old system and merely lukewarm defenders in those who would gain by the new one.

This quote is rather more pointed than it first appears, for it recognises that each new development in atmospheric modelling has threatened the status quo and the balance of importance between those strange scientific bedfellows, chemists and dynamicists. Their conflict has become less pronounced as the most sophisticated general circulation models have been developed: the atmosphere provides enough of a challenge for scientists from both disciplines to pit their wits against.

The fundamental reason for atmospheric circulation is that hot air rises: the temperature difference between the equator and the poles essentially drives weather systems in the lower atmosphere. At the equator, air is effectively 'freeze-dried' as it passes through the relatively high-altitude tropical tropopause and, as we saw in Chapter Four, the stratosphere is uncommonly dry. To some extent, the circulation of the stratosphere mimics that of the lower atmosphere. The strongest absorption of solar radiation by ozone itself takes place over the 'summer' pole (i.e. in the hemisphere that is pointing towards the Sun) and there is large-scale circulation towards the winter pole because of the temperature difference. This basic dynamic motion drives the strong stratospheric 'jets', winds that are either westerlies or easterlies depending on the hemisphere. Stratospheric wind patterns are also distorted by underlying topography which gives rise to large-scale waves within the stratosphere. They serve to transport heat and momentum towards the poles and are formally known as 'planetary-scale eddies', or, more usually, 'planetary waves'. They are more pronounced in the northern hemisphere because there are greater numbers of mountain ranges (such as the Rockies) to distort stratospheric circulation. In short, planetary waves have to be accounted for in any accurate representation of the circulation of the upper atmosphere and its subsequent transportation of ozone.

In many ways, the stratosphere *should* be far easier to model than the rest of the atmosphere below. The bulk of the troposphere is characterised by large-scale motions of vast air parcels that slosh around and cause all manner of complications. But in the stratosphere, most of the atmospheric motions are horizontal and there is very little mass motion upwards. The stratosphere nestles between the troposphere below and the rarefied regions above (known as the mesosphere) where the influence of the solar environment starts to be felt. But Byron Boville, deputy head of the Climate Modelling Section at NCAR, warns: 'If you don't understand the stratosphere by itself, you're not going to understand it when it's coupled to the rest of the atmosphere.'

In truth, a useful representation of how the ozone layer behaves should describe how the chemistry is affected by the large-scale dynamic motions within it. In the 1970s, most of the modelling was performed with the first 1-D models which had traditionally been developed by dynamicists. They considered the upward or downward movement of trace gases in a column which was taken to represent the whole of the atmosphere. At various points along this line, the physics of how those gases radiated heat, or underwent convective motion as a result, was calculated. These representations were necessarily simplistic because of the limited computer power then available. They could hardly tell modellers about variations from season to season, but it was at least a start. If nothing else, it helped focus research activities by showing up areas where there were fundamental deficiencies in both data and understanding.

By the early 1980s, two-dimensional models had been developed by atmospheric chemists. They essentially averaged conditions out along a line of longitude: imagine slicing the Earth along the Greenwich meridian with a circular saw and etching points on the saw if it was left at the centre. 2-D models represented a step forward because they incorporated vertical motions as well as atmospheric flows from north to south within their calculations. So they had to average any atmospheric flows from east to west, and hence often underestimated the underlying picture of ozone transportation within the stratosphere. Nevertheless, 2-D models did start to reproduce the patterns of global ozone distribution as circulated by the dynamics of the stratosphere.

With increased computing power came greater strides in the modelling. For example, by the mid-1980s, there were perhaps at most fifteen grid levels represented in the bottom 20–30 kilometres of the atmosphere. Less than ten years later there were up to forty levels, with the new ones added above the 20-kilometre mark. 'We are starting to look at how we can simulate concentrations of trace species and how that feeds on the dynamics,' said NCAR modeller Byron Boville. Over time, the models have been fine-tuned to include a better understanding of dynamics. It had been thought that planetary waves would act to diffuse the normal variations in ozone abundances, but observations in the real atmosphere have shown this to be incorrect.

These waves are better rendered in three-dimensional models, which can describe their interaction with atmospheric chemistry far more accurately. And though such models have been assembled, they require virtually the whole of the supercomputer's analytical power to run them. Normally Cray computers subdivide the work being done: to run a 3-D general circulation model, however, other users have to be kicked off even with the unparalleled number-crunching capacity now available. Such 3-D models will include over a hundred chemical constituents and thirty to forty grid levels separated by three degrees at the surface. They will also need to consider what is happening where the Sun is rising and setting, at which point – known as the terminator – photochemistry becomes paramount in importance.

Guy Brasseur, head of NCAR's Atmospheric Chemistry Division, has started to assemble the most comprehensive and accurate three-dimensional representation of atmospheric chemistry and dynamics ever. 'We add a number of steps in the computer code which assume that at each grid point chemical reactions will be taking place,' Brasseur explained. 'At each of those 50,000 grid points we look at all the chemical equations taking place.' A 3-D model can take into account the position of the Sun and how its rising and setting will fire off photochemical reactions. This will then alter the concentration of the trace constituents in the stratosphere, which in turn will affect the dynamics. This will then alter the transportation of ozone, which will have a knock-on effect on temperatures elsewhere, since there will be less ozone present to absorb ultraviolet. Keeping track of all these complications needs the full computational ability of a supercomputer.

How Soon Is Now?

A singular handicap remains for even the most accurate models. It is very difficult to obtain a continuous stream of real observations throughout the whole of the stratosphere across the globe. Models attempt to fill in those gaps by careful extrapolation. So, in that sense, they can be thought of as mathematical representations of understanding. New data verify the accuracy of the models; and deficiencies that are thrown up identify new areas of research. A perfect understanding of the stratosphere would thus be based on a perfect representation of all the known factors within it. 'The difference between our models and the real atmosphere helps research by making people search for additional chemical mechanisms which we may have overlooked,' said Guy Brasseur.

However, even the best models cannot reproduce the exact workings of the atmosphere completely. It is no surprise that they failed to predict the Antarctic ozone hole, for example. Even today, the most accurate 3-D models cannot quite match the observed distribution of ozone in the upper atmosphere. Most notorious is the cold-pole problem, where the polar regions are 30 or 40°C too cool at the upper stratosphere: this means too many polar stratospheric clouds are produced by the models, thereby reducing the ozone quite drastically. Ozone concentrations are generally underestimated across the whole of the globe in the upper stratosphere and currently observed mid-latitude ozone losses have yet to be reproduced accurately. 'That is a feature common to all the models,' said Guy Brasseur. 'That means we're either overestimating the loss rates or underestimating the ozone production. Possibly there may be missing chemical mechanisms.'

The scientific basis for modelling the atmosphere owes its origins to a remarkable polymath named Lewis Fry Richardson. His singular contribution to modern science is often overlooked, perhaps because he was employed in the relative obscurity of the UK Meteorological Office, whose predominantly military background has given it a tendency to hide lights under bushels. Not until recently has the full range of his contributions to other sciences – physics, mathematics, psychology and even war studies – been fully appreciated. Yet it was in the field of meteorology that he realised his most crucial and enduring achievement. Single-handedly Richardson outlined the mathematical blueprint upon which all atmospheric models are based, in his book *Weather Prediction by Numerical Process* which was published in 1922.

After studying under J. J. Thomson (the discoverer of the electron) at Cambridge University, Richardson became a jobbing meteorologist within various branches of the UK Met Office. He was appointed director of the Eskdalemuir weather station in 1911, and, while stationed there among the rolling hills of southern Scotland, came up with the germ of an idea that it might be possible to predict how the atmosphere behaved if it could be modelled mathematically. By use of the laws of physics which defined how the atmosphere behaved at any given time or point in space, Richardson realised that it would be possible to work out what its future behaviour might be. At this time, weather forecasting was still very much a hit-or-miss affair because there was little information available upon which to base the forecasts. To be a weather forecaster largely involved remembering how previous weather patterns changed, and the ability to draw maps of atmospheric pressure based on that experience.

Richardson would perhaps have developed his ideas sooner had the First World War not intervened. As a Quaker, he was a conscientious objector and joined the Friends' Ambulance Unit as a driver. It was during this period that he began making notes on mathematical weather forecasting in earnest. Eventually he assembled them into a manuscript which he sent for safekeeping behind the lines in 1917. The unspeakable carnage and horror of the war would for ever after prey on his mind – particularly as several of his former students were killed, or barely survived injuries which were tended in his own ambulance. As a result, Richardson subsequently devoted his considerable abilities to the prediction of warfare, which, mathematically, bore similarities to weather prediction in that it involved any number of ill-defined factors. Richardson became a pioneer in applying mathematics to war studies, though these contributions would not be recognised until long after his death.

His magnum opus on weather forecasting suffered a setback from the sheer absurdity of warfare as his manuscript was mislaid. Only after the Armistice was it found under a pile of coal in Belgium, and, doubtless pondering the curious hand of fate, Richardson polished its contents and prepared the book for publication. The text essentially considers weather readings on 20 May 1910, methodically imposed on a grid across Britain and northern Europe. Each grid was about 200 kilometres across and four layers were added vertically. At each point, values for temperature, water vapour content and wind velocity were assigned. Using the laws of conservation of mass, momentum and energy, it was possible to determine how they would change. Richardson started his analysis at 6 a.m. and considered how the weather would change at six-hour intervals.

The results were astounding. The difference in air pressure was about a hundred times too big, so that Britain would have suffered from hurricane winds had his forecast been correct. Richardson realised that this immense pressure gradient had resulted from the relative coarseness of the grids used and the related fact that only spotty data were available. These two handicaps still face modellers today, although the grids and information available have improved far beyond anything that Richardson would have dared imagine. He also intuitively understood that the larger weather patterns would depend on small-scale physics, so that ultimately, it would be difficult to model large-scale physical effects without knowing what was happening at smaller scales. Years later it would be shown that a weather forecasting model will work only if the time step between each calculation is shorter than the time it takes a sound wave to travel between two grid points.

Most famously, Richardson paraphrased Augustus de Morgan's couplet about fleas to encapsulate the limitations of mathematical weather forecasting:

> Big whirls have little whirls that feed on their velocity,
> And little whirls have lesser whirls and so on to viscosity –
>   in the molecular sense.

In that lyrical appreciation of atmospheric motion, Richardson had essentially realised its chaotic nature. So, despite the fact his forecast didn't work, he had the foresight to know why and, more importantly, said so in his book. The basic kernel of his approach to modelling was sound and this remarkable contribution to science was appropriately appreciated by his peers. Richardson was awarded the highest scientific honour which can be bestowed upon a Briton, Fellowship of the Royal Society.

It was clear that if weather forecasting was to become a reality there would have to be strides in the analysis of the data upon which they would be based. It had taken Richardson something like six weeks to do all the calculations for just one six-hour time period. The only way he could envisage 'operational' forecasting was in the construction of a dedicated 'weather factory', wherein 64,000 people would use slide rules and hand-cranked calculating machines to solve all the equations at each point along the various grids. A sense of what it might have looked like may be had from the following quotation from *Weather Prediction by Numerical Process*:

> Imagine a large hall like a theatre, except that the circles and galleries go right round through the space usually occupied by the stage. The walls of this chamber are painted to form a map of the globe. The ceiling represents the north polar regions, England is in the gallery, the tropics in the upper circle, Australia on the dress circle and the Antarctic in the pit. A myriad computers are at work upon the weather of the part of the map where each sits, but each computer attends only to one equation or part of an equation. The work of each region is coordinated by an official of higher rank. Numerous little 'night signs' display the instantaneous values so that neighbouring computers can read them. Each number is thus displayed in three adjacent zones so as to maintain communication to the North and South on the map. From the floor of the pit a tall pillar rises to half the height of the hall. It carries a large pulpit on its top. In this sits the man in charge of the whole theatre; he is surrounded by several assistants and messengers. One of his duties is to maintain a uniform speed of progress in all parts of the globe. In this respect he is like the conductor of an orchestra in which the instruments are slide-rules and calculating machines. But instead of waving a baton he turns the beam of rosy light upon any region that is running ahead of the rest, and a beam of blue light upon those who are behindhand.

It should be noted that the 'computers' Richardson refers to are the people using the machines rather than the machines themselves. His weather factory would also have obeyed the strict regimentation of English society. Higher officers would oversee the work of lower ranks – essentially mathematical underlings – region by region across the globe. If Richardson's vision had ever been realised, it would probably have appeared similar to the records department where Sam Lowrie worked in Terry Gilliam's film *Brazil*.

Richardson's concept was astoundingly prescient, and for once the cliché of 'years ahead of its time' is totally justifiable. Like most of the best ideas, his book was not a success at the time. Cambridge University Press priced *Weather Prediction by Numerical Process* at a hefty 30 shillings so that very few people ever saw copies (it would be three decades before the initial print run of 750 sold out). Nor was it warmly received by reviewers: the *Manchester Guardian* tartly dismissed Richardson and his ideas for sounding 'like the rhapsodies of an irresponsible visionary'. The anonymous reviewer also added that it would be for 'future generations to consider the desirability of reorganising their forecasting systems in accordance with the new methods. One has little hope of seeing it done unless we can breed a race of millionaires to find the money and of Richardsons to organise the work.'

The result was that most meteorologists nodded politely in Richardson's direction, smiled and went about the business of weather forecasting in the arcane fashion they had done before. Richardson left the Met Office and branched out into other areas of science while his remarkable ideas were effectively forgotten for decades. In many ways this was hardly surprising, for in the years between the two world wars it became fashionable to believe that the human race's ever increasing technological prowess would be more than equal to weather and climate modification. Who would need to predict the weather when you could actually change it? An editorial from the *Boston Herald* in July 1937 was typical of this naively unshakable belief in a brave new world:

> [A] hundred years from now it is quite possible that the forecast of a heat wave will be only of academic interest to most persons. We shall probably be working in air-conditioned

offices and factories, sleeping in homes where we may select our favorite temperature, and travelling in trains, automobiles and airplanes, all of which will be protected from the outside weather. Then only farmers, incurable out-of-door friends and a few sentimental die-hards will even bother to read the weather reports.

The first experiments in weather modification in the 1950s when clouds were seeded with crystals to cause rainfall to little effect showed just how ludicrous a notion weather modification was – and is.

Richardson's work was rediscovered just before his death in 1953. By then, a machine which was rumoured to use up half the electrical power in Philadelphia when switched on would bring weather forecasting that little bit nearer to reality. Named the Electronic Numerical Integrator and Computer, it was a behemoth of an electronic calculating machine and was used by meteorologists at the Massachusetts Institute of Technology to solve the equations which Richardson had outlined thirty years before. Richardson had the pleasure of hearing about this work and outlines of more complex experiments which followed. Nothing short of a revolution took place in meteorology thanks to the subsequent invention of silicon-based electronic chips and the data-gathering ability of satellites. By the end of the 1960s, weather forecasting had become a reality throughout the globe, as eventually would predictions of future climates, using the basic mathematics outlined by Lewis Fry Richardson fifty years before.

In *Changing Places*, the writer David Lodge remarked in 1975 that British weather forecasts sound to the ears of visiting Americans like a bizarre extension of the satire industry as they predict 'every possible combination of weather for the next twenty-four hours without actually committing itself to anything specific'. Today, however, the UK Meteorological Office prides itself on the unprecedented accuracy of its forecasts: for the next twenty-four hours, its forecasters claim, they can achieve 93 per cent accuracy in predicting the weather in your part of the world. Despite the public opprobrium after they failed to warn of the great storm of October

How Soon Is Now?

1987, Met Office forecasters based in Bracknell are still among the best in the world. When storms lashed Europe in January 1990, they more than made up for their error of three years before by correctly predicting their passage days before they had even formed out in the Atlantic (the chief forecaster on duty when Towyn was flooded was, with cruel irony, Mr C. Flood). British forecasters are amused that most US airlines prefer to use their weather forecasts to fly the Atlantic rather than those produced by the US National Weather Bureau.

The place from where these forecasts emanate is appropriately named after Lewis Fry Richardson. It is to be found inside a nondescript building which fits the general tone of Bracknell, an otherwise unloved and featureless ornament of the Berkshire commuter belt. The L. F. Richardson Centre could loosely be described as a forecast factory, although it is hardly as the great pioneer envisaged it for there are at most only a few dozen people employed within. The computing power needed to grind out weather forecasts is provided by the unseen electronics at the heart of a Cray Y-MP elsewhere in the building. When it was opened in 1972 by Prime Minister Edward Heath, he expressly acknowledged Richardson's peculiar vision of five decades before. The civil service, Heath said, would have been quite happy to hire the additional 64,000 people needed to perform manual calculations. 'It is only when I tell [the] Civil Service [to be] cut down by 64,000 people that the problems really arise,' Heath added, to the amusement of the meteorologists in the audience.

Just down the road is a newer building which bears the name of another illustrious British scientist. The Hadley Centre for Climate Prediction is named after the gentleman scientist who in the 1800s explained why the trade winds occurred, a lawyer by the name of George Hadley. (According to one Met Office source, it nearly was called the Climate Research and Prediction Centre, until someone wrote down its acronym.) It was opened in May 1990 by Margaret Thatcher, who had by then undergone a transformation as dramatic as that of Saul on the road to Damascus. Environmental matters, which she had previously scorned, had now attained importance of the vote-gathering, election-winning variety. When she opened the centre, Thatcher declared that its scientists 'will help us look into the future and predict more precisely the changes in our climate'. She added on an uncharacteristic note of caution that the 'very detail of the forecasts may not be quite right. It very rarely is when you are trying to predict the future.' Six months later Thatcher herself was to suffer from an event which she could not have predicted – her forcible ejection from office as part of a Conservative Party coup.

Although the weather forecasters in the Richardson Building and the climate modellers at the Hadley Centre employ the same electronic machines – Cray Y-MPs – they do so for completely different ends. The difference between them is as stark as between the grimy, Ministry of Supply decor in the former and the brighter colours and designer furniture of the Hadley Centre. The weather forecaster is interested in short-term changes and most of the computer power goes into tightening the grids to resolve geographical detail. The climate modeller has a looser vision of the world which is a prerequisite for extrapolating far into the future. This distinction is often lost and is central to misunderstandings about predictions of how our atmosphere may change in the future.

Whereas weather forecasting covers an interval of a few hundred hours, climate forecasting has to look decades hence. While a weather forecast model can accurately predict the weather change over a small area, climate models do so on a far larger scale. The Hadley model represents the United Kingdom by three grid points, so that it effectively isn't represented as a country on a global map. In the mind's eye of a Cray supercomputer, neither Japan nor the United Kingdom is big enough to register. At best, models of future climate change can only give a broad picture of how it will be manifested globally.

Climate models first have to come up with a semblance of today's atmosphere before their users can confidently gallop off into the future. The NCAR climatologist Stephen H. Schneider has an amusing tale of being quizzed by an almost apoplectic Congressman in the 1970s about the first climate-modelling predictions. He could hardly believe his ears when he heard that scientists had been spending millions of tax dollars to show that winters are cold and summers are warm. Schneider replied that if they couldn't prove *that*, then he could hardly stand up in front of Congress and argue that the climate could be changing from the basis of the models used. Schneider and other climatologists often refer to climate models as dirty crystal balls.

Another important distinction between climate and weather forecasts concerns their sensory input. For a weather forecast, the computer's view of the world is updated continuously so that it can gauge how storm systems, for example, will evolve in the space of a few hours. Climate simulations have to be run over several years with little sensory input: the value of their prediction really depends on the underlying accuracy of understanding how the atmosphere behaves from season to season.

Weather forecasts are limited by the inherently chaotic nature of the atmosphere. Predictions beyond fourteen days are worthless because small changes in the atmosphere will accumulate and render the overall forecast totally useless. This is the 'big whirls have little whirls' to which Lewis Fry Richardson alluded, and today it is usually termed the 'butterfly effect'. Its most vivid expression is 'Can the flap of a butterfly's wings today affect the weather elsewhere in a week's time?' The answer is that it is possible although highly unlikely. The point is that effects on such a small scale cannot be modelled and extrapolated in weather forecasts. They set a strict limit on the predictability of future weather patterns. Within that fortnightly time frame, however, weather forecast models can be tested. By plugging in real observations of the weather, the model's accuracy can be gauged against that which actually occurred. If a weather forecast predicts the passage of a storm system at 4 p.m. and the storm passes by at 9 p.m. then it obviously needs fine-tuning.

By comparison, climate models cannot be classed as 'true' or 'false' in the same way. They can be better thought of as being 'useful' or 'useless' in their ability to generate a selected range of facts for a number of possible future atmospheres. Climate models do not encounter the limits imposed by the butterfly effect because of their deliberately vague sense of the world. Nevertheless, some people (usually reactionary politicians and newspaper columnists) express puzzlement that climate models still cannot tell them what the weather will be like on their birthday, say, fifty years hence. 'If we could do that, we'd also be able to tell you how many bloody birthday cards you'd get and what colour shirt you'll be wearing,' said one exasperated modeller interviewed for this book. Climate models can, however, give an indication of how much sunshine or rain you would reasonably expect to find during the month in which your birthday falls half a century from now. So, at best, good climate models can only come up with reasonable values for these sorts of parameters within a meaningful statistical range.

The ultimate test of the validity of a climate forecast is to use the full benefit of hindsight. The same sorts of climate models used to predict the future have been employed to see if they can accurately account for changes in climate over the last few centuries, when relevant meteorological data is plugged in. Experience thus far has shown their predictions are severely lacking: for example, when weather data from the end of the last century is employed in climate prediction models and time steps are run forward to today, they invariably predict that average global temperatures will increase by 5°C. In reality, the figure is nearer to one degree. Yet this failure should not be judged too harshly, for the good reason that weather records that stretch that far back in time are few and far between. Observations are also biased to sites across the northern hemisphere, which sets obvious restrictions on the global outcome.

Most recently, the improved efficacy of past and future climate predictions has been demonstrated by work carried out by Martin Hoffert of New York University and Curt Covey of the Lawrence Livermore Laboratory in California. They used data from geological records from 20,000 years ago and 100 million years ago to test if they could predict what actually happened to the climate in between times. Their models of these primordial atmospheres were designed to respond to observed changes in solar radiation, and the subsequent build-up of carbon dioxide as indicated by air bubbles trapped in polar ice sheets. In both cases, Hoffert and Covey's model predicted that average temperatures at the Earth's surface would increase by about 2°C owing to a doubling of CO_2 levels. This prediction tallies with the recorded change in climate and this has bolstered belief in the ability of models to predict change. Current projections for global change suggest a similar temperature increase due to the doubling of CO_2 towards the end of the twenty-first century.

Any intellectual comfort to be gained by our ability to predict future climate change has to be weighed against nature's capacity to spring surprises. And indeed, recent developments in modelling our future atmosphere reveal the disturbing way in which global warming and ozone depletion will serve to antagonise each other.

It is a supreme paradox that both CFS's and ozone are greenhouse gases. Ozone depletion will effectively feed on global warming. Although CFC-induced ozone loss in the stratosphere will serve to offset the global warming effect of the chlorofluorocarbons themselves, the inexorable long-term build-up of other greenhouse gases will aggravate ozone depletion in the future. It is just one of many paradoxes that have to be accounted for as supercomputers gamely attempt to diagnose the symptoms of global change.

The greenhouse effect always gets a bad press, yet without it life would not be possible on the third planet from the Sun. For the constant stream of radiation from the Sun which reaches the Earth's surface is re-radiated at infrared wavelengths and trapped by greenhouse gases. The current cocktail of atmospheric gases ensures that the surface temperature is 30°C warmer than it would otherwise be. The net result is that, on the whole, the Earth is temperate enough for most life forms to survive. But the cumulative effects of human activity – most notably in the rise of carbon dioxide – indicate that the average global temperature may increase by 1– 2°C by the year 2040. By comparison, the globally averaged ambient temperature increased by roughly half a degree over the preceding century.

Much of the blame for this rise can be laid at the door of carbon dioxide. It is estimated that the average concentration of CO_2 was around 280 parts per million by volume before the start of the industrial age. Now it has nudged 350 ppm and is increasing by roughly 1 per cent per annum. However, the fact of this rise is clearer than its possible effects in the future. Carbon dioxide is just one of a handful of greenhouse gases and is certainly not the most efficient at trapping heat: that distinction is held by water vapour, which has mostly accounted for the warming that keeps our planet temperate. After that, the heat-trapping abilities of ozone and CFCs are the most prominent – something Jim Lovelock pointed out as far back as the 1970s. While obtaining infrared spectra of the planet Mars with a large ground-based telescope in the Pyrenees, he was startled to find that absorption lines due to the CFCs in our atmosphere could be seen quite markedly.

And this introduces a tortoise-and-hare aspect to the emerging picture of global warming. Compared to the heating effect of a molecule of carbon dioxide, the most common chlorofluorocarbons, CFC-11 and -12, are 5,800 times more efficient, and ozone is 2,000 times so. But at nearly 350 parts per million, carbon dioxide effectively swamps the effect contributed by the roughly 10 parts per million for ozone and 0.0005 parts per million for CFC-11 and -12. So, although ozone and CFCs have a greater potential to contribute to the greenhouse effect, there is much more carbon dioxide available to cause the greatest overall effect. During the late 1980s, it was assumed that the direct greenhouse contribution from CFCs would far outweigh the indirect cooling which results from their destruction of stratospheric ozone. Conventional wisdom was that the percentage contributions towards the greenhouse effect in today's atmosphere (excluding water) would be as follows:

carbon dioxide	55 per cent
methane	15 per cent
CFC-11 & -12	17 per cent
other CFCs	7 per cent
nitrous oxides	6 per cent

(It is also clear that the rise of tropospheric ozone will play a smaller, but increasingly important, part in heating the Earth's atmosphere at lower altitudes. There will also be an additional indirect greenhouse effect from water vapour: as temperatures increase, so will the amount of water evaporating from the oceans.)

Belatedly, however, climatologists have realised that stratospheric ozone destruction will be greatly amplified in its effect on climate. Quite simply, most ozone depletion occurs at the lower stratosphere where it is at its coldest. As the ozone is destroyed, much less heat is absorbed at ultraviolet wavelengths there, and this enhanced cooling more or less offsets the rise by CFCs. The ozone loss at this altitude has tended to decrease the average global temperature, so overall it has a negative effect on global warming. Keith Shine of Reading University and V. Ramaswamy of Princeton have looked at the issue for both the Intergovernmental Panel on Climate Change and the United Nations Environment Programme and concluded in a *Nature* paper in February 1992 that 'the net decadal contribution of CFCs to the greenhouse climate forcing is substantially less than previously estimated'. Bob Watson of NASA – who also coordinated the IPCC panel on greenhouse gases and aerosols – put it more forcefully: 'We've overpredicted by a factor of about a quarter. So, for example, predictions of global warming over the 1960s and 1970s should have been 25 per cent less. That's a slight, but not insignificant, figure.'

The question of ozone depletion is just one of innumerable complications raised by global warming generally. For the moment, the emerging 'signal' of the onset of the greenhouse effect remains hidden within the 'noise' of the year-to-year variability of climate. Recently, however, ozone researchers have revealed that our climate is rather more sensitive to change than had previously been suspected. This will have far-reaching consequences in the subtle, inextricable linkage between ozone depletion and global warming in the future.

The crystal balls that are supercomputers remain murky because the climate models run on them often have to make bland assumptions about many of the factors – or parameters – upon which they are based. It is a fundamental axiom in climate research that computer models are only as good as the assumptions which are built into them. In the parlance of climate modellers, this is known as 'parameterisation' and involves making large-scale assumptions about factors that are just too difficult or complicated to be modelled with any degree of certainty or accuracy. Of all the problems faced by climate modellers, the way in which clouds and the oceans interact with climate present the most fundamental handicaps.

Clouds are by far the most perplexing. If whole countries slip through the grid, it is hardly surprising that individual cloud decks cannot be accounted for in climate models. In the early 1990s, a comparison of how eighteen general-circulation models of the atmosphere 'handled' a doubling of carbon dioxide produced a stark result. If all the models assumed that there was a cloud-free world, they all agreed upon the subsequent rise in global temperature despite the different assumptions built into them. When clouds were introduced, however, the numbers ranged between 1.5°C and 4.5°C. Each model portrayed feedbacks – such as how much light was reflected back to space or how much was re-reflected back by higher-altitude clouds – in fundamentally different ways. Different paramaterisation results in very different future atmospheres. In the late 1980s, for example, the UK Met Office predicted that a doubling of CO_2 would result in a temperature rise of 5.2°C worldwide. Within a couple of years, however, modellers at the Hadley Centre had reduced this estimate to 1.2°C when they were able to discriminate between transitory icy clouds and persistent water clouds in an improved version of the model. Previously they had treated clouds en masse and had arbitrarily assumed that they would all have the same reflective properties.

The oceans, too, complicate matters by absorbing carbon dioxide and redistributing heat over decades in ways that are not well understood. As nearly three-quarters of the Earth's surface is covered by oceans, they have a fundamental impact on our climate. In a sense, the surface of the seas act as a global 'thermostat setting' for the reservoir of heat contained below them. The efficiency of the oceans in absorbing or redistributing heat may act to delay the onset of global warming. The problem is that such information is not yet available: climate modellers often have to make educated guesses about how the oceans will change because of that incomplete knowledge. The World Ocean Circulation Experiment, which started in 1990 and will last for five years, is a concerted effort to learn more about these uncertainties by the use of satellites and dedicated survey ships.

Also important in parameterisation is the way in which vegetation absorbs or reflects radiation at the Earth's surface. Although assumptions have to be made, they are often done on minimal experimental basis, so during the 1990s there are large field experiments with wildly varying vegetation all around the world to investigate this problem more closely.

The net result, then, is that the climate modeller is faced with the intellectual equivalent of the tasks which vexed Hercules. A variety of interrelated and often inconsistent factors make the diagnosis of global warming far from certain. In time, as the inconsistencies are ironed out and a more detailed, seamless global coverage from satellites becomes available, the picture will doubtless come into focus. 'What these problems show beyond all doubt is that we have to get the first few pieces of the jigsaw in their place before we can complete the rest,' said Bob Watson with commendable clarity.

And despite their inherent flaws and idiosyncrasies, the only possible way of manipulating these myriad pieces into a coherent whole is to use the unparalleled power of supercomputers.

Just how effectively global warming may aggravate ozone depletion is a nightmare that has been played out on a supercomputer. In November 1992, just days before government representatives met in Copenhagen to work out new controls on future CFC emissions, the scientific community was given a timely reminder of how ozone depletion would be enhanced if carbon dioxide levels doubled. British researchers using one of the Met Office Crays showed that an Arctic ozone hole is ever more likely in the future. John Austin of the Met Office, Neal Butchart at the Hadley Centre and Keith Shine of Reading University used the standard 3-D stratospheric model to predict the future state of the ozone layer over the northern hemisphere. Their results made the cover of *Nature*.

They modified the Met Office model by effectively throwing away the troposphere so that more detailed chemistry involving chlorine and bromine compounds could be plugged in. 'We exchanged computer time we'd normally devote to the troposphere to the chemistry to see what might happen,' Keith Shine said. The researchers received a considerable fillip for when they used current values for CO_2 and CFCs, the model predicted the patterns of ozone loss in the lower stratosphere seen during the most recent Arctic winters. In the mid-1990s, CO_2 rates were increasing by roughly 1 per cent per annum, and although the Earth Summit in Rio in the summer of 1992 agreed that emissions by 2000 should have been cut back to 1980 levels, it is clear that a 60 per cent cut in emissions would be required to prevent climate change. Austin et al. looked at an atmosphere where carbon dioxide would have reached 660 parts per million and chlorine loading would have stabilised at 3 ppm.

As the greenhouse effect takes hold in this doubled-CO_2 world, the stratosphere cools as the troposphere warms up. This is because

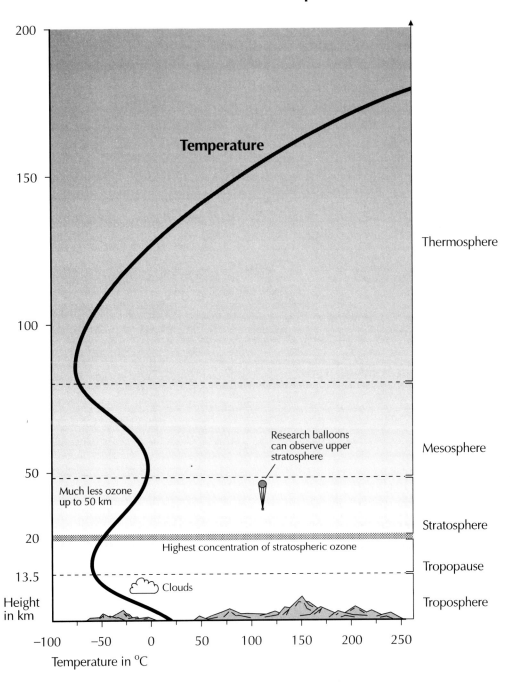

The Ozone Layer and Ultraviolet Radiation

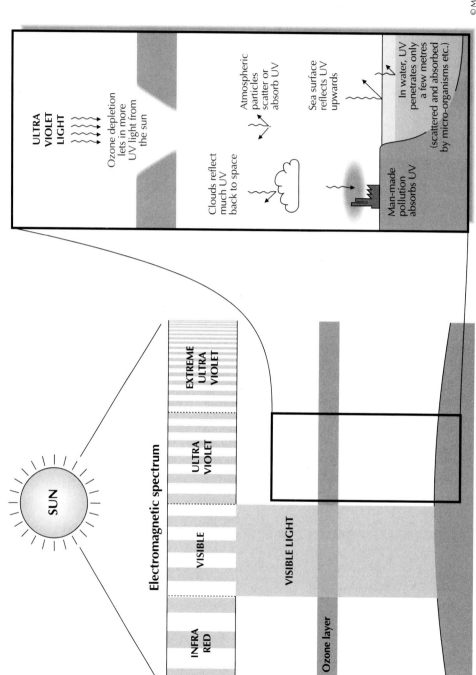

Prophet In The Wilderness
Twenty years ago, Sherry Rowland discovered how the ozone layer might be at risk from CFC's. Until the discovery of the ozone hole a decade later, his predictions were often ignored although he has now been vindicated.
(Kerry Klagman)

Voice Of Reason
British-born Bob Watson has been a key player in the ozone issue, ensuring that the complexities of the science were translated into language understandable by diplomats and politicians.
(Courtesy Elizabeth Watson)

The Man Most Likely To
Joe Farman (above) has spent all his working life measuring ozone in Antarctica. In the early 1980s, his team at the British Antarctic Survey found evidence that two-thirds of the ozone over the south pole was disappearing each spring.
(Nick Sack, 1990)

The Sky Is The Limit
Susan Solomon of the NOAA Aeronomy Lab in Boulder determined how so much ozone is lost over the poles. She is seen here on the first US National Ozone Expedition in 1986 at McMurdo Sound.
(Courtesy Susan Solomon)

Simply Sampling The Stratosphere
Large scientific balloons remain the most cost-effective way to make direct observations of the ozone layer. Seen here is one of the larger balloons launched out of Kiruna in the spring of 1992.
(European Ozone Research Coordination Unit)

We Come In Peace
Two of NASA's three ER-2 aircraft flying over the Golden Gate Bridge, close to their base south of San Francisco. The ER-2 remains the workhorse of NASA-led campaigns to investigate the ozone layer.
(NASA)

Call Me Snoopy
The UK Met Office uses a modified Hercules transporter (above) to investigate the increasingly important question of tropospheric ozone. Because of its brightly coloured nose probe, it is called Snoopy. *(The Met. Office)*

Up And Running
In September 1991, NASA's Upper Atmosphere Research Satellite was deployed in orbit from the Space Shuttle. Despite technical problems, the $700 million satellite has revolutionized studies of the ozone layer from space with a new generation of instruments. *(NASA)*

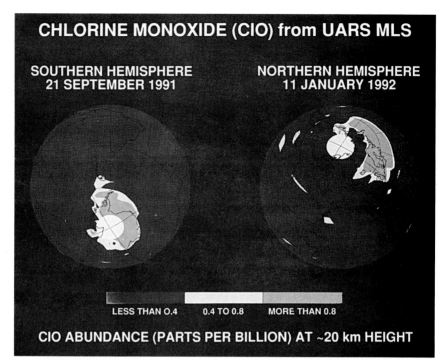

North And South

The Microwave Limb Sounder aboard UARS has mapped the extent of chlorine monoxide - the smoking gun in ozone depletion - from space. As seen above, the amount of chlorine monoxide in the northern hemisphere was similar to that over Antarctica when the ozone hole is at its greatest extent. This worrying observation led to the February 3rd press conference discussed in Chapter 6.

In the austral winter of 1992, the Microwave Limb Sounder revealed hitherto unexpected details of ozone loss over the south pole leading to the formation of the ozone hole. As shown below by the white areas on the globes, the ozone loss started early on in the winter season and is correlated with the presence of chlorine monoxide. *(Joe Waters, NASA JPL)*

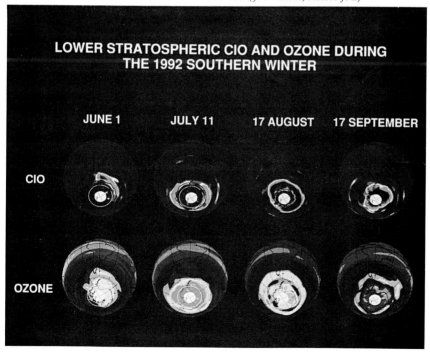

Down Under
Increases in skin cancer will largely be averted by changes in social behaviour. As this photograph above shows, sunbathers on Bondi Beach are a thing of the past, as the Australian populus at large has learned to stay out of the sun.
(The Science Photo Library)

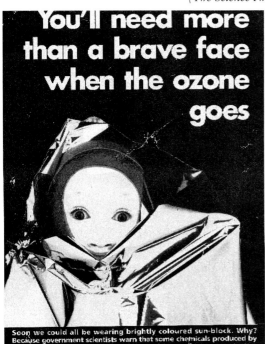

Environmental Overkill
Unlike the measured and preventative campaign in Australia, Greenpeace in Britain have used scare tactics to alert the population to the dangers of ozone depletion. The campaign poster shown here was widely criticised for being over the top.
(Greenpeace)

the atmosphere as a whole has to maintain thermal equilibrium – the amount of radiation coming in must be equal to that being re-radiated out. As a result, Austin et al. concluded that 'it might be expected that up to 20 per cent of winters will produce Arctic ozone holes for doubled CO_2 conditions. For the remaining 80 per cent of winters, it is unlikely that doubling of CO_2 will suppress the occurrence of a stratospheric warming sufficiently to produce an Arctic ozone hole, although even in these conditions ozone depletion could be enhanced.'

The timing of that warming could not be predicted with any accuracy, for the Met Office model did not consider how weather patterns might change in the future. 'When we double CO_2 we don't know if mid-latitude storms will get stronger or weaker,' Shine said. According to some authorities, the thermal difference between the poles and equator will lessen, so these storms will become weaker and the stratosphere will be kept cooler because it will be less affected by the warming from wave activity forced from below. On the other hand, there is a school of thought that suggests increased temperatures will bring about increases in wave activity as storm systems become more active owing to greater evaporation from the oceans. That would tend to warm the stratosphere and thereby lessen the likelihood of ozone depletion. Future work in these areas will help resolve this not insignificant matter.

Although this greenhouse world may be decades away, and a doubling of CO_2 might be averted, the effects of a cooling stratosphere have already become apparent. Observations already recorded in the 1990s over Antarctica have revealed that there is a reduction in stratospheric temperature associated with the formation of the ozone hole. As ozone is depleted, less ultraviolet radiation is absorbed which normally would keep the stratosphere warm. Formations of polar stratospheric clouds have become more extensive and the ozone hole has increased in area.

John Austin and his colleagues believe that polar stratospheric clouds will cover greater areas of the Arctic regions in the future. They concluded their *Nature* paper by noting: 'It is therefore likely that a combination of conditions will occur during the next fifty years which will give rise to an Arctic ozone hole unless both CO_2 and chlorine emissions are curbed.' These findings portend a disturbing trend. It had been vaguely comforting to think that global warming's deleterious effects might be decades away, by which time chlorine loading would have passed its peak. But the fate of the ozone layer in the greenhouse world of the future may not be so certain.

Chilean fishermen were the first to notice it. Around Christmastime, they observed that the regular coastal currents to which they had long been accustomed had changed. The strong offshore current was warmer than usual and had resulted in smaller catches of fish. They named this warm current after the infant Jesus as 'the boy child', or El Niño, because of its coincidence with the nativity. Little did they suspect that this current was one small part of a climatic anomaly which centuries later would underpin many theories of global change and ozone depletion. It is yet another complication in a seemingly endless catalogue of phenomena which the climate modeller will ignore at his or her peril.

El Niño events, as they are termed, occur roughly every four to seven years when the circulation of both the oceans and atmosphere in the central equatorial Pacific change. They are caused by an unexplained warming of water near to the ocean surface which markedly alters the local climate. Normally winds and ocean currents travel towards Asia in the central Pacific, but when an El Niño occurs they turn around towards the East, possibly owing to an accumulation of warmer water which then sloshes back in the opposite direction. The warm current which brushes past the Chilean coast is not the end of El Niño's Pacific-wide influence, for it wreaks climatic damage across much of the world. Global atmospheric circulation and weather patterns are so severely disrupted that deserts in California, Peru and Ecuador have been flooded and droughts have occurred in the Far East and Australia. The start of the El Niño event in October 1991 was heralded on the West Coast of the United States by severe storms which caused many millions of dollars' worth of damage to property. El Niño also caused forest fires in Indonesia, Australia, Borneo and Sumatra, with similar consequences for their local economies.

The effects of El Niño are all the more surprising because the oceans are normally perceived to exert a steadying influence on climate. Because the oceans trap heat more effectively than the land, they can be thought of as a vast thermal reservoir and their thermal inertia — their tendency to keep hold of that heat — has to some extent delayed the onset of global warming. In the thermostat analogy, the temperature of the ocean surface could be thought of as its 'setting'. An increase in warming at the Earth's surface would be mirrored by a rise in sea surface temperatures. Although evidence of this has yet to emerge globally, localised changes in sea surface temperatures can occur and bring a marked climate effect.

El Niño events are among the most persistent, but there are many other, less extensive changes in sea surface temperatures which are clouding the emerging picture of global change. These changes can occur quite unexpectedly, revealing just how little is known about the current setting of the global thermostat. After 1976, for example, atmospheric circulation patterns changed over the northern Pacific and continued for over a decade. Most noticeably it ushered in storms and large waves along the coast of California. For Walter Komhyr of NOAA's Climate Monitoring and Diagnostic Laboratory in downtown Boulder, it provided clues to solving a mystery of ozone depletion.

A Canadian who drives a four-wheel jeep perhaps too recklessly for one nearing retirement, Komhyr has been involved in ozone measurements since the International Geophysical Year. Like Joe Farman a continent away, Komhyr assumed responsibility for NOAA's Dobson spectrophotometers operated at twenty-five sites around the world, including its South Pole station. It was this background knowledge of the Dobson record that led him to postulate whether a change in sea surface temperatures could affect the ozone hole over Antarctica. He was curious why purely chemical methods could not explain some of the observed trends in ozone depletion as recorded by Nimbus 7 and the Dobson network. Gas-phase photochemistry alone could not account for the rather large seasonal decreases in ozone at middle and high latitudes. 'It appeared to me that some other phenomenon was contributing to the observed changes,' he said. 'And, almost on cue, nature provided a clue as to what might be happening during 1988.'

In June of that year, spectrophotometers at the NOAA 'clean air' monitoring station on Samoa recorded total ozone levels which suddenly increased to values not seen there since the mid-1970s. A few months later, the same thing happened at Mauna Loa in Hawaii. And at the South Pole, the NOAA station recorded a relatively shallow hole compared to previous years. Curiouser and curiouser, Komhyr thought. He then recalled some work by Jerome Namias of the US National Weather Bureau linking anomalous atmospheric circulation patterns over North America to anomalous sea surface temperatures in the north Pacific. Could there be a similar relationship between sea circulation and the ozone measurements that had been made at the South Pole?

The next step was to compare ozone measurements and sea surface temperatures in the eastern equatorial Pacific over a twenty-five-year period. He found that between 1962 and 1975 the Pacific cooled and ozone levels increased. When the waters warmed from 1976 to 1988, the ozone decreased. According to an analysis which he performed together with other NOAA colleagues, sea surface temperatures in the eastern equatorial Pacific appear to influence the interaction between equatorial winds and mid-latitude planetary waves which are several thousand kilometres in extent. They normally carry momentum and heat towards the poles and can obviously disrupt weather patterns. 'These observations suggest that the downward trend in ozone observed over the globe in recent years may have been at least meteorologically induced,' Komhyr said.

When his findings were published in the November 1991 issue of the *Canadian Journal of Physics*, many atmospheric chemists pronounced themselves sceptical. But at a conference on ozone held in Williamsburg, Virginia, the following June, E. A. Jadin of the Central Aerological Observatory in Moscow presented a paper entitled 'Long-term total ozone variations over Europe and sea surface temperature changes in the Atlantic' with similar evidence for a relation between the two. Though this correlation may be important so far as long-term trends are concerned, it is, Komhyr admitted, only of secondary importance so far as the cause of the Antarctic ozone hole is concerned. He was quick to point out that these dynamic effects are much less important than the role of the heterogeneous chemistry taking place on polar stratospheric clouds. Yet it is nonetheless an interesting conundrum among the many presented by ozone depletion worldwide.

Pieces in a Puzzle

The events of the summer of 1816 were a remarkable augury for dramatic climatic change. For the 'year without a summer', as it was later immortalised, ingrained itself in the public psyche thanks to two celebrated works of literature which it directly inspired. The greatest literati of the day were holidaying on Lake Como in northern Italy, where they found that the weather was unaccountably 'ungenial'. As a result, Lord Byron, Percy Bysshe Shelley and his bride to be, Mary Wollstonecraft (daughter of the feminist of the same name and the anarchist William Godwin), would gather by the fire and try to outdo each other in telling ghost stories (and if Ken Russell's film *Gothic* is to be believed, engage in weird forms of sex and have even weirder dreams about sex with nuns – but that's hardly new for Ken Russell).

Byron was certainly more melancholic than usual, and the dreadful weather inspired his deeply disquieting '*Darkness*' quoted at the start of this chapter. Out of those fireside stories, Mary Shelley would eventually complete *Frankenstein*, perhaps the most accomplished warning about technological advance ever written. It too is replete with references to gloomy weather, which became a key element in its subsequent cinematic adaptions. In fact, widespread misery was the most enduring feature of 1816 and one contemporary chronicle crisply summarised it as a time of 'hunger, want, sickness, death, manufacturing unemployment, trade stagnation and calamitous weather'. It would not be too much of an exaggeration to say that the weather went completely haywire, and other historical accounts record that it snowed in Madras, the New England corn crop failed, there were large numbers of beggars in the streets in cities the world over as well as repeated food riots in Britain and France. The final death toll from 'the year without a summer' was in excess of a thousand lives, at a conservative estimate.

Although it was not known at the time, these calamitous events were the direct climatic consequence of one of the most violent volcanic eruptions in recent times. The Indonesian volcano Tambora had erupted the previous summer and the estimated 50 cubic kilometres of dust and ash it injected into the stratosphere shielded the Earth's surface as its spread across the globe. Small wonder that the Sun was rarely seen in the months which followed, and that it remained unseasonably cold for the rest of that decade. Tambora is the most remarkable example of how our climate can be changed over many years by volcanic eruptions. And, as has been found in subsequent years, volcanoes do not have to be nearly so cataclysmic to cause an effect worldwide.

When one of Tambora's neighbours erupted with smaller force in 1883, there were palpable effects on climate across the globe. In the months following Krakatoa's explosion, newspaper headlines could be read at night from light reflected by streetlights across London. Atmospheric haziness and often brilliant sunsets around the world were unequivocal evidence for the global spread of the dust and gas from Krakatoa. In the decades after the eruptions of both Indonesian volcanoes, the Earth's surface was cooled by the effect of their dust clouds. Fortunately, volcanic explosions in more recent times have been firecrackers by comparison, but even so their impact on temperature records has been well established.

The eruption of yet another Indonesian volcano, Mount Agung, in 1963, gave important insights into a far more subtle volcanic influence on the stratosphere. In the early 1960s, Christian Junge, then with the US Air Force Cambridge Research Laboratories, had discovered a layer of aerosol particles in the stratosphere which now bears his name. The Junge Layer is composed of sulphuric acid droplets, 0.1 micron in diameter, and is within roughly the same altitude range as the ozone layer. These aerosols tend to remain suspended there, hardly mixing with the bulk of the atmosphere below. Almost on cue, the eruption of Agung revealed that the particles it injected upwards – usually termed aerosols – replenished the Junge layer and heated the stratosphere by 2–5°C as a result. Subsequent modelling revealed that this aerosol layer absorbs infrared radiation and scatters solar radiation back to space: effectively it acts to warm the stratosphere and cool the troposphere below by shielding it from the incoming solar radiation.

Subsequent eruptions have helped refine models of how volcanoes can affect climate. Not all volcanoes, however, have as marked effect on the stratosphere as Mount Agung did. The eruption of Mount St Helens in 1980 produced an awful lot of ash, much of which was injected into the stratosphere, but it didn't remain there for long enough to bring about a long-term climatic effect. It is now known that the sulphur content of the volcanic material – not the dust and ash – results in enhanced climate forcing from long-lived sulphuric acid aerosols. Mount St Helens was comparatively poor in sulphur and that is why it had little climatic effect.

Volcanic aerosols may also have an important effect on ozone. Susan Solomon and David Hofmann of NOAA proposed in 1989 that such particles would significantly aid and abet ozone depletion by providing surfaces on which catalytic reactions could take place. Sulphur-based aerosols would serve much the same purpose as polar stratospheric clouds do in the heterogeneous chemistry which causes the Antarctic ozone hole. Oxides of nitrogen would be transformed into nitric acid and precipitate out, thereby removing the natural safety mechanism which the atmosphere exhibits towards the ozone layer. Volcanic aerosols would produce a double whammy: simultaneously, the most abundant form of chlorine in the stratosphere, chlorine nitrate, would be converted into a more reactive form. This liberated chlorine could then take part in ozone-destroying chain reactions. Although these reactions would not be as efficient as they would be on polar stratosphere clouds, they could still cause significant ozone depletion across whole swathes of the warmer latitudes away from the poles. Solomon and Hofmann cited as evidence the marked drop in ozone levels over the northern hemisphere (an average of 10 per cent) after the eruption of the Mexican volcano El Chichón in April 1982. Once again, it was the sulphur content of its aerosols which 'drove' many of the chemical reactions involved.

In November 1990, Guy Brasseur of the National Center for Atmospheric Research in Boulder used a 2-D model to show that another large volcanic explosion would intensify worldwide ozone depletion because of the increased chlorine loading of the stratosphere since the eruption of El Chichón. If the aerosols were sulphur-rich and injected into the stratosphere over the equatorial regions (where they would then circulate around the globe hand in hand with the natural dynamic spread of ozone), then, Brasseur calculated, ozone depletion over the mid-latitudes could reach 10 per cent. Almost on cue, Mount Pinatubo in the Philippines erupted six months after his paper appeared in *Nature*. Aerosols from the explosion were injected into the equatorial stratosphere, and quickly spread around the rest of the globe over the subsequent winter.

How Soon Is Now?

As we have seen, it helped prime the stratosphere for marked ozone depletion over the Arctic winter of 1991–92. Over the long term, its effects on stratospheric chemistry have not been as marked as Brasseur had suggested, but significant nevertheless. More immediately, and most spectacularly, the dust from Mount Pinatubo demonstrated just how sensitive our climate is to volcanic influences. The explosion in the Philippines would soon be described as 'the greatest geophysical event of this century' by one of the few climatologists brave enough to make specific predictions about its short-term influence on climate.

In upper Manhattan, where 112th Street crosses the northeast corner of Broadway, there is a nondescript building. At street level is a café of the greasy-spoon variety which is frequented by students from the nearby Columbia University or office workers in need of cheap and cheerful nourishment. This café too would merit little attention had Suzanne Vega not immortalised it in her song 'Tom's Diner'. And the rest of the building above is, for climate scientists, equally notorious for it plays host to the NASA Goddard Institute of Space Science and its staff of twenty climatologists directed by James Hansen.

Hansen himself achieved a great deal of notoriety in 1988 when he attested at a Congressional hearing that he was 99 per cent certain that the greenhouse effect was already taking hold. Very few colleagues have allowed him to forget that statement, which some characterise as brave and bold, others as ridiculous and mischievous. Whatever its merits, Jim Hansen is one of the few climate modellers who will actually make predictions. True to form, after the eruption of Mount Pinatubo, he was one of the few climatologists in the world to calculate the effect it would have on the Earth's climate in the months which followed.

Ostensibly part of NASA's Goddard Space Flight Center, GISS has been associated with Columbia University since its formation in the early 1960s. The idea behind establishing a theoretical research centre in New York City was to encourage intellectual cross-fertilisation outside the confines of a NASA establishment. And indeed, the Goddard Institute is more relaxed and informal than the other NASA centre which shares its name. In late June 1992, for example, a xeroxed poster announced a forthcoming talk entitled 'As the World (Barely) Turns'. Underneath it continued: 'Abstract: Comparative planetologists dream of being able to twist a few knobs in a model and turning one planet into another. This dream has not yet been realized. Keywords: Superrotation, Venus/Titan, inadequate funding levels, frustration.' Hansen's ostensible bosses just outside the Beltway would doubtless balk at such displays of anti-establishment sentiment.

Given his controversial reputation, one might expect Jim Hansen to be as emotional and outspoken in person. So it is quite a shock to be greeted by a shy, hesitant man with a Midwestern accent, delivered in a style strangely reminiscent of Jimmy Stewart in *It's a Wonderful Life*. There are other surprises within the imposing edifice of GISS. To run its climate models, it employs an Amdahl for which 'supercomputer' would be overstating the case. It was using this 'antiquated calculator' (in the words of one onlooker) that Hansen and his colleagues predicted global warming would increase the average temperature of the Earth's surface by 0.6°C over the next three decades. On the basis of this work he was called before Congress to discuss the greenhouse effect at length in the record-breaking, swelteringly hot summer of 1988. On that occasion, it should be noted, he had to testify as a 'private citizen' rather than incur the wrath of the Reagan administration. The extent of the notoriety this gave him only struck him when he saw a headline in his 'hometown' paper, the *Cedar Rapids Gazette*: IOWAN JAMES HANSEN IS PROPHET OF GLOBAL WARMING.

Hansen's research team at GISS has been considering the effects of what he has aptly termed 'nature's own great climate experiment' – the eruption of Mount Pinatubo. For Jim Hansen it has meant a return to the subject of his master's degree which he obtained from the University of Iowa in 1963. His thesis involved precise measurements of the effect of the aerosol from Mount Agung upon a lunar eclipse shortly thereafter. 'We couldn't really see the Moon with our naked eyes,' he recalled, 'as the atmosphere was very dark because of the dust.' Four years later he joined GISS freshly armed with a doctorate, and while helping establish its pre-eminent role in climate modelling, kept a keen eye on the climatic effects of volcanoes, particularly after the El Chichón eruption in 1982.

For Hansen the subsequent eruption of Pinatubo had the same effect trapdoors have on daemons in children's pantomimes. It would give GISS climatologists the chance to run their computer models and assess the effect of volcanic aerosols on climate. 'Pinatubo was twice as large as El Chichón,' Hansen said. 'If we don't get a signal, then there's something wrong somewhere. If we can successfully model the effects of Pinatubo, we will have a better understanding of global warming,' he added.

Once again, Hansen was willing to make specific predictions. In a paper which appeared in *Geophysical Research Letters* in January 1992, he suggested that Mount Pinatubo would 'delay by several years the time at which global warming becomes generally obvious'. On the GISS Amdahl they ran a number of scenarios (each having slightly different initial atmospheric conditions) to relate the shielding effects of its aerosols to the temperature of the Earth's surface. 'All our models, which take into account varying degrees of greenhouse gases, predict a cooling of 0.5°C by the end of 1992,' Hansen told me a year after the explosion. Further proof of his propensity to stick his neck out came with an even more startling prediction: that a city at the latitude of Moscow could expect to have unusually cold winters, that Omaha would expect to have cooler summers and that in Washington, DC, cherry blossoms would be noticeably delayed in the spring seasons that followed.

His critics sat back, rubbed their hands and could hardly believe his foolhardiness as nature provided a positively Machiavellian complication. No sooner was the dust cloud from Pinatubo spreading throughout the stratosphere than the most recent El Niño event started up towards the end of 1991. Climatically, El Niño and the dust from Mount Pinatubo would tend to work against each other. The rise of the former's warm current in the equatorial Pacific heats the atmosphere above it: the shielding of sunlight by Pinatubo has tended to cool the ocean below. Which effect would win out? Hansen stood by his belief that Pinatubo would be far more important and that the GISS models would be vindicated.

'If this year's El Niño is a typical one, then its effect may get swamped,' Hansen told me in June 1992. 'It looks now like El

Niño has drawn to a close. There are still aerosols so we'll get a clearer test of the volcanic impact.' Though the average global temperatures had dropped towards the end of 1991, there was a steady rise throughout the first half of 1992. Where lesser scientists would have panicked, Hansen remained upbeat and was often heard to remark to colleagues and journalists alike: 'My guess is that this is probably a case where the model is right and the real world is wrong!' But with whatever blind faith he can summon for the efficacy of his predictions, the atmosphere more or less responded as Hansen said it would, by cooling somewhere in the 0.3–0.4°C range.

Ever the pragmatist, Hansen cautioned that he was by no means on a home run yet. The GISS models predicted that global temperatures would rapidly recover from the Pinatubo cooling and towards the end of the 1990s we will experience record temperatures. In a paper his GISS team contributed to *National Geographic Research and Exploration* in the summer of 1993, they declared that their long-term predictions 'will provide a check of fundamental science [and the] degree to which the climate system is presently out of equilibrium with radiative forcing'. In plain English, that means the effect of the Pinatubo aerosol has shown that our climate is even more sensitive to change than had been previously thought. Their certainty came from the fact that the actual temperature recovery after the Pinatubo cooling has followed the uppermost range of values predicted at GISS (a less marked recovery would hint that our climate is much more resilient). 'New record global temperatures in the 1990s, if realized, will be evidence in support of [this] high climate sensitivity,' Hansen et al. concluded in *National Geographic Research and Exploration*.

His detractors beg to differ. The annual variability of weather means that it will be difficult to see if his temperature patterns are followed exactly. A noted greenhouse sceptic, Richard Lindzen of MIT, has gone so far as to say that 'Hansen has pulled a fast one with this volcano'. He claims that any models would show a similar temperature recovery from the Pinatubo cooling to that which Hansen has claimed as his own. And though temperatures within continental landmasses were cooler over the winter of 1992–93, there was no noticeable delay in the cherry blossoms in the US capital in the spring. Only time will tell whether Jim Hansen has stuck his neck too far out on the issue of our climate's increased sensitivity to volcanic explosions and the greenhouse effect.

The undulating downs of Oxfordshire provide an incongruous backdrop for research at the cutting edge of many scientific disciplines. A handful of pleasant, modern buildings pepper a pleasant, quintessentially English landscape, slightly despoiled by the large cooling towers on the horizon near Didcot. The laboratory that bears the name of two of Britain's most famous physicists – and was once memorably described as 'a high tech centre in the leafy English countryside' – contains many surprises. Until recently, the Rutherford Appleton Laboratory was largely eclipsed by the more famous Harwell Laboratory next door, better known to peace campaigners and military scientists alike for its involvement with nuclear research. But today, the establishment which commemorates the physicists who split the atom and determined that radio signals would bounce off the ionosphere is pre-eminent in many fields. RAL's large space research division, for example, has contributed to the Upper Atmosphere Research Satellite and built hardware for both the ISAMS and MLS instruments. Other RAL scientists are involved with nuclear research or materials science and advanced computing with specific applications in the modelling of the stratosphere.

Within the handful of buildings at RAL is another Cray supercomputer which has been used to run models describing global ozone distribution. The Universities' Global Atmospheric Modelling Project is jokingly referred to by one participant as the 'Me Too Modelling Project', for it allows academics in Britain to gain access to a state-of-the-art 3-D general circulation model which accurately portrays both chemistry and dynamics. Formed by a consortium of British universities and funded by the National Environment Research Council, the contributors to the UGAMP project increasingly belie the traditional divisions (and, it must be said, enmities) between dynamicists and chemists, for the good reason that it used the talents of both from the outset. Where many models added chemistry or dynamics at a later stage, both were envisaged as integral parts of the model.

The UGAMP scientists adopted a weather-forecasting model developed by the European Centre for Medium-Range Weather Forecasting in nearby Reading. 'We don't run it in the forecast mode,' said the RAL researcher Lesley Gray, a dynamicist. 'We are interested in things like seasonal cycles so we run the model over a year or more.' The UGAMP scientists will model the distribution of ozone, for example, over a few years with real data to check if their understanding of the atmosphere is essentially correct. They extended the vertical grid within the original ECMWF model from 30 to 90 kilometres to keep tabs on the influences on ozone distribution higher in the atmosphere. The model is able to calculate the relative importance of the dynamic, radiative and chemical factors at work in the stratosphere.

The UGAMP model has led to many developments in the understanding of the subtleties of ozone depletion throughout the world. At mid- to high latitudes, for example, stratospheric air is pulled down behind weather fronts to form what is known as 'tropopause folds'. The UGAMP model has started to resolve these curious events which are important in determining how dynamics can alter ozone levels at any given time. Atmospheric waves have also come under greater scrutiny thanks to the resolving power of the UGAMP model. In the profusion of wave activity in the upper atmosphere, one of the more perplexing is known as the quasi-biennial oscillation (QBO). It manifests itself as a distinct change in stratospheric wind patterns near to the equator and just over every two years (twenty-eight months to be exact) they change direction.

The origins of the QBO remain tantalisingly elusive, but the phenomenon has to be taken into consideration when monitoring the progress of ozone from the equator to the poles. It can cause alterations of between 30 and 40 Dobson units towards the polar regions, 'which is significant when you're trying to perform trends analyses,' said Lesley Gray. 'The signature of the QBO is very distinct. It's just the sort of problem for which you really need 3-D models.' Previously, she had run around twenty 'ten-year runs' of the QBO on a 2-D model, but its exact effects on ozone depletion were smeared out. On the UGAMP model, she was now preparing to 'watch' the evolution and effect of the QBO on global ozone levels with its most accurate and realistic representation ever assembled.

Throughout the rest of the atmosphere, planetary waves behave like the ocean breaking on a beach. As they propagate up towards the poles, they break up into smaller turbulence like the surf found at the seafront. This means that naturally occurring trace gases will be disturbed by the mixing-in of air transported by planetary waves. Overall, this muddies the picture of upper-atmosphere chemistry and in 2-D models these effects would tend to get 'averaged' along a line of longitude so they would effectively be ignored. 3-D models like UGAMP have for the first time attempted to show how parcels of air mix in and are chemically altered across whole portions of the northern hemisphere. This remarkable 'back-tracking' of air parcels was used during the European Arctic Stratosphere Ozone Experiment to plan launches of balloons to search out the most interesting chemistry.

Researchers at the University of Cambridge, led by one of the EASOE coordinators, John Pyle, used the UGAMP model to predict how chlorine monoxide would build up in and around the vortex. His team at Cambridge added a complex scheme to portray 'chemical simulations' at grid points within the stratosphere. Remotely accessing the RAL Cray from Kiruna, they used actual meteorological data to 'initialise' the model and predict how the atmosphere would behave in the immediate hours afterwards. Because of the butterfly effect, such predictions would be useless after a few days. In the immediate hours after the data were entered into the model, however, UGAMP could produce valuable insights into the ozone-destroying chemistry at work within the polar vortex. 'What we saw during the EASOE campaign was a massive amount of chemical processing in the polar vortex in January 1992,' said John Pyle. And the way in which chlorine monoxide filled the vortex when it was at its coldest mimicked actual observations from Joe Waters's MLS instrument aboard the Upper Atmosphere Research Satellite. The ability of the new model to predict data from a space-based instrument was a significant indication of the increased sophistication of both areas of research.

'We want to use the model as a tool to diagnose ozone loss,' Pyle said of the chemistry which can be 'tracked' using the UGAMP model. He admitted that 'we weren't quite ready' to use it operationally in real-time during EASOE, but it has helped in the understanding of the chemistry of how chlorine is activated within the vortex. Computer simulations allow the same experiments to be run again and again with slight alterations in the initial data. By altering concentrations of some radicals or ignoring others altogether, it is possible to determine a picture of the details of the underlying chemistry. The approach helps sharpen understanding of the way the Arctic stratosphere primes itself for ozone depletion.

'From EASOE we have a good picture of the chlorine activation,' said Pyle. Simulations showed that air processed at the edge of the vortex peeled off and mixed in with air at lower latitudes. However, the way in which the chlorine was 'deactivated' at the end of the spring season – when it returned back into reservoir forms – is rather more complex than had been previously thought. As a result of EASOE, Pyle and his colleagues can now address new areas of research to understand better how chlorine is deactivated when the vortex breaks up. This symbiosis of experimental campaign and theoretical modelling is a new and promising development in the understanding of Arctic ozone depletion.

The latest approach to modelling the stratosphere may be found at NASA's Goddard Space Flight Center just outside the Washington Beltway. Data assimilation is a halfway house between weather prediction and evaluating climate changes using supercomputer models of the atmosphere. It takes real meteorological data – such as temperature, wind and moisture – to look at how ozone distribution around the globe changes as the weather does. The 'assimilation' refers to the fact that an inordinate amount of time and effort is used to gather data together to portray accurately the state of the atmosphere as it was at one particular instant. 'We use winds from assimilated data to drive a 3-D transport and chemistry model,' said Dr Richard ('Ricky') Rood, head of the data assimilation centre at Goddard. 'We're the only major data assimilation research group not at a weather prediction centre.'

Data assimilation aims to get around even the best 3-D climate model's 'bias'. The parameters of some observations lead to uncertainties in the representation of the real atmosphere. Specifically, data assimilation aims to circumvent a different sort of bias in traditional weather forecasting models. For example, most weather observations are made at stations located at airports or near to cities, which are warmer than the surrounding countryside. This then leads to a small, yet significant, bias when trying to understand the dynamic circulation high above the Earth's surface. Overall, weather prediction models will use data obtained at a certain time and then predict how those data will change over the space of a few hours. The data assimilation technique takes the data obtained at both times to see how the weather actually changed in between. When these 'assimilated' observations are plugged into the model, they accurately reproduce the dynamic workings of the atmosphere.

'Unlike weather prediction, we're trying to get the best physical and chemical representation of the atmosphere at any one time,' Rood said. The data assimilation technique will model atmospheric winds accurately and then show how ozone is transported by adding chemistry at the grid points. This requires accurate representation not only of horizontal winds, but also of the way in which ozone is transported vertically. In particular, the data assimilation results have shown that the way tropospheric storms tend to propagate around the polar vortex in the northern hemisphere is rather like small flywheels spinning around a larger flywheel. The result has been to pull 'blobs' of chemically processed air out into lower latitudes. The Goddard data assimilation model has shown that these same storms propagating around the vortex will change the temperature in and around it quite dramatically. As a result, polar stratospheric clouds are more likely to form towards the edges of the vortex where it is cooled, and this will increase the likelihood of ozone depletion at mid-latitudes.

Overall, the assimilation technique is very good at describing today's atmosphere in detail. Rood's aim is to use it to extrapolate forward to see how dynamics will change as the atmosphere does. So far, the modellers at the Goddard Space Flight Center have been using a Cray Y-MP into which chemistry has been added at a later stage. The Data Assimilation Office at Goddard received a more powerful Cray C-90 in July 1993 into which more extensive chemistry has been added from the outset. 'When that model is up and running, we'll really be able to let it go,' Ricky Rood said in January 1993.

The ultimate synthesis of all these different supercomputer simulations of the stratosphere is to produce an unbiased, accurate representation of today's and tomorrow's ozone layer. Their basic techniques are subtly different; more than one modeller uses the analogy of climbers approaching a summit from the different faces of the same mountain. The onward march of technology means that before the year 2000, even faster supercomputers will become commercially available. Cary Y-MPs and C-90s are in the 'gigaflop' range – that is, they can perform 10^9 calculations every second. The new generation will be a 'teraflop' machine, a supercomputer capable of making 10^{12} calculations in a second. It will bring with it the most expensive bill for hardware thus far: a conservative estimate is that the big brother of today's machines will have a price tag of around $100 million.

Will faster computers mean more accurate models and better predictions? Not necessarily, warn those involved with the work today. 'There is a misconception that our current difficulties are somehow an engineering problem,' said Byron Boville of NCAR. 'Bigger and better computers will not solve the parameterisation problem. Even if we run a model with a grid every 50 kilometres across compared to the 300 kilometres today, it's still too large to account for cloud processes. Our understanding will still require approximation.' At best, supercomputer modelling of the stratosphere allows researchers to sharpen their understanding of the subtleties of ozone depletion. In turn, that will result in a fine-tuning of those models from which a deeper knowledge will emerge. This symbiosis at the heart of modelling has been – and doubtless will continue to be – a recurring feature of the science of ozone depletion.

For those people who feel slightly threatened by the impersonal precision of the supercomputer perhaps there is some comfort in the crucial importance of personal imprecision in the unravelling of the myriad complexities of ozone depletion. Human insights are needed to interpret the predictions and to understand the limitations of even the most accurate supercomputer visions of the future. And in another sense altogether, supercomputers are no match for human experience. They cannot predict at all how human beings will react when confronted with the latest research into the ozone layer. As the next chapter shows, that's where all the trouble starts.

7

Instant Science

> In order for any event, public or private, to become a major news story [the] coverage of the event must acquire a key ingredient. Without that ingredient, the story drifts and eventually withers. The ingredient is reaction, broad public reaction. Until there is reaction [the] event is like an airplane moving along an endless runway, unable to get up enough speed to take off. Some events are sufficiently momentous to compel substantial and varied reaction from the time they occur until far into the future.
>
> David McClintick, *Indecent Exposure*

JOE WATERS looked out into the bright lights of the auditorium. For the scientist who describes himself as 'a farm boy from Tennessee', the prospect of the press conference was a new experience and one that was all the more nerve-racking because he would shortly present disturbing observations of the Arctic stratosphere from space. Although he had already shown them to a scientific meeting, it was now – as he and his NASA bosses well realised – time to go public. The date was Monday 3 February 1992; the setting was the press auditorium at NASA Headquarters on Pennsylvania Avenue, just across the street from the Smithsonian Museum.

How Soon Is Now?

Waters was sharing the podium with colleagues from the Airborne Arctic Stratosphere Experiment II and together they would alert the reporters filling the auditorium to the possibility of something dramatic: the highest ever levels of chlorine monoxide recorded anywhere in the Earth's atmosphere and the possibility of an Arctic ozone hole. 'By early January we were seeing as much ClO over populated areas of Europe and Russia as we had ever seen in the depths of the Antarctic ozone hole,' Waters said. 'We had a duty to the public to let them know.'

Controversy would soon envelop them and explode throughout the US media. The 'Hole Over Kennebunkport Scare', as it soon became known, would eventually widen to take in accusations of bad faith, scaremongering, apocalyptic reporting and even fraudulence in research. These criticisms would incubate over the next eighteen months and eventually burst into the media as the 'Ozone Backlash'.

These controversies are worrisome and startling symptoms of how politicised ozone research has become, particularly in the United States. Though similar results were reported by scientists involved with the European Arctic Stratospheric Ozone Experiment on that same first Monday of February 1992, they did not produce nearly as much fuss and furore. Yet many of those same criticisms are equally applicable on the European side of the Atlantic. So how should disturbing scientific observations of the ozone layer be presented to the public? Ozone depletion is a complicated issue which does not lend itself to the crisp compression required by newpaper sub-editors or the brevity of the sound bite. Widespread ignorance of scientific method and the work of scientists in general has led to misconceptions and distortions in the reporting of the subject.

This chapter will closely examine the dilemmas raised by conducting scientific research in the fishbowl of media attention. The need to present instant results is one of many issues in ozone research which have broader implications in science generally. In many ways, the events after the 3 February press conference are entirely symptomatic of the difficulties many scientists have had to face. With no sense of the impending deluge of criticisms, it was Joe Waters' particular misfortune that it was to be his first experience of 'instant science'.

* * *

There is a saying in Maine that there is no point in speaking unless you can improve on silence. For months afterwards, the hundred or more scientists working on the second Arctic Airborne Stratospheric Expedition would often reflect on that homespun wisdom. Out at the airport in Bangor, planning, executing and analysing the 'quick look' data from the flights of the ER-2 and the DC-8 had left little time for reflection.

Each new piece of data from the aircraft campaign was spelling an increasingly disturbing picture of conditions in the stratosphere. Measurements of ClO had inexorably started to rise to levels never seen before in the north. By the end of January 1992 nearly 80 per cent of the chlorine in the stratosphere had been converted to this active form, while passive reservoirs of chlorine – in particular, hydrogen chloride and chlorine nitrate – were virtually non-existent. Measurements in and around the vortex showed that the Arctic stratosphere was being primed for large-scale ozone loss. If the stratosphere stayed as cold as it then was, dramatic ozone loss could result when the Sun returned in February. The dilemma for the scientists out at Bangor was how and when to announce their results.

In many ways, the second AASE was always going to be more controversial than the first. NASA's stratospheric research programme can afford to mount large aircraft campaigns (costing up to $10 million) only every other year, if that. Some of the US stratospheric research community had felt that a return to Antarctica would have been scientifically more valid. After all, the south polar vortex is better defined and, as a result, presents fewer difficulties in interpreting the relative contributions from the dynamics and chemistry which conspire to destroy ozone. No matter how intense the Arctic investigation, it was argued, conditions in the north would be too complicated to unravel the chemistry from the dynamics. On the other hand, the first-order questions had been answered in Antarctica, and another Arctic mission would – in Jim Anderson's telling phrase – 'break away from the lunacy of thinking that the ozone problem was somehow fixed'.

There were compelling reasons for heading north. The previous Arctic campaign had shown elevated ClO levels occurring over large portions of the northern hemisphere. A roll of the meteorological dice in the form of pronounced cold in the stratosphere could thus result in prolonged ozone loss. Another important consideration had come from analysis of ozone trends from Nimbus 7 and the ground-based Dobson network. There was a distinct late-winter, mid-latitude ozone loss over the northern hemisphere. Was it caused by encroachment of the same sort of chemical processing which had previously been seen over the Arctic regions? Clearly this was no longer an abstract or remote question for the unwitting populations of the United States and Europe below.

One participant in the extensive planning discussions offered a powerful argument: 'I wonder what the American public's reaction would be if we went south and a hole opened over the north.' As manager of NASA's stratospheric research efforts, Bob Watson asked Adrian Tuck of NOAA to present the scientific case for going south and Jim Anderson that for returning north. At extensive review meetings at NASA Headquarters early in 1991, it became clear that the north would win out. 'There's no doubt that some perception of politics played a role in choosing the north,' said Adrian Tuck. 'Most of the voters live in the northern hemisphere and they will want to know is ozone depletion going to affect them. I don't think anyone would have thought that in their wildest dreams the results would affect national policy in a matter of days.'

The validity of their eventual decision was graphically demonstrated on the first ER-2 flight out of Fairbanks in October. Levels of the oxides of nitrogen which normally react with active chlorine to render it harmless were much lower than had been expected. As the winter progressed, evidence emerged that most of the nitrogen species had been converted into nitric acid, which forms the heart of polar stratospheric clouds. 'If nitric acid is tied up in PSCs,' Jim Anderson explained, 'it cannot be photolysed nor made available to chew on high levels of ClO.' As if on cue, after the mission had moved to Bangor, ClO levels started to increase 'before our very eyes,' said Anderson.

He, at least, had instantly grasped the severity of the situation. ClO levels over the Arctic at the start of 1992 were higher than had ever been seen over the Antarctic in late September. In other words, the north had reached a comparable stage in chemical processing nearly three months earlier than the south.

In the new year, the aircraft flights showed that there were quantifiable links to the Pinatubo aerosol. On 12 January, AASE II measurements down to 22°S also showed relatively high levels of reactive chlorine and reduced nitrogen oxides. Then, on 20 January, came the biggest shock of all. The polar vortex had extended as far south as Bangor and when the ER-2 took off it started to notice elevated levels of ClO almost straight away. One of the 'profound simplifications' of the ClO measurements is that they provide an index of how much ozone could be destroyed in sunlight. The destruction of ozone is quadratic with ClO: if you increase ClO by 50 per cent, then ozone loss will double. With measurements as high as 1,700 ppt of ClO in the lower stratosphere, all the ozone present there could be lost within three weeks if temperatures remained the same. It was clear that conditions had reached a dangerous precipice.

The global perspective from the Upper Atmosphere Research Satellite served to confirm the severity of the situation. By early January, high levels of ClO were seen over large areas of northern Europe and Eurasia. But further measurements were hamstrung by the orbit of the UARS. On 14 January, the spacecraft would have to yaw south to avoid the Sun encroaching upon its instruments. 'We knew that we couldn't get any more observations for over a month,' Waters said. 'But if these levels stayed around when the Sun returned, one would expect to destroy 1 or 2 per cent of the total ozone a day.'

So by the next time UARS looked north it might well be seeing a pronounced hole in the Arctic ozone layer whose genesis would have been faithfully recorded by the aircraft missions. Clearly it was time to let the general public know.

For its participants, the 3 February press conference would resemble the Kurosawa film *Rashòmon* in that none of them remembers it in exactly the same way. Perhaps it didn't help that none of the scientists on the podium at NASA Headquarters had ever taken part in a press conference before. Even Jim Anderson, an old hand at Congressional and Senate testimonies, was nervous. 'I can tell you that it's certainly very different from teaching at Harvard,' he said.

Compared to the UARS mission, the aircraft campaign was easy to grasp, particularly because – for the first time – it was based within the boundaries of the United States. The previous campaigns out of Stavanger and Punta Arenas had been much more difficult for the press to follow because of their remoteness. Psychologically, at least, Bangor make it easier for journalists to find information and obtain quotes from mission scientists. Throughout the winter of 1991 and into the new year, reporters had been repeatedly phoning in to find out the progress of the campaign. Increasingly, there was the risk that one reporter might scoop the others from some unguarded comment.

One person's Pulitzer Prize was another federal employee's nightmare: it could lead to accusations of favouritism, which, as a government agency, NASA had to avoid. To avert that possibility, Jim Anderson and the manager of the AASE II at NASA headquarters, Michael Kurylo, felt that their preliminary findings should be announced to the media. Kurylo's bosses, including Bob Watson, agreed that a press conference should be called. The NASA high-ups also decided that the MLS results should be announced at the same time, and Joe Waters agreed to fly in from California to present them. Releasing both satellite and aircraft data together, it was felt, would emphasise their wholly different contributions and make it clear that the sum of the parts was very much greater than the whole. Both were needed to weave together a seamless global picture of conditions in the stratosphere. It was an unusual step, to be sure, as the MLS instrument was still being validated and the AASE II was only halfway through. But it seemed warranted because of the worrying observations across much of the northern hemisphere.

That Monday lunchtime, the press auditorium at NASA Headquarter was full and buzzing with excitement. Reporters were handed two press releases (one each for the AASE II results and those from MLS) accompanied by a more detailed fact sheet written by the scientists themselves. This document justified the unusual step of presenting the data from Bangor while the campaign was still in progress: 'Although the mission is not yet complete, preliminary results show the potential for greater ozone depletion than had been previously expected, suggesting the need for a timely release of these early data.'

Viewed on videotape, the conference does not seem to strike a particularly portentous tone. Chaired by Mike Kurylo, Jim Anderson and Joe Waters presented their respective data with overhead projections and fairly low-key graphics. David Fahey of the NOAA Aeronomy Lab was also on hand to elucidate the finer details of the nitrogen oxide measurements which he undertook on the ER-2. In tone and manner, it should be said, they were reasoned, and they repeatedly mentioned the preliminary nature of their observations. From the tenor of later criticisms one has the impression that they had somehow been ridiculously messianic in their fervour. On the strength of the 3 February press conference, the scientists involved would probably have been equally as dispassionate had they been announcing the Second Coming of the Lord.

In the endless analyses afterwards, some scientists (Jim Anderson included) were disturbed that the AASE II press release was headed 'Scientists Say Arctic "Ozone Hole" Increasingly Likely'. Though it had been cleared with NASA high-ups, some felt it was forcing the hand of journalists too much. In the event, the participants performed verbal gymnastics to avoid using that very phrase – 'Arctic ozone hole' was a concept that was most assiduously avoided. That said, none of them pulled any punches. 'These findings have increased my concern that significant amounts of ozone will be removed this decade,' said Jim Anderson. Mike Kurylo was quick to point out the hemispheric implications of the work by adding there was 'an ever-increasing danger of ozone depletion in the mid-latitudes and Arctic area'. Referring to the nitrogen oxides results, Jim Anderson said: 'The immune system of the atmosphere is weaker than we had suspected before.' Asked how he would rate the alarm over the findings, Anderson suggested that if they were scored out of ten, 'my personal impression is, it's a good solid eight'. That was the nearest any of them came to venturing anything which could be construed as an opinion.

From the evidence provided thus far, it was suggested by Mike Kurylo, if the stratosphere remained as cold as it had been in late January, then 1 to 2 per cent of ozone in the lower stratosphere would be stripped away when the Sun returned by the end of February. In the reports which followed these caveats were ignored, thereby giving the impression that an ozone hole had already opened. With the wire services first off the mark, this was the erroneous message that went out across the world. For Reuters, Deborah Zabarenko's story explicitly stated: 'The NASA data indicates ozone depletion as severe as 30 or 40 per cent over the Northern Hemisphere and the presence of ozone-destroying chlorine monoxide as far south as Mexico, southern Texas, Florida and north Africa.' The results were front-page headlines the next day all around the world and were also the lead story on CNN bulletins for the subsequent thirty-six hours. Some of the television reports, it should be said, mistook Joe Waters's blob of ClO for an ozone hole, as it shared more or less the same false colours used to present data from TOMS to which the visual media were long since accustomed.

As was almost inevitable, trenchant editorials followed which called for the accelerated timetables of phasing out ozone-depleting chemicals. The *Washington Post*'s was entitled 'THE VANISHING OZONE LAYER' and declared that 'the protective ozone layer in the sky is being destroyed faster than even the pessimists had expected'. The *New York Times* weighed in with an editorial whose title left no doubt where blame was to be apportioned: 'THE OZONE hole over mr bush's head'. And to add to the clamour the next issue of *Time* had on its cover a photograph of a hole burned into an image of a cloudy sky and the legend: 'VANISHING OZONE – THE DANGER COMES CLOSER'. Within the first paragraph, *Time*'s cover-story writer Michael Lemonick declared: 'No longer is the threat just to our future; the threat is here and now. Ground zero is not just the South Pole anymore: ozone holes could soon open over heavily populated regions in the northern hemisphere as well as the southern.'

A mile away on Capitol Hill, politicians were quick to respond. On 6 February, the US Senate voted 96–0 to pass legislation to bring forward the complete phasing-out of CFCs from the year 2000 to the end of 1995. Senator Albert Gore – then being touted as a possible Democratic contender for the presidency – goaded the incumbent with a wounding, eminently quotable sound bite: 'George Bush has kept his hands over his ears and closed his eyes. Now that there's the prospect of a hole over Kennebunkport, perhaps Bush will comply with the law.'

The very next day Bush used the platform of his formal declaration for the presidency to announce new legislation. Days before the electoral primary in New Hampshire, he declared that CFC production would be phased out five years before the end of the century. Almost inevitably Bush was said to have been railroaded into declaring this new legislation as a pre-election sop, very much against the judgement and free-market ethos of his party. But it would be erroneous to conclude that the 3 February press conference had forced the hand of the federal government.

As a matter of course, Bob Watson always ensures that policymakers get to see important results twenty-four hours before they are released. Though he was out of town that crucial Monday, his office made sure that the President's science adviser D. Allan Bromley, as well as politicians like Senator Gore, received advance warning of the results. NASA was acutely sensitive about making sure that nobody could get any political mileage out of its ozone-related research. As a result, federal officials and politicians of all persuasions had had time to digest the data and its implications before it hit the newsstands.

In any case, an accelerated timetable for phasing out CFCs in the United States was already being discussed within the executive branch of government. Three months previously, a biennial review of the state of stratospheric ozone research undertaken by the United Nations Environmental Programme and the World Meteorological Organisation had concluded that, worldwide, ozone depletion was occurring at a far greater rate than had previously been suspected. It was clear that rapid action would be required to minimise the chlorine loading of the stratosphere before the century was out. As one of the coordinators of that report, Bob Watson had been instrumental in ensuring the White House was fully appraised that only reduced CFC emissions would stop the situation from deteriorating in the near future. The case was unlike the White House position on global warming and CO_2 emissions; all the President's men were agreed that something had to – and should – be done. 'February 3rd may have accelerated the phase-out timetable,' Watson said, 'but it was close to being presented anyway.'

Nevertheless, certain right wing ideologues whose power still swayed in the White House were disgruntled. Chief amongst them was Bush's Chief Of Staff, John Sununu, who was a confirmed Greenhouse Agnostic. At that time it was still unclear whether Bush would attend that summer's Earth Summit in Rio, and Republican party nerves were stretched tight over the prospect of a Gore candidacy. The net result was that the AASE II results started to attain a significance for George Bush's re-election prospects as the release of Willie Horton had done for Michael Dukakis four years earlier. Battle lines were being drawn.

By comparison, the mid-term results from the European Arctic Stratospheric Experiment caused much less of a stir. It was no coincidence that the Kiruna-based campaign issued its findings on the same day – and, of course they echoed those from Bangor. Though there was no formal coordination between the campaigns, there was a great deal of exchange of information and gossip, usually in the form of electronic mail. On its flights in and out of Stavanger, the NASA DC-8 would fly through air masses which were almost simultaneously being investigated by balloons flown out of Kiruna. Wherever possible, aircraft and balloon scientists tried to coincide their observations in space and time.

Similarly, British participation in the UARS Microwave Limb Sounder experiment via scientists at the Universities of Edinburgh and Heriot-Watt also ensured that its global perspective on the growth of the chlorine monoxide was well known. Theorists on both sides of the Atlantic were astounded by the similarity of the predictions from the EASOE computer models and observations from the recently launched satellite. John Pyle's theoretical modelling using the British universities' Cray at the Rutherford Appleton Lab had generated a worrying blob of ClO over northern Europe which was uncannily similar to that seen from space.

The EASOE campaign confirmed all the major findings from the US aircraft campaign. Balloon profiles had repeatedly shown the suppression of the oxides of nitrogen throughout the lower stratosphere, extending to altitudes where the Pinatubo aerosols had settled. Starting in the new year, the first ever balloon measurements from Kiruna of chlorine dioxide (OClO) had shown large concentrations below 22 kilometres. Altogether, it added to the impression that 'the atmosphere in the northern hemisphere is highly perturbed', in the slightly formal wording of a press release issued by the European Ozone Research Coordinating Unit.

Perhaps one of the reasons that reporting in Europe was more muted was that most of the leading European researchers were still at ESRANGE in Kiruna and altogether more difficult to reach by telephone. Most science correspondents, then, had only the press release to work from: the precisely honed words of the agreed statement were less open to interpretation. It explicitly mentioned Pinatubo as the culprit for much of the chemical processing of the Arctic stratosphere: 'Into an atmosphere already heavily loaded with man-made chloride compounds, large amounts of aerosol have been injected following the eruption of Mount Pinatubo.' The EASOE results suggested there was a tenfold increase in stratospheric aerosol compared with recent years across middle and high latitudes. The nearest the press release came to apportioning blame was in a vague, yet reasoned statement: 'The calculated concentrations of active chlorine are large enough to be causing ozone destruction over a wide area.'

The next day's broadsheets in Britain reported the story rather less dramatically than one might expect. Even the headlines included suitable caveats: 'OZONE LAYER FACES THREAT OF DEPLETION OVER ARCTIC' (*Daily Telegraph*), 'SCIENTISTS WARN OF OZONE "HOLE" OVER ARCTIC' (*Independent*), 'THREAT TO ARCTIC OZONE LAYER POSES HEALTH RISK IN EUROPE' (*Financial Times*) and 'OZONE FEARS BRING WINTER SUN WARNING' (*Guardian*). The *Guardian*'s story mentioned that scientists had warned that Britons should wear sunglasses during sunny days over the next few weeks. Quite who these scientists were was not clear as none of the EASOE scientists ever remember saying that. This quibble was the nearest to controversy that the European reporting came.

At first sight, the notion of northern Europeans having to wear sunglasses in springtime would seem to lend itself to tabloid treatment. Initially, the tabloids were rather more preoccupied with the ongoing saga of HRH The Duchess of York's questionable dalliances with Texans of varying respectability. A few days later, however, the *Daily Mail* found an angle destined to attract the attention of the island race. Entitled 'SHOW YOUR CAT YOU

CARE – PAINT HIS NOSE WITH SUNCREAM', it chimed with the curious British habit of relating important events to domestic animals. Suzanne O'Shea's story warned that cats with pink noses and white fur were particularly at risk from increased levels of ultraviolet radiation. It chronicled the sorry tale of Kipper, a white cat from Bristol, whose right ear had had to be amputated. His owner, a Mrs Susan Davies, was quoted as saying 'he loved the Sun'.

Although the cat story didn't mention the EASOE by name, it was clearly implied that new scientific results about ozone depletion had intensified concern. For ozone researchers who have felt themselves 'burned' by the press, it is little consolation that cats can't be quoted out of context.

In the event, the northern hemisphere was saved from an Arctic ozone hole because stratospheric temperatures increased at the end of January. Though this warming amounted to only a few degrees, it was enough to inhibit the formation of polar stratospheric clouds by the freezing-out of nitric acid at $-77°C$. Later analysis by the NOAA National Meteorological Center in Washington, DC, showed that from mid-December to mid-January there were thirty-nine days when temperatures were low enough for such clouds to have formed; an average winter, statistically speaking, would see sixty-eight such days. A region of high pressure had sat over the north Atlantic for most of February, warming the stratosphere. Within the vortex, ER-2 measurements revealed that ClO levels had fallen to 1,000 ppt by mid-February when the Sun returned.

Nevertheless, criticism of how NASA had released its results was inevitable. As in all large undertakings, there were mumblings of dissent from within the scientific community involved in the AASE II campaign directly. Some questioned the wisdom of presenting aircraft and satellite results on the same stage. It was entirely natural for NASA managers to combine the *in situ* data from the former with the polar perspective of the latter. But one AASE II participant noted that NASA's desire to show off all its new toys worked to its overall disadvantage: 'The satellite smears out all the secrets of the distribution. You need to know the dynamics before you can figure out the chemistry. Mixing the satellite results in with the aircraft results didn't serve either well.'

On the other hand, there was a forceful argument for combining both data sets, as they did reinforce the overall conclusion by confirming each other's observations. 'It would have been unbelievably stupid not to have done so,' said Bob Watson. 'Otherwise you'd have had results coming out from the aircraft people and UARS and others at different times.'

Some of the AASE II participants had urged caution in releasing results early. Later press reports painted a picture of Jim Anderson as having somehow gone over their heads and courted the press. It is a charge that Anderson refutes, as do all the scientists interviewed for this book who were in Bangor for much of that time. In most cases, scientists were happy for Anderson to be their spokesman for the simple reason that he was good at explaining detailed chemistry to non specialists. At the 3 February press conference, for example, David Fahey did not speak at all, having agreed to let Anderson present his results at the start. 'I was quite happy to do that,' he said later. 'If anything, I think my silence counted more against me in the long run.' But it did lead some observers to conclude that Anderson wanted to do all the talking.

Another, rather more serious criticism voiced by some AASE II scientists themselves was soon picked up by the media: science by press release. Releasing interim data before it has been peer-reviewed is always a contentious practice. Journals like *Science*, *Nature* and the *Journal of Geophysical Research* are the preferred forums of the ozone debate. By going public before its findings had been peer-reviewed, the AASE II scientists found themselves at the mercy of the media. At the end of February, the *Wall Street Journal* weighed in with a huffily written editorial entitled 'PRESS-RELEASE OZONE HOLE'. The basic tone of its criticisms was that NASA had been wrong and that, bizarrely, it should have anticipated that the stratosphere would warm.

'Before agreeing that the sky is falling, we'd like to see more time spent on the data,' spoke the *Journal*'s editorial writers, 'starting with the likelihood that NASA now believes there won't be any ozone hole this year.' And it also argued that using the 20 January measurement as a basis for a press conference was wrong (although, as we have seen, it wasn't). 'Arguing the destruction of the ozone layer on the basis of one day's, or a few weeks' data, is a bit like announcing the comeback of retail on evidence that takes Neiman Marcus's receipts from the day before Christmas and annualizes them,' the editorial ran.

As mission scientist, Jim Anderson felt honour bound to reply, as did President Bush's science adviser Allan Bromley. Although the newspaper chose not to publish Bromley's letter, it could hardly ignore Anderson's, which was nothing short of a *tour de force* and systematically refuted all the *Wall Street Journal*'s criticisms. 'The word "alarming" was not used in our briefing,' Anderson declared, 'but what level of ozone thinning would elicit your concern?' And the nearest he came to an emotional outburst was in adding: 'To equate the six-year, peer-reviewed research process exploring ozone in the polar stratosphere with the analysis of one day of receipts at Neiman Marcus is your editorial prerogative.'

The criticism of science by press release was less applicable to the data from the Upper Atmosphere Research Satellite. Joe Waters had presented his MLS observations to the American Meteorological Society two weeks before the 3 February press conference. After his presentation, he had discussed them at length with colleagues not intimately involved with the UARS project. And for good measure, Waters and his JPL colleagues had presented five papers which had discussed the rise of ClO across sizeable chunks of the northern hemisphere. 'Because we had already presented the results to a scientific meeting,' Waters said, 'I was quite happy to submit a press release to NASA Headquarters.'

The *Wall Street Journal* also complained that the effects of Mount Pinatubo were downplayed. In fact the AASE-II briefing document written by the scientists explicitly discussed Pinatubo and cited it as an underlying reason for mounting their campaign. Though it was not as prominent in the NASA press releases as those from EASOE, the US scientists did discuss the effects of Pinatubo during their presentations and at length in the extensive question-and-answer session which followed. If the press were more concerned about an Arctic ozone hole and chose to ignore the Pinatubo results, that was hardly the scientists' fault.

Much was also later made of the fact that no ozone measurements were presented on 3 February. Because the Arctic stratosphere was still mostly in darkness, little ozone had been destroyed by the ClO and, indeed, observed ozone levels had fluctuated well within the bounds of natural dynamical variability. In early January ozone levels had risen slightly, only to fall significantly by the end of the month. When word of this earlier rise got around, it provided the less discerning journalists with ample evidence for conspiracy theories that the sky was not about to fall in, so to speak.

Ironically, the MLS team decided not to present their ozone observations on 3 February because they were still preliminary and the relative effect of dynamics on the chemistry had yet to be ascertained. 'We didn't want to release those data,' Joe Waters said, 'as we didn't know if it was a dynamical blocking effect or a chemical loss.' It would be nearly a year before the answer to this puzzle would appear in print in a scientific journal, by which time, of course, it was old news.

And, ultimately, news values were the root of the problems which followed the 3 February press conference. Many of the first-order questions in ozone research have been answered and as a result, it has become increasingly difficult to interest the media with what some editors would probably perceive as tedious minutiae. The dismay felt by AASE II scientists on seeing a press release announcing an increasingly likely Arctic ozone hole has to be reconciled with trying to interest reporters. 'Whether we like it or not,' Jim Anderson reflected, 'our work is newsworthy.'

Although he was the focus of much of the subsequent furore, he said that he would repeat the 3 February briefing if another aircraft campaign found similar results. 'I wouldn't change the content of the briefing,' he said, 'as we have a duty to inform the public.'

The final end-of-experiment results from EASOE were released before NASA did the same with the AASE II. Because of the Europe-wide involvement, they were formally released at a press conference at the European Commission in Brussels on Tuesday 7 April 1992. But to ensure that national correspondents could get quotes, there were press conferences held in the various European capitals the day before on a strictly embargo basis. The London press conference was held at the Institute of Electrical Engineers in

Birdcage Walk just by Big Ben with Neil Harris, Joe Farman and Tony Cox (from the National Environment Research Council) in attendance.

The message was simple: high levels of ozone-destroying chlorine compounds had been observed. Low concentrations of nitrogen oxides had been measured throughout the campaign, believed to be the result of the Pinatubo aerosol. Ground-based observations of ozone tied in with aircraft and balloon measurements which showed beyond all doubt that stratospheric ozone levels were noticeably down. On a month-by-month basis, the results were:

November	slightly low
December	10–20% down
January	10–20% down
February	5–15% down
March	down (data still being analysed)

The scientists went into some detail about the unusual weather conditions which had prevailed. Joe Farman pointed out that it had been known since Gordon Dobson's time that high-pressure systems cause a thinning-out of the ozone layer. He used the graphic analogy of a stream passing over a rock. 'Where the water passes over the rock, the water thins out and goes faster,' he explained. 'It's the same with ozone. That's why you have a thinning out.' Nevertheless, the end results from the campaign came down in blaming the chemistry: 'The mid- and high-latitude ozone values in January and February are lower than would be expected from simple extrapolation of recent analyses of ground-based and satellite measurements. However, no feature was seen in the northern hemisphere this winter which could appropriately be called an ozone hole.' That was the message which graced the front pages of most newspapers the next day.

By this time, it was clear that the future of the Arctic ozone layer was not immediately threatened. The same, however, could not be said for John Major's government. Three days before a British general election, the EASOE scientists had assiduously avoided any form of scaremongering or straying into the realms of policy. Almost inevitably, that same morning Greenpeace had staged a protest climb of the Houses of Parliament to draw attention to the results. The arrest of a handful of activists made a few inside pages in brief, if at all.

NASA's reporting of the final AASE II results on 30 April 1992 was eclipsed by riots in Los Angeles. In any case, after the criticisms of the 3 February press conference, participants this time around crossed their t's and dotted their i's. Part of the more formal mission summary stated: 'If temperatures had remained cold into the third week of February, which has occurred several times in the last decade, greater amounts of ozone would have been destroyed. The loss of ozone in winter 1991/1992, while significant, should not be described as an "ozone hole", a term to denote the sharp transition to dramatically suppressed ozone levels over Antarctica.' Once again, the MLS results shared centre stage with those from the aircraft campaign. From low Earth orbit, it too had observed that the stratosphere had considerably warmed by the next time it was pointing north and this had averted the large-scale development of an ozone hole over the North Pole.

Nevertheless, all participants agreed, it had been a close-run thing. 'This tells me that conditions in the upper atmosphere are in a very delicate balance,' Joe Waters said. 'With so much chlorine in the stratosphere, a slight temperature difference can make an enormous difference in the potential for ozone depletion.' ClO levels had subsequently decreased; how and why would take the best part of the year to analyse. But the implications of the first ever satellite observations of chlorine chemistry were obvious, as Waters made clear to me when I interviewed him a month later. 'By the end of February, the Earth looked like it was a different planet than in January with respect to ClO,' he said. 'In future years, that blob of ClO is going to be a fixture of the northern hemisphere.'

As the flames and mayhem subsided in Los Angeles, some newspapers tried to turn the heat on NASA. Typical of the criticism was an editorial in the 7 May 1992 *Washington Times* entitled 'NASA CRIES WOLF ON OZONE'. It began by heaping scorn with a rhetorical question: 'Remember the ozone hole that was going to sauté Kennebunkport this year?' Its writers continued sotto voce:

How Soon Is Now?

By checking the January data a little more closely or perhaps by checking at all, the agency could have spared everyone the panic of February 3 – except, of course, that some people in the agency apparently were eager to create the panic of February 3, for reasons having nothing to do with science and everything to do with their ideological environmentalism. As it is, it would be nice if the next time NASA cries 'wolf', fewer journalists, politicians and citizens heed the warning like sheep.

That final sentence is a strange comment from a newspaper which is largely owned by the Reverend Sun Myung Moon.

Of all the criticisms levelled at NASA following the second Airborne Arctic Stratosphere Experiment, the most persistent has been identified as the 800-lb gorilla effect. Because the space agency has been charged by the US Congress to monitor the state of the stratosphere, it is usually the lead agency in long aircraft campaigns like the AASE II. Although these experiments also include the participation of other federal agencies, such as NOAA and the National Science Foundation, responsibility for their efficient conduct and trouble-free logistics is laid at NASA's door. Because of this lead status there is a sense that anyone who criticises the space agency does so at their peril. To use an often quoted analogy: if you find yourself in a room with an 800-lb gorilla, you either stay out of its way or go along with it.

This could, of course, be said of any large agency which has the means to dole out research funds. NASA's stratospheric research is considerably better managed than many other government-funded scientific programmes, in the experience of this writer. Because of its inherently political nature, however, the normal infighting and bickering which accompanies most research is more acute in stratospheric science. NASA has in fact been sensitive to the needs and criticisms of the stratospheric research community. After the first airborne campaign out of Punta Arenas in 1987, a number of excluded researchers felt that the AAOE had relied too much on the cronies of NASA personnel and the NOAA Aeronomy Lab. So, in the subsequent campaign out of Stavanger in 1989, new faces participated and greater numbers of theorists were included.

Ironically, this led to more marked infighting, prolonged arguments and what one onlooker calls 'dynamic tensions'.

At a time of worldwide recession and burgeoning federal deficit, budgets are tightened so greater numbers of scientists will find that their funding requests are rejected and, as a result, their complaints provide evidence for the sort of conspiracy theories upon which the media often thrive. All researchers could well do with more money, whether they be atmospheric scientists, astronomers, geologists or biologists investigating the mating habits of nematode worms. In repeated surveys of the 'health' of science in the United States, a sense of demoralisation has increasingly become the norm. Stratospheric research merely reflects these larger concerns, but the prominence of the funding issue in ozone science is amplified because of media preoccupations with it.

Against this background, then, the mid-term release of the AASE II results prompted the accusation that NASA scientists deliberately used scaremongering tactics to create an opportunity whereby they would be assured of continued – if not greater – funding for the future. 'If you have a doomsday scenario you get a lot of money,' a disgruntled NOAA researcher told the US magazine *Insight*, which looked at the 3 February press conference two months afterwards. 'Research organizations are in great competition with each other to get the politicians' ears and obtain the necessary resources. If you want money you have to come up with a doomsday scenario.' In other words, if scientists exaggerate their findings, they are more likely to maintain funding levels. In reality, the AASE II scientists would have been hard pressed to have cooked up anything more serious than what was found. 'It's absolutely ridiculous to claim we somehow massaged the results,' Jim Anderson said. And with regard to the funding question, Joe Waters echoed the sentiments of others interviewed for this book: 'I can assure you that this wasn't our motivation.'

There is, however, another dimension to these accusations. It is a topic which has also reached prominence in recent years: the spectre of fraud in American science. Scientific fraud has generated controversies which far outshine those in ozone research, and at its heart lies the pressures on some researchers – particularly in the medical sciences where patents costing billions are at stake –

purposefully to pervert the course of scientific enquiry. A survey carried out by the American Association for the Advancement of Science in early 1992 suggested that fraud may be more prevalent than had been previously suspected. Of the 469 scientists polled, 27 per cent had encountered an average of 2½ cases of suspected fraud in the previous decade. One respondent put it thus: 'I believe the problem has worsened as research has become driven by more outside forces than by true academic need-to-know. Competition between "colleagues" has become extreme, and is based on more economic considerations and media exposure.' Five years earlier, Daniel Koshland Jr, the editor of *Science* (which is published by the AAAS) had previously suggested that science was 99.9999 per cent free from fraud. Koshland argued that the 1992 survey was based on feelings, and probably fed off the protracted fallout from high profile cases of fraud in medicine.

'High profile' is certainly a description of ozone research, and not just because of its prominence in the media. A survey of scientific literature published in 1991 revealed that stratospheric research is, academically-speaking, a hot topic – or rather one aspect of it appears to be. A Washington-based research organisation, the Institute for Scientific Information, looked at a number of scientific disciplines to assess their 'trendiness' by a complicated analysis. It involved the number of cross-references to academic papers in specific fields and the number of citations to those papers in other fields. In the top ten of 'hot topics', atmospheric chemistry of CFCs came ninth, with the top four being medical topics (the hottest topic was laparoscopic cholecystectomy). Small wonder, then, that stratospheric research is a subject in which controversy is an almost continuous companion.

But what of *fraud* in ozone research? Despite the many underlying pressures upon it – both blatant or unspoken – there is, it should be stated, absolutely no evidence for deliberate fraudulence in stratospheric research today. With hindsight, NASA's early release of its data on February 3rd 1992 could be viewed as 'questionable' in view of the controversy which followed in its wake. But the validity of the data upon which that judgement was made is irrefutable. In fact, there is only one documented case of 'questionable' practice in the whole convoluted history of research into the ozone layer.

It occurred over fifteen years ago, and showed, if anything, an overwhelming desire for hegemony by American scientists rather than any desire to pervert the course of scientific enquiry.

While watching a televised race in which his daughter was participating, Jim Lovelock was struck by its aptness as a metaphor for the pressures on American scientists. As someone who has been involved in research on both sides of the Atlantic, Lovelock has had the opportunity to view how it is carried out at first hand. He feels that many researchers in the United States behave like competitors in a race for 'glory, grandeur and prizes'. His daughter Chris was an Olympic-standard runner who often participated in televised races; on this particular occasion, she let someone who was fast approaching her overtake. As she obliged, her competitor stamped hard on her foot, thereby pushing Chris Lovelock out of the race. It was, her father says, altogether too reminiscent of his experience of the 'vicious competitiveness' of American science generally.

Although Jim Lovelock believes that there are many outstanding stratospheric scientists in the United State, he feels that the system is scored for personal recognition. 'Unless you hype it, nobody is going to take notice,' he says of a prevalent attitude there. Lovelock has an interesting perspective on American stratospheric science in the immediate aftermath of Rowland and Molina publishing their findings in 1974, when he became a largely unwitting participant in the clamour which followed. He was asked to participate in a review of knowledge on the state of the stratosphere convened by the US National Academy of Sciences. It was, as Lydia Dotto and Harold Schiff so perceptively called it in their seminal record of the time, *The Ozone War*, one of the first acts in 'the incredible Stratospheric Travelling Road Show and Debating Society'.

Because of his pioneering measurements of atmospheric CFCs, Lovelock was called upon to describe how he had made them at a meeting in Snowmass, Colorado, in 1976. With characteristically scrupulous honesty (and, he is the first to admit, innocence) he was candid about their accuracy. Unlike measurements made in a laboratory, those performed in the real atmosphere have no standard against which a comparison could be made. So Lovelock had to make assumptions. By joining electron capture detectors in series, he could compare how his air sample changed as it progressed through each detector in time. It was, he well knew, only accurate to 20 per cent. At the Snowmass meeting, one American participant said that he could make much better measurements than that and boasted that he could determine the presence of CFCs to within 1 per cent accuracy. Other US scientists concurred that they could probably do the same, and both Lovelock and the National Academy review panel were suitably impressed.

The result was that subsequent analyses of clean air samples used to monitor the rise of CFCs would be performed domestically within the United States. Within a few months, however, Lovelock started to receive air samples from across the Atlantic. Some of them contained letters asking him to determine the concentrations of trace gases, which he was quite happy to do. He was slightly puzzled by why his analyses were required, given the supposedly greater accuracy of American laboratories. Only when he started to elicit replies along the lines of 'Yes, we find those numbers too!' did he become suspicious.

So did the US Bureau of Standards (now the National Institute of Standards and Technology) the federal agency charged with checking on the reliability of important scientific measurements. In 1977, the Bureau sent out two air samples to many of the leading American laboratories which claimed greater accuracy than Lovelock. One sample contained clean air, the other that same clean air diluted 30 per cent by nitrogen. The results, as compiled by the Bureau, were astonishing: instead of being within 1 per cent of each other, analyses varied by 350 per cent. In Lovelock's phrase it was clearly evidence for 'collusion and gross-scale lying', though the official report produced on this incident ('Evaluation of Methodology for Analysis of Halocarbons in the Upper Atmosphere', Phase 1, NBSIR 78–1480, March 1978) was rather more diplomatic in its language. Part of its conclusion reads: 'This large error appears to be systematic in nature and probably due to differences in calibrations, and/or sample handling techniques, *although other causes cannot be excluded at this time*' (author's italics added).

Today, Lovelock simply says that 'one must draw one's own conclusions' from this strange episode. And rather than drawing any pleasure or personal vindication, he feels concern for the state of health of American science and the peculiar pressures incumbent upon it.

The undoubted Cinderella in stratospheric research is as low-tech and inexpensive as others are glamorous, state-of-the-art and costly. Despite repeatedly demonstrating their usefulness and the incomparable value of their priceless data archive, the ground-based network of Dobson instruments has always had low status in ozone studies. Had they not been in place, however, it is doubtful that both the Antarctic ozone hole and the trends in northern hemisphere ozone would have been so readily discovered. Today, the venerable instruments have been supplemented by the grandiloquent Network for the Detection of Stratospheric Change, which involves laser-based instruments, such as Lidars, that return information independently of the Dobsons and satellites to maintain a continuous observational record. More than sixty-five scientists from fifteen countries are involved in this relatively informal network, which began operations in January 1991.

If nothing else, the fact the ozone hole was picked up by fairly antiquated equipment proves the adage that a good observation is a good observation no matter where it came from. And it is probably fair to say that had Joe Farman not discovered the ozone hole, the British government might well have ceased to fund the Dobson instruments operated by the British Antarctic Survey out at Halley Bay. The circumstances behind the scenes at the time of Farman's discovery provide an illustration of the way funding bodies tend to view basic, low key research projects in a system scored for large, prestigious projects. These attitudes are not entirely restricted to the British Isles.

British taxpayers' money which is earmarked for research is funnelled through a number of governmental funding bodies that oversee the various research disciplines. The British Antarctic Survey comes under the aegis of the Natural Environmental Research Council, or NERC. Every so often, the ruling committees and chairpersons of organisations like NERC survey the work which it funds in its various institutions across the length and breadth of the land. It is an excuse for the great and good to see at first hand how well 'their' money is being spent (and, usually, the opportunity for an excellent lunch somewhere along the way). Joe Farman has good reason to shudder when he thinks of one particular visit of the NERC council to the British Antarctic Survey.

In the early 1980s, Farman was well aware that his ozone work was under threat from cutbacks. In vain, he had tried to make the point that if BAS stopped making ozone measurements, it wouldn't actually save any money since it was such a small part of his work. His salary and that of his research assistants would still have to be paid while they pursued other geophysical monitoring activities in Antarctica. His words largely fell on deaf ears. In the various reviews of scientific research prompted by the Thatcher government, a sense of 'value for money' had started to permeate its way into the arbitrary decision-making processes of funding bodies like NERC. One rule of thumb which supposedly indicated the value of research was drawn from the number of citations of work in the general scientific literature. For this reason there were grumblings within BAS that Farman had not been publishing the results of much of his recent work.

At the start of 1984, a visit by the new NERC chairman was in the offing. The distinguished cosmologist Sir Hermann Bondi was well versed in the mysterious ways of government, having previously been the chief scientist at the Ministry of Defence, and had also been a director general of the European Space Research Organisation. More poignantly for Joe Farman, however, Bondi had been something of a legend at Cambridge when he was an undergraduate there in the 1950s. He was delighted to have the opportunity of letting the great man in on his group's secret – the springtime loss of ozone over Antarctica. On the appointed day, Farman eagerly presented the data from Halley Bay as irrefutable evidence of the value of small science.

Yet it elicited a strange response: Bondi asked why on Earth would anybody want to make such measurements? Farman patiently explained that systematic observations were the only sensible method of demonstrating a change in global ozone levels. To this Bondi exclaimed, imperiously: 'Oh, you've been making them for posterity! What's posterity ever done for you?' Farman wisely said nothing and there the matter was dropped.

Instant Science

The physicist Ernest Rutherford once dismissed all science as physics or train spotting, and in that sense, Farman is unavowedly a train spotter. He is keen to emphasise that ostensibly dull, systematic observations are fundamental for scientific progress. He likes to paraphrase the words of Lord Rothschild, a former head of the UK government think tank, that there should be room for scientists 'to play the fool'. There should be an inbuilt flexibility in the grand scheme of things for scientists to pursue avenues of research whose value has not been presupposed. Despite the lessons which should have been learned from his discovery of the Antarctic ozone hole, the tendency for small science to be ignored continues as complex, large-scale projects are an increasing preoccupation.

'No great advances ever came out of a committee planning ten years ahead,' Joe Farman said by way of diagnosing the problem. He remains concerned about the way in which large-scale projects that are 'trendy' get in the way of 'good science'. In a piece he wrote for the *Independent*'s science page in April 1990 after his last professional visit to Antarctica, he aimed a broadside at the proponents of big science. In particular, he was concerned about the way in which governmental money was being directed to build bases and enhance the infrastructure for Antarctic science at the expense of renewing simple scientific equipment. His concluding words serve as a rationale for little science:

> Is it cynicism that makes me wonder why many of these projects should be regarded as urgent when much remains to be learned from simple observations? I hope that Antarctic Science will not become a summer-only business and that young people will continue to want to work there throughout the year, seeking to understand. I am sure they will, but will the funding agencies continue to support this long-term science? The outlook is not good: at current levels of support, it will take several years to get new instruments into the field.

Characteristically, Farman did not mention one of the reasons why he had made that last trip down to the South Pole. One of his tasks was to locate a particular rock on the ice shelf upon which French scientists had made a note of the average sea level in 1906.

How Soon Is Now?

Farman and his colleagues located the rock in question but found that today's average sea level had not changed since earlier in the century. So it was that the man who had discovered the ozone hole was not to be the first to find incontrovertible evidence for the onset of global warming.

Personal recognition has been a constant feature of Sherry Rowland's life since that shocking December morning in 1973 when he and Mario Molina discovered that the Earth's fragile ozone shield was at risk from chlorofluorocarbons. Unlike many other prominent ozone scientists who are decidedly ambivalent towards the media, Rowland has used it well and has been fairly outspoken in his dealings with the press. As a better part of his reputation was staked on the validity of his hypothesis, Rowland understandably had a vested interest in presenting it as favourably as possible. It is easy to forget that during the first few years of the ozone debate, very little was known about the stratosphere. Unlike many of his fellow researchers, Rowland could – and did – speculate. 'Federal employees cannot express opinions,' he commented. 'I can.'

So it is hardly surprising that throughout the wilderness years of the 1970s he often found himself publicly at odds with the giant chemical industry. 'When you find there is a problem with ozone and CFCs,' he said with considerable understatement, 'and you suggest production of the latter should be stopped, you are going to draw fire.' This caused him to take on the aspect of a martyr figure, as exemplified by the self-conscious title of one profile in a US science magazine during the late 1980s: THE MAN WHO KNEW TOO MUCH. And though Rowland would take issue with couching his participation in such melodramatic terms, he did make a telling point in a rather more sober profile conducted by Paul Brodeur in the *New Yorker* in 1986: 'Chemists have tended to feel stigmatized by all the adverse publicity that has surrounded their profession in recent years. Their reaction to environmental problems caused by chemicals [is] frequently a defensive withdrawal from public involvement. Many of them are convinced that such problems are either nonexistent or grossly exaggerated.'

Rowland certainly did not withdraw from the debate. His critics counter that he went too far in the opposite direction by unashamedly straying into advocacy and (for scientists) the uncharted waters of attempting to dictate public policy. Where other researchers would assiduously avoid commenting on what (often draconian) measures would be required to forestall the ozone problem, Rowland would talk about them to the press. 'I think I'm labelled more of an advocate than I actually am,' he told me. He maintained that he did not actively court the media and that his only 'policy' was simply to pick up the telephone and answer reporters' questions. 'After all, the public paid for my research and it has a right to know about it,' he said by way of explanation.

Some scientists who attract publicity – whether by accident or design – become what can only be termed media whores. They can no longer function as working scientists. Proof that Sherry Rowland's research interests have not suffered as a result of intense media interest in his work is in the number of research papers which he has written throughout his career. During the first twenty-five years of his work as a research scientist, he published 170 papers (the Rowland–Molina hypothesis was number 171). In the subsequent nineteen years, he published a further 150, and commented that this 'represents not much change in rate one way or the other'.

His peers have certainly recognised his abilities and eloquence in addressing broader issues in science. In 1992, he was elected to the presidency of the American Association for the Advancement of Science and used the platform of his address to its annual meeting in Boston in February 1993 to urge the need for greater communication of science based on his own experiences. After twenty years of answering telephone calls from reporters on a daily basis, he could certainly speak from experience. The week before, he told his audience, he had been asked over the phone if he knew where it would be possible to obtain tapes of UFOs. Though tempted to joke or give his caller someone else's telephone number, he didn't, for they might well have got the impression that there really were people who had them. Rowland has learned that there is little room in the ozone issue for jesting, however misplaced.

One symptom of the problems which all scientists face, Rowland suggested, was 'simply the enormous growth of scientific knowledge itself and its accompanying tendency toward narrower and narrower specialities'. Because we live in a media age, he argued, we are bombarded by messages whether we like it or not. If scientists shy away from publicising their work, then non-specialists are increasingly likely to learn 'about these advances through another, less scientific route'. For good measure, he emphatically spelled out the dangers of ignoring the media: '[Outside] the mainstream exist other magazines and books which often have specialised points of view, ideologies and axes to grind, and frequently their proponents are unwilling or unable to give the attention to detail for full and accurate understanding.'

In many ways, Rowland has acted as a lightning rod for much of the controversy and criticism in the ozone issue. Yet he has a forceful justification for his participation, particularly if one considers what might have happened had worldwide controls on CFC production come into effect sooner. He surmises that even if, say, just worldwide aerosol uses had been curbed in 1976, it is debatable whether an Antarctic ozone hole would have opened up within a decade. Certainly for chlorine loading of the stratosphere to have been minimised it would have required pre-emptive action on a virtually worldwide scale. Yet it was only two years after the discovery of the ozone hole that globally binding legislation was put into place.

As it is, Rowland is thankful that he was – in the time-honoured way of most scientific breakthroughs – the right person at the right time. 'I often wonder how it would it have played out if Molina and I hadn't done what we did. What would have happened if nobody had been looking at chlorine when Farman's paper came out? They wouldn't have put CFCs in the first drafts nor would there have been the experimental equipment available. It might have taken ten to fifteen years just to figure it out.'

Susan Solomon of the NOAA Aeronomy Lab has avoided scientific advocacy at all costs, but because of her singular contributions to ozone research, she has often been at the mercy of press attention. As someone who has achieved pre-eminence at a relatively young age, she has also been the recipient of honours normally bestowed on older scientists. In May 1992, for example, she was elected to the US National Academy of Sciences, and, as a result, became its youngest member at thirty-six.

Her gender has also resulted in special burdens and responsibilities of its own. On Susan Solomon's computer is a small figure of Dorothy from *The Wizard of Oz*, a metaphor for the role in which female scientists find themselves. A survey of female researchers by *Science* magazine (in which she was featured) in March 1992 examined in detail the appalling gender bias which still persists in most scientific disciplines. Overall, 'Women in Science' concluded that '[subtle] obstacles and unconscious assumptions are taking the place of explicit sexism. Amongst those obstacles is the expectation – sometimes internalized by women as well as by men – that women will not have the same level of success as their male colleagues.'

Research chemistry still has a long way to go. Of the working chemists surveyed by the US National Science Foundation in 1988, for example, only 18 per cent were female. Though the later *Science* survey found that chemistry was more open compared to some disciplines (mathematics being the most sexist), there were still problems for women who wished to work in the field. One female chemist – who for obvious reasons was quoted anonymously – said: 'I was accused of not being serious about graduate school because I was having a family.' Many university chemistry departments still have no permanently tenured female staff, and in a brave attempt to force change, each year the American Chemical Society produces a 'dirty dozen' of universities where there are still no female academics.

For her part, Solomon said: 'I don't think in my career I've ever faced significant discrimination.' Like many of the chemists questioned in the *Science* survey, she is keen to emphasise the positive side to her work (an aspect which many female scientists pointed out in a follow-on survey a year later). Solomon feels that perhaps the most daunting aspect of scientific research for a woman is that its standard for successful behaviour is predominantly male. Scientists, generally speaking, are encouraged to be totally logical, forceful and completely dispassionate. Intuition is something that isn't very highly regarded. Although she was quick to point out that this is a gross simplification, intuition is most predominantly a female trait. 'To be a good scientist you don't have to be afraid of your intuition,' she said. 'A whole lot of scientists are afraid of it.'

Solomon has also become a role model for women entering or wanting to enter research. She feels ambivalent about this and has reservations about being thrust into such a position. 'There's a fine line between being a model and a mould,' she said. 'All women scientists shouldn't have to be like Marie Curie and that's the real danger.' She is nevertheless keen to encourage young people to enter science, but because of her myriad commitments (the curse of a respected researcher is to sit on innumerable committees), she can talk to only local colleges and has to decline many more invitations to address women's colleges throughout the United States. 'It's painful to have to say no to about 90 per cent of everything,' she admitted.

Another small figurine on Susan Soloman's filing cabinet is the Good Witch of the North, which she has christened Bob Watson. 'It seemed appropriate,' she said, 'because he appears at irregular intervals and fixes things.'

On the way to that First National Ozone Expedition in 1986, a reporter in New Zealand asked Solomon what she considers to be the single dumbest question in her career. Had she noticed that she was the only woman in a group of dozen men? She replied sarcastically: 'You're right! They *are* all men! I hadn't noticed before!' 'The problem was he took it at face value,' she told me shaking her head at a tribulation to which only female researchers would ever have to be subjected.

The Messianic Fervour of environmental evangelism seems to have taken a back seat to the vicissitudes of economic recession. If in the late 1980s one could hardly escape platitudes from any (and, seemingly all) two-bit pop singers, soap-opera stars or publicity-seeking politicians, then the early 1990s saw regular reports on the precarious state of funding for environmental pressure groups. As the pendulum of popular conception has swung in the opposite direction, it has given rise to the single most peculiar aspect of the ozone debate in all its strange contortions over the past two decades: the Ozone Backlash. Even by the standards of the curious events which are chronicled in these pages, it is downright bizarre and preposterous, leading *Science* magazine to comment in June 1993 – 'Welcome back to the ozone wars, which many scientists believed were long settled.'

In the fully fledged ozone wars of the 1970s, industry and

environmentalists clashed in the media while scientists sat uneasily on the sidelines. The Ozone Backlash of the 1990s is curious in that *all* participants have been cast as villains. In essence, it is a skirmish promulgated by a vociferous minority who view ozone depletion as a myth. They cannot be ignored, despite their often self-evident foolishness, for their numerous pronouncements can seem credible to people with little knowledge of the complexity of the ozone issue.

The tenor of the Backlash's collective criticisms, polemics and wrong-headed rantings rank from the mind-bogglingly ridiculous to those at the fringes of speciousness. Its proponents can be categorised into three vague associations: free-market devotees who see the ozone problem as a scare to diminish capitalistic free enterprise; those to whom environmental controls are anathema and who regard ozone depletion as a weapon in the armoury of eco-terrorism; and those who charge there is evidence of large-scale scientific fraud. Curious bedfellows, to be sure, all of whom rely solely on the media and voraciously feed off its less discerning representatives.

The animating genius behind this movement is a portly, reactionary US chat-show host whom *Newsweek* has described as 'a radio blowtorch'. As Rush Limbaugh has a best-selling book (*The Way Things Ought to Be*) in which he describes ozone research as 'poppycock', his wide audience of many millions of people in the United States and elsewhere may also have come to this conclusion. For Limbaugh, at least, the 3 February 1992 press conference was all the evidence he needed to question the motives of ozone scientists, whom he collectively likes to call 'the ozone priesthood'. To him, the NASA press conference was 'a scam' and he asks quite plaintively: 'What could be more natural than for the National Aeronautics and Space Administration, with the space program winding down, to say that because we have this unusual amount of chlorine in the atmosphere we need funding?'

His polemics have as their basis the work of two people who have at least some modicum of scientific training. Where Limbaugh and fellow ideologues can be dismissed as knee-jerk reactionaries (or for that matter, just jerks), the authors of a book entitled *The Holes in the Ozone Scare: The Scientific Evidence That The Sky Isn't Falling* are much less difficult to ignore. For Rogelio A. Moduro and Ralf Schauerhammer have collected what appears – at first sight – to be a plausible series of critiques of the validity of ozone science. Moduro is a graduate geologist who is an editor of a magazine called *21st Century Science and Technology*, which is also linked with the right-wing political activist Lyndon La Rouche, currently serving fifteen years' imprisonment for tax evasion. It is not exactly a mainstream publication and neither are its opinions (for example, it also champions cold fusion).

The Holes in the Ozone Scare could be charitably termed 'The Complete And Unabridged Conspiracy Theorist's Guide to the Ozone Issue'. We must be thankful that it stops short of implicating environmentalists in the assassination of John F. Kennedy, but elsewhere, it goes a few stages further into the realms of wild speculation. Its express aim, in the words of its authors, is nothing less than 'the overthrow of the murderous environmentalist regime now ruling our schools, government institutions, and media'. Nobody, least of all the chemical industry, escapes the scorn of Moduro and Schauerhammer: they even suggest that the 'global chemical cartels [financed] and supported the movement to ban CFCs'.

Their book is highly entertaining, but for all the wrong reasons. Its conclusion, for example, is a chapter entitled 'Great Projects to Transform the Globe', which breathlessly suggests that the world needs more population, that central Europe should be linked by a vast rail grid to provide resources for the United States, that North America should be crisscrossed with magnetically levitated railways, and that the world's water shortage could be solved in one fell swoop by arbitrarily cutting canals or causing floods throughout the Third World. After ploughing through these bizarre offerings, I rubbed my eyes, half expecting to see sows floating in the stratosphere.

It would be easy to dismiss this most arrant of nonsense were it not for a slightly sinister side effect: that even those who are well versed in science have been taken in by the selective facts which Moduro and Schauerhammer have published. 'What I am most concerned with,' Moduro told *Science* in June 1993, 'is that scientists who have been presenting an opposing view have a public forum, the ability to present their work to the public.' Though he claims that his book sets out to air some of these 'opposing' findings, it fails to report that much of the evidence chronicled within it has been superseded by more recent research.

This fundamental axiom of scientific research seems to have passed the Backlash polemicists by. The body of scientific knowledge is peppered with the scar tissue of false conclusions and wrong leads, but they are largely corrected in the subsequent checks and balances of scientific enquiry. That science is largely a self-correcting enterprise is something for which Moduro and his cromies have no appreciation. If a theory is not supported by experimental evidence, it is discarded or amended according to subsequent observations.

Today's insight isn't necessarily tomorrow's paradigm shift: many of the arguments held against the 'ozone priesthood' have been demolished by more recent observations. 'In the course of scientific enquiry,' Sherry Rowland said, 'errors are made. What Moduro and others have done is collected all the errors.' And, indeed, by this purposeful assembly of redundant conclusions and largely superseded theories, *The Holes in the Ozone Scare* presents a weirdly believable portrait of the iniquities of stratospheric research. For his part, Moduro contends that scientists have 'systematically ignored all the massive research which debunks elements of their theory'. It is a claim that cannot be substantiated as demonstrated by three of their most pervasive criticisms of ozone depletion: that CFCs are too heavy to reach the stratosphere; that Gordon Dobson stumbled upon an ozone hole in 1956; and, most importantly of all, that the observed ozone depletion can be explained away by natural sources of chlorine. These three tenets of the Ozone Backlash can quite easily be demolished under closer scrutiny.

Chlorofluorocarbons are indeed heavier than air – about five times heavier, in the case of CFC-11. But molecular weight presents no restriction on the efficiency of molecules to be mixed throughout the atmosphere by the large-scale motions within it. When Rowland and Molina first proposed their chemical mechanism twenty years ago, there was no unequivocal evidence that CFCs had been found in the stratosphere. At that time, however, there was evidence from sounding rocket experiments far into the mesosphere that two relatively inert molecules (krypton-84 and neon-20) were observed in the same ratio as they appeared near to the ground. So although the former was much heavier than the latter, these observations indicated to Rowland and Molina that a molecule's weight did not preclude its ability to reach the uppermost reaches of our atmosphere.

Subsequently, many hundreds of balloon flights and scores of ER-2 flights have observed that CFCs and more than a dozen halocarbons are present in the stratosphere. The many hundreds of peer-reviewed papers in which these observations have been recorded seems to have passed Moduro and Schauerhammer by. As Sherry Rowland pointed out in his AAAS address in February 1993, it tells more about the critics themselves and 'the reluctance which many have to accept the results of observation or even to bother determining whether any pertinent observations have been made'.

The circumstances surrounding the first ozone measurements in Antarctica were chronicled at the end of Chapter Two. As we saw, Gordon Dobson himself was excited by the findings of much lower ozone levels than had been expected. Instead of the 450 Dobson units, they hovered around the 300 DU mark. *The Holes in the Ozone Scare* goes so far to carry the relevant chapter from the second edition of Dobson's seminal *Exploring the Atmosphere* (part of which was quoted from the first edition of the book on page 50). What Moduro and Schauerhammer fail to explain is that Dobson is discussing the Antarctic *vortex* and not the ozone hole. Proof, if it were needed, of the development of the latter comes from the observations in recent years that ozone regularly drops below 125 Dobson units, which 'is at the limits of what some of our instruments can detect', to quote Joe Farman directly. When asked about the so-called hole in 1956, he shook his head and uttered one word: 'Ridiculous!' And Farman should know, as he made many of those original measurements himself.

The question of natural sources of chlorine is rather more complex but worthy of detailed examination. The sources of naturally occurring chlorine are sea water, biomass burning, ocean biota – and, most importantly of all, volcanoes. A cursory glance at their chlorine content is impressive: 600 million tonnes (sea water): 8.4 million tonnes (biomass): 5 million tonnes (biota) and 36 million tonnes (volcanoes). Compared to the annual emission of some 750,000 tonnes of CFCs, they may seem overwhelmingly important. What Moduro et al. fail to explain is that these natural sources of chlorine are soluble, so that they get rained out of the atmosphere before they go very far. CFCs, on the other hand, are inert as they rise upwards until they reach the stratosphere, at which point they are photodissociated. Those seemingly meagre 750,000 tonnes reach the stratosphere unimpeded, where they then start to strip away the ozone layer in chain reactions.

If stratospheric chlorine did come from sea water, then there should also be equally enhanced levels of the sodium from the salt (NaCl); if it originated with the biomass burning or ocean biota, there should be greater abundances of methyl chloride. There aren't: and in the case of the latter, it would account for only perhaps 5 per cent of the total chlorine loading observed today. In *The Holes in the Ozone Scare*, Moduro and Schauerhammer quite arbitrarily use earlier estimates for biomass burning which produce nearly twenty times more methyl chloride than is now known to result from this source. Proof of human responsibility in ozone depletion comes from measurements of the concentrations of hydrogen fluoride, which undisputably result from the breakdown of CFCs in the stratosphere. They have increased on average by a factor of five over the last two decades.

Volcanoes in various shapes, sizes and locations are cited as another 'overlooked' cause for ozone depletion. Most notorious is the supposed starring role of Mount Erebus, which is an active volcano a few miles upwind of McMurdo Sound in Antarctica. It is rather more famous as the crash site for an Air New Zealand DC-10 which claimed 260 lives in November 1979 (the greatest death toll in one single aircraft accident). Moduro and Schauerhammer make much of the fact that Erebus 'is constantly blowing out a huge cloud of chlorine and other volcanic gases'. What they fail to point out is that it does not erupt explosively and is several kilometres below the stratosphere. *In situ* measurements have revealed that gases from steam vents reach perhaps as high as half a kilometre. Moduro and Schauerhammer also cite that some thousand tonnes of chlorine are emitted per day by Erebus, based on research which was published in 1985. Subsequently the same vulcanologists have reported that only 15,000 tonnes are emitted per *year*. This fact is ignored in *Holes in the Ozone Scare*.

By far the most misused data from volcanoes is that from the eruption of Mount Augustine in Alaska in 1976. David Johnston, a US Geological Survey geologist, proposed in a paper which was published in *Science* in July 1980 that as much as 175,000 tonnes of hydrogen chloride might have been injected into the stratosphere from this one eruption. Most recently, however, observations by aircraft which have flown through even the most chlorine-rich eruptions have shown that most HCl is rained out in the lower atmosphere or else scavenged near to the volcanic calderas by steam which is also often produced. After the eruption of Mount Pinatubo, for example, NCAR researchers William Mankin and Michael Coffey measured the HCl ejected upwards from aircraft and found that there was little change in its background levels in the stratosphere.

This point was overlooked by perhaps the most exotic ornament of the Ozone Backlash, the late Dr Dixy Lee Ray. Originally trained as a marine biologist, Dr Lee Ray headed the Atomic Energy Commission from 1972 to 1977, and thereafter was governor of Washington State for five years. She was well known as a vociferous critic of the ecological movement before she passed away at the end of December 1993. It was during her gubernatorial tenure that an event occurred which more or less galvanised her crusade against what she termed 'environmental fascists' and 'technological luddites'.

It could almost be said that the eruption of Mount St Helens in Washington State was the defining moment of her life, for as she relates in her book *Trashing the Planet*, she 'received my lesson in humility, my respect for the size and vast power of natural forces on May 18, 1980'. Thereafter she took to calling it 'my volcano', although this is a moot point: the geologist David Johnston probably has a greater proprietal claim for the very sad reason that he was killed by the explosion. He had remained perilously close to the volcano before it erupted while he was taking scientific measurements.

Dr Lee Ray seemed not to have realised that not all volcanoes erupt 'equally' as it were, for although Mount St Helens injected much dust and ash into the stratosphere, it had little discernible effect on the chemistry of the ozone layer. In her book, she took the speculation in Johnston's paper on sources of HCl at face value and ascribed them to the Mount Augustine eruption. Johnston had also calculated from the geological record that the explosion of Bishop Tuff in California 700,000 years ago may have injected 289 million tonnes of HCl into the stratosphere. Dr Ray took this figure as fact and erroneously attributed it to Mount Augustine. To add insult to injury, Rush Limbaugh has then blithely taken it and ascribed it to the amount of HCl spewed into the stratosphere by the eruption of Mount Pinatubo! For good measure, Limbaugh describes Dixy Lee Ray's writings as 'the most footnoted, documented book' he has ever come across. Perhaps it only proves the veracity of George Bernard Shaw's comment that half-truths are the most dangerous of all, an observation which could be made of the Backlash as a whole.

Dr Lee Ray was an indefatigable speaker on the subject of 'environmental myths' and all that they entail. In her book, she gamely tackled acid rain, nuclear waste and the greenhouse effect, suggesting they pose little or no threat and are instruments of fashionable panic orchestrated by those who are 'anti-development, anti-progress, anti-business, anti-established institutions, and above all, anti-capitalism'. Small wonder that few scientists would be publicly drawn into debating with the venerable Dr Lee Ray, although collectively, they did once famously blow a raspberry in her general direction – and with musical accompaniment to boot.

At the end of a conference on ozone at Williamsburg, Virginia, in June 1992, delegates sang a song with suitably tongue-in-cheek lyrics. Just in case anybody missed the point, they were conducted by a meteorologist wearing a confederate uniform. The chorus ran thus:

> We're just concerned about low ozone
> and the harm that it does to the Earth's protection,
> and we're OK, Miss Dixy Ray
> You have your faith that the Earth's forgiving
> Of all that mankind keeps on emitting
> But we're just concerned about low ozone.

They also satirised many of the funding criticisms discussed earlier in the following verse:

> I've heard them say that Dixy Lee Ray thinks we're worth
> nothing at all.
> For everyone knows that, just like the snows,
> Freons fall back to the ground.
> We do science for gain,
> and students we train,
> to make sure the money flows fast.
> So we can besport
> at those winter resorts
> Kiruna, McMurdo and Maine.

Dr Lee Ray's reaction was not recorded.

One persistent critic of the 'ozone priesthood' was one of its leading lights many years ago. Unlike most of the critics of the vast body of knowledge in stratospheric science, S. Fred Singer has interesting credentials to participate in the debate. He was present at the dinner in the home of James Van Allen where the International Geophysical Year was proposed in 1952, and shortly thereafter came up with a forerunner of the backscatter technique for the measurement of ozone from space which has subsequently been made routine by TOMS. His veering off the main track of research occurred in the 1970s when he was involved in the issue of the impact of supersonic aircraft on the ozone layer. At that time he served in various capacities for the US Department of Transport and the Environmental Protection Agency and has subsequently become concerned that ozone scientists are now preoccupied with funding issues.

In a piece he wrote for the *National Review* in 1989 (entitled 'My Adventures in the Ozone Layer'), Singer admitted that his scientific colleagues had behaved 'honourably' throughout the ozone debate. But, pointedly, he added in the same article that 'egos and ambitions collided with facts, leading to a temptation to ignore the facts. Politicians had no hesitation in manipulating science. And the media had a field day.' Singer has had no hesitation in using the media himself, and has been played off against some of the other participants in the Ozone Backlash. But, quite unlike many of the critics of ozone science, he does at least attend scientific meetings and is open-minded on some of the aspects which they entail.

Fred Singer's interest in the issue of ozone depletion began when he considered the increasingly important role of methane on ozone chemistry. In the early 1970s, its increase had started to be quantified by atmospheric scientists throughout the world. Singer realised that its photodissociation in the stratosphere would begin to humidify the ozone layer and calculated that half the methane seemed to originate from artifical sources. As a result, he predicted that it would therefore increase during the following years and also calculated that the water which would result would be roughly the same as that from a supersonic aircraft fleet. But few journals wanted to publish his findings, and one colleague advised him not to get involved in the issue at all.

Subsequently, Singer watched the ferocious debates on the ozone issue with head-shaking concern at the way in which its complexities were glossed over for public consumption. His *National Review* article was an attempt to lay bare some of the complexities, though he did include the 'ozone hole was there in 1956' rhetoric. He is now associated with the Science and Environmental Policy Project in Washington, DC, which has analysed the misuse of scientific data and the problems of assessing risk in the face of scientific uncertainty.

'The ozone issue cannot be decided on ideological grounds,' Singer told me. 'CFCs must be an important source of chlorine, but we must investigate other natural sources.' He was willing to admit that much of the writings of Moduro et al. are red herrings, but believes more work should be done without presupposing answers. He pointed out that much of the trends work has been performed over just one eleven-year cycle of solar activity, from which he thinks it is not possible to get a true picture of the underlying global change in stratospheric ozone levels. And on the matter of the Antarctic ozone hole, Singer believes more work should be done on the peculiar dynamics of the surrounding region. The hole's sudden development, he thinks, has more to do with cold temperatures in the stratosphere than even enhanced chlorine loading. 'The Antarctic ozone hole is triggered by stratospheric temperatures rather than by chlorine,' Fred Singer said. 'Chlorine is no longer the controlling factor.'

How Soon Is Now?

The body of evidence in the ozone debate points unequivocally to CFCs as the root cause of ozone depletion. It is the direct consequence of human activity that has altered what had formerly provided pervasive protection against harmful radiation from the Sun. Many questions remain about the exact details of ozone depletion, but enough is known to piece together a coherent portrait of the science involved. Ignorance of the scientific method, disdain by scientists for the media, deliberate stirring by some reporters and conspiracy theories espoused by polemicists all play their part in garbling this underlying coherency.

Scientists endeavour to be objective, but, as one US scientist interviewed for this book said: 'We live in the age of the sound bite and there is so much pressure to say things in short, snappy sentences. I know it sounds like an excuse, but the subject is incredibly complicated. There is a danger with journalists constantly looking over our shoulders and the pressure to publish results. We must never lose objectivity.' The furore over the 3 February 1992 press conference has made it more difficult to publish disturbing results in the future. As the clock inexorably ticks towards the maximum chlorine loading of the stratosphere, this dilemma will have to be faced. Silence must not become the price of objectivity.

8

Deadly Springtime

> A systematic review of medical records from all the ophthalmic and dermatologic physicians in Punta Arenas was performed, looking for the occurrence of acute sentinel medical events in relation to the ozone hole. Comparison periods from recent years and seasons in which no ozone hole has been reported were used. No convincing evidence of a temporal association of acute ophthalmologic disease with the ozone hole was found.
>
> *Ocular and Dermatologic Health Effects of Ultraviolet Exposure from the Ozone Hole in Southern Chile: A Pilot Project*, Report submitted to the Environmental Protection Agency of the United States and the Ministry of Health of Chile, 1993

OF all the possible dangers from ozone depletion, the most serious are those to human health. As the ozone layer above our heads strips away, more ultraviolet radiation will reach the surface of the Earth. The danger begins where our vision ends: as its name suggests, ultraviolet light is found beyond the blue portion of the rainbow spread of colours comprised by visible sunlight to which our eyes are sensitive. Ultraviolet light is much more dangerous than visible light because of an immutable fact concerning the electromagnetic spectrum. With decreasing wavelengths, each photon of light imparts more energy to the material which has the misfortune of absorbing it.

So how real are the dangers? Data records are incomplete and anything approaching a consensus has yet to emerge. Unlike the chemistry of the stratosphere, which has been investigated for two decades now, the biological effects of ozone depletion are a research topic to which adequate resources have only just been devoted.

Circumstantially, skin cancers have been increasing throughout the world, but this has more to do with changes in social behaviour than anything else. Current levels of skin cancer incidence have little to do with the ozone hole and their future increase may be alleviated by changing lifestyles. Yet there is little room for complacency. While human beings can avoid prolonged exposure to sunshine, the same cannot be said of plants, crops and many aquatic life forms. The economic consequences of destroying these flora and fauna represent the bottom line of ozone depletion.

In the summer of 1991, newspaper reports from southern Chile about blind animals and deformed plants signalled to many people that ozone depletion was already taking its toll. For this book, I visited Punta Arenas to interview local people about the problems which many believe they now face. Most of the evidence turned out to be exaggerated or downright false, a sad example of how the truth has been distorted by environmental pressure groups for their own ends. But the fact that footballing youths were badly sunburned in southern Chile in October 1992 (equivalent to it happening to Britons or Canadians in March) should not be dismissed out of hand. Backlash polemicists are equally culpable for suggesting that there will be no dangers in the future.

The cumulative effect of ozone depletion in the twenty-first century will doubtless bring unpleasant surprises. Just because there haven't yet been any discernible health effects doesn't mean that there won't be in the future.

The first time technological advance threatened Punta Arenas it was clear cut. The opening of a shipping route through a far-off isthmus would almost immediately reduce the city's pre-eminence as a busy port for ships passing through the Straits of Magellan.

Deadly Springtime

Probably without realising it, the engineers who excavated the Panama Canal effectively consigned Punta Arenas to anonymity at the southernmost tip of South America. Though the livelihood of many of its citizens was threatened, fate intervened: a gold rush in Tierra del Fuego kept the local economy buoyant for many years. In time, many of the frustrated gold diggers found work as herders or farmhands in the rolling countryside which surrounds the city.

Seventy years after the opening of the Panama Canal in 1920, the onward march of technology has put the citizens of the world's most southerly city at risk again. This time the threat is more insidious, promising to affect not just the economy of Chile's Magallanes region but the health and well-being of its inhabitants. The citizens of Punta Arenas have long ago accepted the uncomfortable fact that they are nearer than anyone else to the Antarctic ozone hole and have thus braced themselves for the first signs of damage from the unwanted spectacle of El Agujero each spring.

A visitor to this busy, teeming city will be struck by its similarity to a large Scottish sea port, for the oil platforms which pepper the Straits of Magellan are vaguely reminiscent of Aberdeen. Only the brightly coloured roofs of houses, many of which are little more than tin shacks, reveal that you are in the southern hemisphere. The occasional car engines misfiring or soccer posters in shop windows serve to remind you that Punta Arenas is indubitably part of South America.

The city itself is surrounded by pleasant, rolling hills which shield it from the ferocious winds in this part of the world. In Spanish Punta Arenas means 'sandy point'. For it was the only place along the bleak coastline north of Cape Horn where Ferdinand Magellan and his ships could moor before they attempted the journey across the Pacific in 1520. A large statue of the explorer who founded the city stands in its main square and casts an imperturbable glance towards the waters which bear his name. In time, as others followed in Magellan's pioneering tracks, the pleasant bay developed as a coaling station for ships heading east into the Pacific or returning back home to Europe. By the nineteenth century, Punta Arenas was the home of the Great White Fleet, and it is a measure of the city's development that it boasted electric lighting twenty years before Santiago did.

How Soon Is Now?

Today, the world's most southerly city is once more a thriving port with some 125,000 inhabitants; it is now the focus of Chile's burgeoning oil and methanol industries. The government of this part of Chile has actively sought the participation of foreign oil companies in offshore drilling and the development of oil fields on Tierra del Fuego. When the weather is clear, the outline of the island can just about be made out where the sky meets the sea in the cold waters across the Straits of Magellan.

Each spring, the inhabitants of Punta Arenas have become accustomed to legions of scientists and adventurers who make their way down to 'the ice', the unseen continent of Antarctica some thousand kilometres distant. The city is still the nearest point on Earth from which those intrepid enough to want to venture south can do so. It was for this very reason that *los Americanos* came here in the austral spring of 1987 and proved that humans were responsible for the wholesale stripping away of the ozone layer. Of this problem the population of the city needs little reminder, for the visitor to Punta Arenas cannot fail to notice the numbers of people who wear sunglasses and cover their heads with shawls and scarves. Around lunchtime, in particular, the streets fill with hordes of schoolchildren either on their way home or hurrying to afternoon classes. They look, in the telling phrase of one teacher, 'like the blind-club day out'.

It is not an uncommon sight in this part of South America to find signposts pointing vaguely in the direction of Europe which cite impossibly large distances to improbable destinations in the former Yugoslavia. Quite why anyone a continent away would yearn for a country which is self-destructing may seem like another curious conundrum in this part of the world. The answer is that an estimated 40 per cent of the population in and around Punta Arenas originates from Croatia, their grandfathers lured to Tierra del Fuego by the promise of the gold rush in the 1920s. When the gold ran out, many Croatian families stayed and invited their relatives in Europe to escape the uncertainty and turmoil of their homeland – which, generations later, has once again erupted. Today, the natural melancholy of these east European émigrés is a stark counterpoint to the exuberance of the darker-skinned indigenous Chileans.

Bedrich Magas Kusak was one of the more recent arrivals; he came here with his family when he was seven in 1959. A tall, pale man with a quick smile, he is a lecturer in engineering at the Universidad de Magallanes in Punta Arenas. Like the hardest-working of academics, he has the pokiest of offices and shares it with a colleague. Bedrich is usually to be found seated upon a cushion on the floor behind the door. He cuts a colourful figure with his students, and is more amenable than most tutors. 'Everyone should behave like a student at university and that includes the lecturers,' he said. 'That way knowledge will flow more freely.'

It was this almost childlike appetite for knowledge that led him to become one of the first southern Chileans to visit Antarctica. The University of Magallanes was keen to make its facilities – both laboratories and their attendant personnel – available to the many international vessels which passed through Punta Arenas en route for the pole. When the German research vessel *Polarstern* docked there in the austral spring of 1984, the university provided back-up technical assistance for its month-long biological investigation mission to Antarctica. Bedrich's technical skills ensured he was invited aboard, an experience which he likens to a journey to the Moon. 'There is a special feeling you get from going to the Antarctic which I find difficult to put into words,' he says. 'You feel the loneliness but also the power out there in the expanse of whiteness. It's depressing and elating at the same time. You don't ever feel the same again.' When the ship returned to port, he remained on board for another two days.

Over the next few years, Bedrich helped out whenever he could when polar expeditions needed equipment to be fixed or manufactured at the last minute. As a result of his interest in Antarctic research, he joined the American Geophysical Union and read the results of the first National Ozone Expedition from McMurdo Sound in 1986. Up to this point, he had had a vague interest in environmental issues generally, but it would take a phone call from the US embassy in Santiago to transform it into a professional preoccupation with ozone depletion. That call came in April 1987 and its purpose was to find a Chilean scientist who could act as a local liaison for the Antarctic Airborne Ozone Experiment. Without hesitation, Bedrich offered to help.

How Soon Is Now?

An advance party from NASA and the State Department would shortly be making a visit to Punta Arenas to determine the logistics for the forthcoming campaign. A few days later, Bedrich and the president of the university met Estelle Condon and Bob Watson at the Hotel Cabo de Hornos. His obvious technical knowledge and eagerness to help impressed the NASA team. They were relieved to have found so helpful a local link man for the project, and Bedrich turned out to be a godsend for the American scientists who would be holed up in the freezing aircraft hangars for days on end. Many of them remember the tall, pale, lab-coated local who uncomplainingly sorted out their most niggling of problems. When an important telecommunications link with the United States went down, Bedrich spent seventy hours tracing the problem to a misconnection with a feed to a satellite over Mexico. After he had spent a week fixing some other obscure electrical equipment, Bedrich told Bob Watson to 'forget it' when he was told to submit an invoice. If there were any problems, the AAOE scientists soon learned that a quick call to Bedrich Magas would sort things out.

The citizens of Punta Arenas were at first intrigued by the presence of *los Americanos* who had booked all the rooms at the Cabo de Hornos hotel. El Agujero seemed like an abstract concept, far removed from their everyday concerns, and after the novelty of their arrival had worn off, most of the locals forgot about their presence and what they were up to. But Bedrich Magas, at least, formed an unshakable opinion of the seriousness of the results from the Airborne Antarctic Ozone Experiment. 'I quickly got the idea that something big was happening,' he said of his conversations with American scientists, usually in the bar of the Cabo de Hornos hotel. 'If the chlorine hypothesis was correct,' he added, 'then the time scales of the Montreal Protocol would be a joke. Good intentions were not enough.'

The locals warn visitors to Punta Arenas that they can often experience weather typical of all four seasons in the space of one day. The winds can whip up quite suddenly, only to die minutes later. Some of the trees hereabouts bear permanent testimony to the wind as they have been literally bent sideways. So it is hardly surprising, then, that in such an unseasonable place, odd events would be reported and that, in time, they would increase in number and weirdness. The very remoteness and relative inaccessibility of southern Chile makes it an ideal setting for supermarket tabloids to chronicle sightings of Elvis, villages being ravaged by dragons or, indeed, monster babies mutating from the ozone hole above.

Press reports of odd goings-on in Punta Arenas have to be weighed against the tendency for human beings to ascribe all sorts of unrelated events to pandemic *dei ex machinae*. Something similar has occurred in Britain, where severe storms in recent years have been falsely ascribed to the onset of global warming. The hottest summer this century in the United States, 1988, was also taken as proof of the arrival of the greenhouse effect. Popular misconceptions often have no basis in reality and the exercise of caution is warranted. This was something Bedrich Magas realised as he became intrigued by reports of damage which had been directly blamed on the ozone hole.

These started to circulate locally in the spring of 1990. Fishermen off the coast of Tierra del Fuego were finding blind fish in their catches. Farmers wondered why rabbits could be caught so easily, and then found that many of them appeared to have cataracts. And then ranchers started to find sheep, cattle and horses going blind, while other farmers reported their animals were calving at unusual times, mirrored by reports of the unseasonal blossoming of plants. Bedrich was sceptical, but felt that they urgently required further investigation, particularly as the ozone hole was growing deeper each year, and had started to encroach upon South America at the turn of the decade.

In the spring of 1990, a local ecological defence group known as IDDEA funded a small study at Bedrich's behest. From September to the end of November, they examined a statistically significant sample of about a thousand rabbits to discern any strange health effects, but found nothing unusual. In parallel, throughout the whole of that spring season, the environmentalists meticulously examined the eyes of rabbits which had been killed at a local slaughterhouse. A range of eye infections were found, ranging from slight milkiness to complete whiteness in some of the samples. At the end of August, 15 per cent of the rabbit eyes sampled had shown obvious signs of damage: by the end of November, that percentage had doubled.

Oscar Riquelme, president of IDDEA, is a photographer by trade, and had spent much of his time faithfully photographing a hundred pairs of eyes each day. A sample of the most badly affected eyes were kept in formaldehyde and sent to the Veterinary Faculty of the Austral University of Moldavia for analysis. A zoologist, a veterinary doctor and a photobiologist examined the samples at the start of 1991. According to Riquelme, they concluded that they showed signs of corneal damage probably caused by cataracts. However, the scientists were not able to determine conclusively whether the damage had been caused by increased levels of ultraviolet radiation associated with the ozone hole or not.

Finance for esoteric research projects is not easy to come by in southern Chile, and so it would be over a year before another study could be carried out. In January and February of 1992, Greenpeace in Santiago sponsored a field trip in the countryside surrounding Punta Arenas to investigate reports of blind sheep. At an *estancia* on Tierra del Fuego where the greatest number of cases had been reported, Riquelme once again took photographs. Greenpeace also videotaped many of these afflicted sheep and the footage was later shown on ABC News in the United States and the BBC's *Horizon* in Britain. One of the most badly affected sheep was chosen to be killed and both its eyes were sent for analysis. The first was sent to the Chilean Agriculture and Farming Service in Santiago and 'nothing more was heard about them', according to Riquelme. The other was sent to the Laboratory for Cytological Analysis in Santiago, where biochemical investigations revealed the presence of an endemic virus. Was it possible that the ozone layer could have caused it? The biochemists could not say for sure.

A reporter from the ABC affiliate in San Francisco, Channel 7 KGO, travelled to Punta Arenas to investigate these reports at roughly the same time. Brian Hackney's findings were transmitted over four consecutive nights in the third week of February 1992 as part of the 'Bay Area Focus' segment on the 6 p.m. news. To the list of odd reports, Hackney added that flamingoes, guanacos and Magellanic penguins had been been breeding out of season. 'None of the evidence for all of this is scientific, though,' Hackney commented in his voice-over. 'Most of it is rumours from sheep farmers who live hundreds of miles from the city.' And, indeed, the reporter and his crew travelled some 300 kilometres down to

Deadly Springtime

Tierra del Fuego to another *estancia* to find perhaps twenty out of 2,000 sheep afflicted by eye problems.

After one sheep was chosen to be slaughtered, Hackney took one of its eyes back to the West Coast for analysis by veterinary scientists at the University of California, Davis. An extensive dissection revealed that there was a white lesion on its cornea, and Dr Ned Buyukmihci of the Veterinary Science Department stated outright that it was 'quite clear it's not a cataract'. From biochemical analyses of micro-organisms within the cornea it was not possible to determine whether a virus had caused the problems or, indeed, whether immunosuppression had taken place. The latter aspect is one of the most worrying aspects of the whole health question concerning ozone depletion.

The most immediate outcome of Hackney's excellent reports – which came just two weeks after the infamous NASA press conference of 3 February 1992 – was a press release from Greenpeace in the United States on ozone depletion. It was, to say the very least, economical with the truth. The environmental group claimed that there were 'flocks of sheep with cataracts' and 'a fivefold increase in skin cancer' in the Magallanes region of southern Chile. Quite where this information had come from is anybody's guess, but in fairness, in Hackney's final broadcast on the issue, he reported that Greenpeace had amended these points in a subsequent update to the original press release.

Yet many of these unsubstantiated reports and myths persist. It comes as little surprise to learn that the most sensational concern farmhands who are going blind.

The Gibbons family have kept an *estancia* on the Otway Sound, north of Punta Arenas, since 1904. To get there takes a two-hour drive from the city along a road that gets progressively more difficult to navigate, winding across a bleak and desolate landscape. On the way one passes through farmland belonging to twenty consecutive farms. The Campo Gibbons is the last, sitting on a headland along from a crescent-shaped bay, with a spectacular view across an inlet beyond which are snow-capped mountains. The wind assaults the senses in both its constant howling and the remarkable vista of trees bent sidewards.

How Soon Is Now?

John Gibbons has managed the ranch since he graduated in economics from the University of Chile in Santiago in 1985. His time there made him realise soon enough 'that I would hate working in an office', so he quite gladly returned to the family *estancia* which had been established by his grandfather in the 1920s. The farmhouse looks like something Norman Rockwell could have painted, and inside, it is comfortably furnished and bedecked with attractive wood panelling. It provides solid protection against the elements outside. Apart from telephone wires and modern cars, there is very little to suggest that this region has changed since the farmhouse was originally built.

As well as the Croatians, this part of South America boasts many farmers of Celtic origin. Indeed, John's grandfather had been lured from his native Ireland by the chimera of gold and, as befits one who would travel across the world in search of a fortune, his life story is replete with the stuff of legends. *Granpapa* Gibbons worked in a goldmine for two years before it went bankrupt and thereafter started ranching with two Scotsmen. They underwent many adventures, including being attacked by native Indians in Argentina where they were moving 500 horses across country. He settled back in the Magallanes region as a ranchhand, eventually becoming a manager of his own ranch. His son inherited the *estancia*, passing it on to John and his twin brother Oscar, a lawyer who lives in Punta Arenas.

Having grown up on the ranch, John noticed some changes after his three years in Santiago. Swarms of African egrets that were common during his childhood no longer came to roost. Weather patterns seemed to change: during the winters of his childhood memory, it seemed to snow continuously. 'Now the snow seldom stays for more than a few days,' he told me with bemusement. When other farmers started to report blind sheep, John didn't really take much notice. 'There are always sheep in these parts going blind,' he said. In any case, the Campo Gibbons did not keep its own sheep, and it had some 1,200 cattle by the end of the 1980s. Nevertheless, by 1989 about fifty of the cattle started to go blind each year, though, oddly, most were only affected in one eye.

In a culture which virtually coined the concept of 'macho', it is hardly surprising that few of the other *estancia* owners discussed the eye problems in their bovine charges. 'If you complain, then you'll get branded as being pessimistic or plain weird,' John Gibbons said. But the farmhands who move from *estancia* to *estancia* each season noticed among themselves that they were suffering from health problems. The Campo Gibbons employs six regular workers and another eight 'casuals' during the summer. They normally spend ten hours outside during the summer months without sunglasses or the benefits of sun cream, the result of low wages combined with characteristic Magallanes machismo.

In the austral summer of 1990–91, many of the workers started to notice they were burning more easily. 'Your face would go red during the day and it would hurt at night,' John Gibbons recalled. 'But then the next day your face would be white again. Now that really was strange.' As summer turned to autumn, some of the permanent farmhands started to suffer intense pains in their eyes.

Walter Ulloa is a dark-skinned farmhand who had moved to Punta Arenas in search of work from northern Chile. That summer of 1991, he started to experience bad headaches and began to lose sight in his left eye. A deep cataract formed and he has been to see a doctor three times since then to find out why. 'He couldn't do anything and he didn't know what had gone wrong,' Ulloa said. 'He did say that I should be operated on.' In Chile, health care is something of a luxury and Ulloa cannot afford the simple operation to remove the cataract. It is a measure of the facilities available that he had bathed his eye in cooking oil when he first experienced the excruciating pains.

Other workers in the Gibbons employ have reported skin problems, severe headaches and prolonged eye disorders, though not as severe as Ulloa's. 'The government isn't taking this problem seriously at all,' said John Gibbons. He admitted that he was worried for the future, particularly since his wife Carolina gave birth to a baby girl, Maria-Paz, in the spring of 1992. 'I am very worried about the health problems which she might have to face,' he said. But, on reflection, he did not feel he would have to abandon the family home in search of a safer place to live. 'After all, the ozone problem is getting worse around the globe,' he said. 'It is just our misfortune to have experienced it first.'

* * *

How Soon Is Now?

The city which owes its direct origins to the greatest Portuguese navigator of his generation has always brought sailors from foreign lands. In the early years of the twentieth century – before the opening of the Panama Canal – three ships a day passed through the docks. In some places, health warnings for cholera dating from that time can still be found, although the signs have not been treated well by the elements. Like most seaports it has a brisk business in the oldest trade in the world: it is said that Punta Arenas boasts the highest number of prostitutes per head of population in South America. For men who are misguided enough to use the excuse that they will soon be spending many months in Antarctica, the city's easily noticeable red-light district offers some cheap comfort. Regular visitors to the city estimate that there are over eighty *whiskerias*, some of which have no health controls.

Given the increasing numbers of reports of strange occurrences, one might expect to see health warnings about the ozone hole. But the reports represent a strange mixture of hearsay and wishful thinking which the local government has decided to ignore. There are two obvious reasons mitigating against the development of noticeable health effects in southern Chile. The first is that the ozone hole hasn't been around long enough for cumulative effects to have developed yet. The second is that although the ozone layer is at its thinnest over Punta Arenas in October, the Sun is at a low angle in the sky. The amount of light reaching the surface is markedly less than that received at lower latitudes during the summer months where the Sun is higher in the sky and more intense sunlight will reach the surface. So during the summertime, greater amounts of ultraviolet radiation will penetrate as far as the surface in the tropics than *ever* would in Punta Arenas. If there was any noticeable damage to sheep in Magallanes, there should literally be an epidemic of blind sheep in, say, Natal in Brazil.

In early 1992, however, the Magallanes Health Ministry made available a few pamphlets for schools on the possible dangers of increased ultraviolet radiation. There were not enough funds available to reprint them. The local government believes that the region is not at risk and its minister for agriculture, Sr Fernando Baeriswyl, told the *Independent* newspaper in April 1992 that 'far more research is needed before we link this phenomenon to illness in animals'. Environmental activists in Punta Arenas are sure it is because the regional government is trying to develop the Magallanes region, and obviously wants to avoid any bad publicity. Though a number of holiday tours have been cancelled, there seems to have been little impact on the economy. Press reports that property prices have slumped in the region are incorrect.

Bedrich Magas has found himself in the role of an environmental Cassandra, having been one of the few local scientists to take the ozone issue seriously. Like Sherry Rowland a hemisphere away, he has been thrust into the limelight because he understands the complicated science which underscores the issue. After the AAOE scientists left Punta Arenas in October 1987, he started to lecture local Scouts, farming and trade associations about the possible impact of Antarctic ozone depletion on Chile. His desire to provide education was motivated by the strange fact that he had seen a photograph of part of a circular piece of ER-2 test equipment described as an ozone hole in a local newspaper. In time, as word spread about 'the ozone man' at the university, the local media would beat a path to his cramped office in the Engineering Building whenever they needed further information. He became an interpreter of the complexities of ozone research for the general public, and, as a result, came to the attention of the local government.

It is a curious thought that if this had happened a decade before, Bedrich might well have disappeared in a football stadium in Santiago. But by the late 1980s, a wind of change had swept through Chilean politics. Virtually the last act of General Pinochet before he stepped down as president in 1990 was to ratify the Montreal Protocol. Following the first democratic elections that month, the new government promised action on the ozone issue – but nothing came of it, at least in terms of additional funding for Chilean scientists. The visit of the *New York Times* journalist Nathaniel Nash in the austral winter of 1991 led to the first unwanted international publicity for the Magallanes region. 'UNEASE GROWS UNDER THE OZONE HOLE' was the headline for his lead story in the paper's environment page on 23 July. The strapheading beneath it was 'In Southern Chile sunglasses and sunblock are in style'.

In the next couple of years, nearly forty other foreign journalists visited the area. The first that Britons heard about it was in November 1991 when the *Financial Times* ran a story entitled 'RABBITS BLINDED BY THE OZONE HOLE'. As recently as January 1993, the *Wall Street Journal* had a front-page story entitled 'SHEEP AND TREES ARE ACTING STRANGELY AT "END OF THE WORLD"', which quoted some of the apocryphal stories of the problems caused by ozone depletion. The net result of all this publicity is that the government of Magallanes has had a serious image problem on its hands, thanks to the ever helpful Bedrich Magas, who has assisted all the journalists from Nathaniel Nash onward.

He was soon branded an 'eco-terrorist' and these accusations intensified when he started to organise protests against a wood-chip project on Tierra del Fuego. A Japanese logging company was planning to cut down forests and ship the wood across the Pacific. On the ozone issue, however, he is more phlegmatic. 'As a scientist, I can help look at the problems posed by the ozone hole,' he said by way of justification. Though he has been portrayed as a fanatic, he is simply trying to get the local government interested in an issue which it has tended to ignore.

He is also well aware that many of the reports in the media are scaremongering and have led to a deliberately false picture. The citizens in Magallanes are not fighting an epidemic of skin cancers, nor are they battling against the ravages of 'bug-eyed bunnies', as *Newsweek* described them in its inimitable style. But he simply cannot believe that some of the weirder occurrences are coincidental. And he provides a clear and obvious rationale for further research into the subject. 'In Magallanes, we are living in the situation today which the northern hemisphere mid-latitudes will find themselves in ten years,' he said. 'The incubation of health factors should be studied as soon as possible so that we can apply the results and help prevent problems in the northern hemisphere.'

The scientific basis for damage to human health from increased exposure to ultraviolet radiation is reasonably well known. Physicists split the ultraviolet band into three groups according to wavelength and label them, for convenience sake, A, B and C. The B in UV-B originates from the fact it is 'biologically active'.

Deadly Springtime

The letters A and C were added later for consistency. The normal unit of measurement is the nanometre, defined as a millionth of a millimetre, and will be used here. Visible light may be found between 700 and 400 nanometres, the latter comprising the violet part of the spectrum. The ultraviolet spectrum is composed as follows:

UV-A 320–400 nanometres
UV-B 280–320 nanometres
UV-C less than 280 nanometres.

The most dangerous part of the ultraviolet spectrum has the shortest wavelength, but luckily, UV-C is almost totally absorbed by stratospheric ozone and will only ever be experienced by people who find themselves exposed to arc-welding or specialised laboratory equipment such as tungsten quartz lamps. UV-A had been thought to be relatively harmless but it has been implicated in skin damage and the ageing process. In particular, UV-A is used in most sunbeds and can harm the deep basal layers of the skin where cells are renewed. Skin fragility syndrome, an unpleasant blistering, has been reported among people who use sunbeds more than three times a week.

Although most UV-B is absorbed by the ozone layer, enough reaches the surface to cause noticeable effect. As any sunbather foolish enough not to use sun cream will notice, the first effects occur almost straight away as a distinct reddening of the skin, technically referred to as erythema. Similarly, snowblindness (keratitis) will also be experienced by those without eye protection. Erythema results from the release of a chemical which is triggered by increased levels of UV-B and causes the subcutaneous blood vessels to dilate, which produces more blood flow near to the skin surface. Scaly red spots known as solar keratoses (or, more commonly, sunspots) sometimes occur and are another dermatological warning of too much exposure to the Sun. All these effects are transitory and the skin reverts back to its normal state.

Prolonged or repeated exposure to the Sun, however, can lead to cumulative damage to what is, after all, the largest single organ of the body. To some extent, our skin is 'designed' for the latitude at which we live: it affords us a basic protection against levels of sunlight we would reasonably expect to experience.

How Soon Is Now?

The outermost layers of the epidermis contain pigment cells, or melanocytes, which produce melanin, the skin's natural tanning shield. Most fair-skinned people cannot produce sufficient melanin to afford protection against increased levels of sunlight. And though dark-skinned people have greater built-in protection, even they will exhibit signs of progressive skin damage. Anyone who has seen outdoor workers in, say, Mediterranean countries will be surprised to see the extent to which people who do not otherwise look old have wrinkled, blotchy skin which has lost its elasticity.

Cancers of the skin can be classed into three basic types from the depth within the skin layer where they are formed. The most destructive are associated with the production of melanin: malignant melanomas have a marked propensity to metastasise and spread to affect other organs of the body. Melanomas generally look like moles or birthmarks and can grow appreciably in a matter of weeks. For some reason which is not well understood, they appear most commonly on men's backs and on women's legs, presumably the most intermittently exposed body surfaces. There is also a marked correlation between outbreaks of malignant melanoma in later life and episodes of intense sunburn in childhood.

There are two other forms of skin cancer which are easier to treat. Within the epidermis are squamous cells near the top and basal cells at the bottom. Basal cell cancer is sometimes called 'rodent ulcer' because it literally eats away at the skin most commonly found on the face and neck. This form of cancer rarely spreads internally. Squamous cell cancers can also spread like melanomas but they are easier to treat. They usually appear as lumps, sores or scaly patches of skin which do not readily heal.

Exposure to sunlight is the single most important factor in the development of these cancers of the skin. An increase in their numbers due to Ozone depletion would take many years to manifest itself in the population at large. But the cases of badly sunburned teenagers in October 1992, at a time when the Sun was low in the sky, is significant. One would not reasonably expect to encounter erythema in Punta Arenas in springtime, as the local dermatologist was only too aware.

It is already growing dark and yet the surgery is still crowded. At first glance you might not expect that a grubby-looking apartment block on one of Punta Arenas's lesser thoroughfares, the Romulo Correa, would house a dermatology clinic. Yet climb the stairs to Apartmento 32, and inside you'll the find the forced quiet of suffering and self-pity which is instantly recognisable as a doctor's waiting room anywhere in the world. As always, Dr Jaime Abarca will be lucky to get home before 9 p.m., for he is the only dermatologist in the region and has his work cut out for him. In recent years, as concerns about ozone depletion have intensified, so has his workload.

Originally from northern Chile, Abarca came here after he graduated from the University of Chile in Santiago in 1986. It was through hearing about Bedrich Magas and his experiences with the NASA aircraft campaign that he became interested in the possible effects of the ozone problem. In December 1987, he won a six-day scholarship given by the American Academy of Dermatology to attend its annual meeting in Washington, DC. He was able to talk at length about the biological implications of ozone depletion with leading dermatologists in the field. Most recently he has been on a three-month sabbatical at the Mayo Clinic in Rochester, New York State, to learn new and improved techniques in dermatology. 'Since I'm not in a university or academic centre, it's important for me to learn the latest developments in treatment,' he says.

In the late 1980s, there were eight to twelve new cases each year of skin cancer in the Magallanes region. This was roughly the statistical average that would be expected for a population of around 100,000 at the same latitude south as Britain is north. As he spends half his time at the local hospital, Abarca has tried to ensure that there has been a systematic 'epidemiological surveillance' of these cases. If there were any glaring anomalies which might have resulted from ozone depletion, he believes he would have detected them in their earliest stages of development. Yet Abarca is hampered because there are no complete records of skin-cancer incidences before 1987, and, despite his interest in the subject, it remains a low priority for the regional government. In 1989, he asked the Ministry of Health to support his study financially, but, though the medical authorities agreed that it was a worthwhile project, no funds were forthcoming. Abarca has tried to maintain ongoing studies of the past cancer incidences by checking up on outpatients.

Lately, Abarca has become used to patients blaming each and every strange skin blemish on the ozone hole. In the six springs that he has been in Punta Arenas, the number of patients reporting ailments has dramatically increased. 'I'd say that about 90 per cent of the cases are not photo-induced but result from allergies to things like pollen,' he said. By the start of the 1990s, Abarca had about forty patients who had chronic, though benign, photodermatoses. These included the usual sun-related symptoms such as polymorphus eruptions and blotches which were largely restricted to their faces, ears and hands. 'Over the last two years, most of them have had more severe, more intense symptoms,' he told me in October 1992. In 1991, as the first reporters descended on Punta Arenas, Abarca had yet to be convinced that the ozone hole could affect the health of the population at large, but a year later he said: 'There is no doubt ozone depletion will cause increases in skin cancer. The only doubt is when and where they will occur.'

In November 1992, a team of half a dozen researchers from Johns Hopkins University in Baltimore visited Chile to perform a pilot study on the effects of increased ozone depletion, particularly in the Magallanes region. Funded by both the US Environmental Protection Agency and the national Ministry of Health in Chile, the study was to assess the seriousness of the situation and suggest the best way to monitor systematically future trends in health effects. 'We'd read the reports in the press about bug-eyed bunnies and so forth,' said ophthalmologist Dr Oliver Schein, the team leader. 'We wanted to investigate the reports of acute disease and do some planning for a long-term study.' In 1985 and 1990, his group had already performed clinical studies into the effects of prolonged exposure to sunlight by fishermen off the coastal waters of Chesapeake Bay to determine if there had been any notable health damage. The only possible effects seemed to be an above-average incidence in cataracts. By far the most important factor in such eyesight deterioration was age, as would be expected for repeated and cumulative exposure to extra sunshine reflected off the ocean surface.

The Johns Hopkins doctors spent nearly a month in Chile and

Deadly Springtime

more than a week in Punta Arenas, where they interviewed Jaime Abarca and Bedrich Magas, among others, at some length. They listened to the fanciful stories about health effects and investigated for themselves animals at remote *estancias* and medical records in local hospitals. Schein and his colleagues are not convinced that there have been any noticeable health effects thus far. 'The bottom line is that there is no evidence for acute health effects on humans and animals,' he told me when the final report which chronicled their findings was submitted to the Environmental Protection Agency six months later. 'Let's say that there are grains of truth around the ozone hole,' Dr Schein wryly commented on some of the more fanciful reports from southern Chile.

Of the animal effects, more than 200 sheep were examined at five *estancias*, and all of them exhibited some eye problems. About three-quarters had experienced 'pink-eye' since the late 1980s, and around 13 per cent appeared to suffer from cataracts of a variety which the Johns Hopkins team suggest would not lead to permanent blindness. Some of the sheep examined exhibited keratoconjunctivitis – a Sun-related eye problem – but this seemed to be consistent with an infection known as *Chlamydia psittaci*, for which the JHU doctors reported that 'there is no known relationship between this infection and solar exposure'. A small sample of hares and rabbits were examined in Puerto Natales, some 250 kilometres north of Punta Arenas, and many had eye problems which seemed to be associated with a common pathogen called *Pasteurella multocida*. The team also examined a herd of thirty Hereford cattle near Punta Arenas and found an unusually high rate of conjunctival or eyelid squamous-cell carcinoma, a tumour which has been related to chronic UV-B exposure in the United States.

For the human health effects, the doctors examined the records of Dr Abarca and local ophthalmologists. Their report concluded: 'Comparison periods from recent years and seasons in which no ozone hole has been reported were used. No convincing evidence of a temporal association of acute ophthalmic disease with the ozone hole was found.' Dr Schein himself added: 'We could see no difference whatsoever for Sun-related diseases between time of peak exposure to ultraviolet or other times.' The only dermatological curiosity was a greater incidence of warts (verrucae) closer to the times when the ozone hole occurred, but 'the significance of this finding is uncertain', the doctors concluded.

On the issue of the sunburned teenagers who reported to Abarca's surgery, Dr Schein was less convinced by the symptoms reported. 'It is possible that the Sun could cause sunburn at that time,' he conceded, but overall believed that this was not as significant as Dr Abarca seemed to think. 'We have actual measurements of ultraviolet incidence when the ozone is at its thinnest and at most it's only 9 per cent higher than normal.'

So, on the evidence of the austral spring of 1992, at least, ozone depletion in southern Chile seems unlikely to have caused any health effects in the population at large. The report which Schein and his colleagues submitted to the EPA forcefully concludes on the subject of exposure to UV-B:

> Given the levels of excess exposure seen in 1992, the effect on cumulative annual exposure is minimal. If one were certain that this level of exposure would remain constant or decrease over the next decade, then it would certainly be unlikely that this level of exposure would be associated with a significant increase in chronic dermatologic or ocular disease. At this level of ozone hole-related UV-B exposure, it is clear that an individual living in Punta Arenas would receive a greater contribution to his annual UV-B exposure by travelling to Santiago in the summer than by remaining in Punta Arenas during the period of ozone-related exposure.

Nevertheless, Dr Oliver Schein acknowledged that there is still a basis for concern for if ultraviolet levels continue to increase – and for greater periods in the springtime – it is entirely possible that chronic effects would become apparent in the twenty-first century. 'The only way to establish this is to do a long-term study,' he said and recommended exactly that to the US EPA and the Chilean Ministry of Health. As part of their pilot study, the JHU doctors examined a small sample of fishermen, shepherds and hospital workers for signs of ocular and dermatological damage as well as immunosuppression. No differences were noted between the outdoor workers and those who spend most of their time indoors,

There are similar ambiguities in the documented rise of skin cancers and observed levels of UV-B reaching the ground throughout the rest of the world. But that does not mean there is room for complacency, particularly for the Chileans who live in closest proximity to the annual ozone hole. 'Nobody knows for sure what is going to happen in the future,' said Dr Jaime Abarca. 'We know we will have to live with this situation for a hundred years or more. I think it's going to get a lot worse before it gets any better.'

There can be few more emotive words in the English language than cancer. From the first time that James McDonald of the University of Arizona made a public connection with ozone depletion and increased skin cancer incidences back in the early 1970s, it has tended to be the focus for the most worrying prognoses about the issue. In a subject already brimming with red herrings and paradoxes, the health implications of ozone depletion are imbued with an additional capacity to shock because of their emotional impact. In many ways they are even more confusing than many complexities raised in stratospheric chemistry for at face value they seem decidedly ominous. It is already well established that skin cancers have been on the increase throughout the world, and estimates of future incidences have, to some extent, dramatically reflected these worrisome statistics.

The prognoses in recent years have become ever more apocalyptic. In the late 1980s, for example, the United Nations Environment Programme (UNEP) reported that for every 1 per cent decrease in global ozone levels, 70,000 new cases of skin cancer could result. There would also be increases in blindness caused by cataracts – as many as 100,000 new cases with each per cent depletion of ozone across the world. In April 1991, the US Environmental Protection Agency reported that there may be as many as 200,000 more skin-cancer deaths over the next fifty years in the United States alone as a result of today's ozone trends being extrapolated forward. A further UNEP report released in December 1991 concluded that a sustained 10 per cent decrease in ozone would be accompanied by 'considerably in excess of' 300,000 cases of skin cancer and 'up to 1.75 million cases of eye cataracts throughout the world every year'.

So it would seem that nothing short of a skin cancer epidemic is expected. Yet these analyses have an unstated social element which is quite often overlooked: they are 'worst-case scenarios' which assume that the human population at large will not change its lifestyles. Unlike dinosaurs or dodos, for example, human beings have a capacity to alter their habits quite markedly in the face of increased danger. Quite simply, if the population at large assiduously avoids exposure to the Sun at all times (or is at least sensible in dress and attitudes), then the skin cancer outbreaks detailed above will be largely avoided. Though ozone depletion will undoubtedly exacerbate matters, sensible behaviour will avert an epidemic of skin cancer.

There is some hope that health warnings are being heeded in Australia, for example, where skin cancer incidences have doubled in recent years. But there is an unstated social element to this statistic which has got very little to do with the development of the ozone hole. This increase in skin cancers is restricted to the light-skinned people who moved to the sunnier antipodean climes in the 1950s. Many thousands of Britons took assisted passages to Australia and are now paying the price for living in a country for which their skin was not 'designed'.

Worldwide, skin cancers are clearly on the increase, but, once again, there is a hidden social dimension. Suntans are linked with perceptions of healthy living, affluence and leisure. In the 1940s, tuberculosis sufferers were told that 'heliotherapy' was an excellent cure and that the golden rays of the Sun were life-enhancing. 'Brown is beautiful' became an opportune slogan for the purveyors of suntan creams and an underlying reason why skin cancers have increased since the war. 'A whiter shade of pale' should perhaps now be the aim, for although many sun-cream manufacturers express concern and point out the dangers of Sun exposure, they are simply exploiting the situation by racking up excessive profits on the highest-factor creams.

Some medical authorities have stated outright that a two-week holiday in the Sun has the greater potential for harm than the cumulative effects of ozone loss in the decades ahead. For pale-skinned natives of northern Europe and the United States, a two-week intensive burst of sunshine is the most dangerous thing they could ever experience, dermatologically speaking. If the skin cells are repeatedly damaged by exposure without being given time to repair themselves in between, the skin 'forgets' how to repair itself and malformations result, thereby causing a cancerous growth. Indeed, the wordwide average 3–5 per cent increase in malignant melanomas since the early 1960s can be directly attributed to holidays in the Sun. In the 1980s, for example, Swedish doctors revealed a remarkable statistical correlation between availability of cheap package tours to Spain and Portugal and the outbreak of skin cancers. In 1993, the Department of Health in Scotland reported an increase of 80 per cent in malignant melanomas over a decade for the same reason. In the UK there were 3,500 cases of malignant melanoma in 1990, compared to 1,682 in 1974. There were 1,300 deaths, compared to 743 in 1974. There is also some evidence that if children are badly sunburned under the age of twenty they will suffer from repeated outbreaks of melanomic skin cancer later in life.

Critics of the science of ozone depletion have focused on this fact as 'proof' that there is no risk associated with it and that there will not be greater incidences of skin cancers in the future. Our old friend Dr Dixy Lee Ray, for example, liked to point out that a 4–5 per cent increase in ultraviolet (which prompted the apocalyptic EPA figure quoted above) is nothing compared to the 22 per cent increase one would expect to experience as one travelled from Washington, DC, to southern Florida. She also pointed out that a journey from either pole to the equator results in an increase of ultraviolet of 5,000 per cent. (Doubtless, critics of ozone depletion will also use the Johns Hopkins finding about southern Chileans moving to Santiago.)

Yet this line of reasoning fails to take account of the fact that ozone depletion will lead to a *permanent* increase in the ultraviolet levels being experienced on the ground. Skin which is 'designed' for one latitude today may find itself being exposed to levels of ultraviolet which are normally encountered nearer the equator tomorrow. In terms of the possible dangers, for many people it would be the equivalent of moving to a sunnier clime against which their skin would have insufficient protection. The only hope is that human beings will have the sense to spend as little time in the Sun as they possibly can.

* * *

How Soon Is Now?

Anyone visiting Bondi Beach in recent years could be forgiven for thinking that the beach bum was an endangered species. Where once splendidly muscled torsos would be on display, few native Australians now sunbathe or sit around without hats, towels or clothes covering their bodies. The threat of skin cancers, fanned by concern over ozone depletion, has reached prominence in the media – and with good reason, for skin cancers account for about half of all the cancers in Australia. It has even been suggested by some medical authorities that two out of three Australians can reasonably expect to suffer from non-melanocytic cancers during their lifetime.

In Tasmania, the state nearest to Antarctica, the incidence of skin cancers doubled between 1983 and 1993. Worst hit is Queensland, once described as 'the sunshine state' in holiday brochures around the world, but is now trying to shrug off the less appealing epithet of 'the skin cancer capital of the world'. Medical authorities believe that 7,000 more cases of melanoma will be reported each year and that 140,000 people will suffer from basal-cell or squamous-cell carcinomas in the 1990s. And in early 1993, the Australian Royal Society for the Prevention of Cruelty to Animals noted that vets in Queensland had reported a sharp increase in skin cancers in lighter-skinned animals, particularly cats around their ears, perhaps the least protected part of their anatomy. Previously there had been negligible figures of feline skin cancers in this part of the world.

As a result of these and other statistics, there is now a well-established educational campaign on the dangers of exposure to sunshine in the Australian media. School principals have distributed literature on an overall 'Sunsmart policy' which exhorts children to avoid the Sun at all times. Two of its most famous slogans are: 'Between 11 and 3, slip under a tree' and 'Slip, Slap, Slop – Slip on a Shirt, Slap on a hat, Slop on some sunscreen'.

There is also another extensive campaign in the media known as 'Me No Fry', which, as its literature declares, is specifically targeted at adolescents 'who represent a particularly appropriate group for intervention to increase the use of Sun protection measures . . . High exposure to the Sun at 10-24 years of age has been shown to be a strong risk factor for the development of superficial spreading melanomas when compared with a consistently low-to-moderate residential exposure throughout life.' At most risk are freckled adolescents with fair hair and skin. Statistically, they are forty times more likely to contract melanomic cancers than their dark-skinned and -haired friends.

Some schools in Australia already provide factor 15+ sun cream, as do many lifeguards on beaches. It is significant that sponsors of the 'Me No Fry' campaign are surfing organisations. There are also travelling clinics on some beaches, advertised as 'Spot Stops', where dermatologists will examine suspicious blemishes free of charge. Similar facilities are provided in New Zealand, and many schools there have brought their lunch hours forward to 10.30 a.m. to avoid exposure to maximum solar radiation by their pupils. Children are actively encouraged to wear T-shirts and hats at all times, and a special 'Anti-Suntan' clothing range is being marketed by the Cancer Council of New South Wales. The council reports that the results from all these various activities are encouraging and telephone surveys indicate that sunburn is much less prevalent than it used to be in all sections of the populace.

Prevention is most of the battle against skin cancer, as the medical authorities in other countries have started to realise. A 1992 report from the UK Medical Research Council, which considered the medical effects of global change in detail, concludes as follows: 'Aggressive health education campaigns need to be developed and evaluated for their effectiveness in getting these messages across: to date this has only been attempted in Australia.' Risk can obviously be reduced by the complete avoidance of sunbathing, limiting exposure and covering the skin at all times. A hat is recommended, particularly one of the wide-brimmed variety which provides the wearer with a 'personal shade zone' as he or she walks around. Anyone still wanting to sunbathe should most definitely avoid the two hours either side of local noon in the sunniest climes.

As for other remedial measures, various medical authorities advise that caution should be exercised when it comes to sunscreens, skin creams and sunglasses. In particular, anyone tempted to smother themselves in suncream and merely lie back thinking they have adequate protection is advised to think again. Many sun creams state a 'Sun Protection Factor', a concept introduced by Piz Buin in the early 1970s. (Interestingly, Piz Buin adverts in Australia show models with much lighter skin colours than in their European adverts.) The SPF gives a rough idea of how much protection is afforded against UV-B by multiplying the factor stated by ten to give the number of minutes which will elapse before most skin types will start to burn. The SPF number, however, does not take into consideration the cumulative effects of UV-A, although a number of manufacturers have introduced a star system to indicate protection against UV-A. Yet there is often no consistency in the way that the factors are evaluated and tested by different manufacturers.

People are advised to exercise caution and look at the packaging – if they can afford the cost, that is. The profit margins on sun creams are about 50 per cent and the expense of the highest SPF is prohibitive for a lot of holidaymakers. It is estimated that in Great Britain alone, sun creams account for some £110 million of revenue, a market which is growing at 4 per cent per annum. And though Boots (which accounts for just under 50 per cent of the market share) joined forces with the Cancer Research Campaign to promote 'sensible Sun behaviour' during the summer of 1993, three of its guidelines suggest the purchasing of sunglasses, sun creams and sun hats – all of which (surprise, surprise) they stock. Many of the claims made by manufacturers bring a new meaning to the phrase 'protection racket'.

Medical authorities may have an uphill struggle in persuading the public at large that a tanned skin is a damaged skin. A survey was carried out by the BBC television science programme *QED* in conjunction with *Good Housekeeping* magazine in early 1993. Of the people polled, only one in thirty preferred 'pale' to 'tanned' as their appearance; merely one in twenty avoided sunbathing in the middle of the day; and only one in six used a high-factor sunscreen. To get the message across, most leading dermatologists allude to the lyrics of Noël Coward: mad dogs and Englishmen should not go out into the midday Sun. And a slightly more downmarket – yet vivid – slogan for holidaymakers may be found on a sign outside a bar frequented by Britons in Benidorm: 'AVOID SKIN CANCER – Bevvy & Telly All Day!'

History provides a telling precedent for a change in lifestyle. In the eighteenth century, the well-to-do would gladly powder their faces with whitening powders made from toxic chemicals. The resultant poisoning of their nervous systems (for which 'lingering death' is the only suitable description) led to untold numbers of madness and premature deaths. Future generations will probably view our twentieth century obsession with 'heliotherapy' in similar disbelief.

Increased ultraviolet flux promises far greater impact than the disappearance of beach bums from places like Bondi, Malibu or Saint Tropez. For it will bring about far worse health effects which are not racially discriminating. All racial types are at risk from increased incidences of cataracts, which are formed when the crystalline lens of the eye becomes clouded and thus blurs vision. Prolonged exposure to ultraviolet light plays a major role in the formation of cataracts, and, to quote the Johns Hopkins ophthalmologist Dr Oliver Schein, 'there's more we don't know than we really know at the moment'. That said, it is well established that there are three basic types of cataract, each depending on whereabouts they are formed within the eyeball (the cornea, the lens or the retina) and it is also known that all three are more likely to form with increasing age. 'The cataract story is similar to non-melanomic cancers,' Dr Schein said. 'Cumulative exposure is by far the single most important cause.'

His study at Johns Hopkins University of East Coast fishermen showed that those who wore glasses or regularly wore hats significantly lowered their chances of being afflicted by cataracts. And while most ophthalmologists advise that ocular protection is advisable, beware the many claims of some sunglass advertising. People who wear spectacles, for example, will be protected from about 90 per cent of the incident UV-B upon their eyes: at best, sunglasses will increase that to perhaps 93 or 94 per cent. In other words, there is no point in buying horrendously expensive sunglasses in the belief that you will be afforded total protection. Most ophthalmologists suggest that people should avoid active Sun-seeking, particularly during the summer months.

As ozone depletion continues, infectious diseases may also increase because UV radiation represses the natural immunity responses in human beings. A UNEP report from December 1991 concludes that sustained ozone depletion 'places all of the world's populations at risk ... including possible increases in the incidence or severity of infectious diseases'. Most people are aware that cold sores – herpes simplex – are more prevalent during the summer. As a result of prolonged UV-B increases, diseases like measles, herpes and tuberculosis may become more prevalent and virulent. This also means that vaccinations will lose their potency. When antigens – substances foreign to the body which stimulate its defences – are injected through human skin that has been exposed to ultraviolet radiation, they seem to induce tolerance to disease rather than protection against it.

Increased exposure to ultraviolet light causes damage to DNA, the complex molecules which contain genetic information. Ultraviolet radiation contains just enough energy to break the delicate molecular structure of DNA molecules. A cell with damaged DNA cannot function properly because the genetic information contained within it is scrambled. One school of thought suggests that immune suppression starts when DNA within the cell absorbs UV-B. Others believe that it is due to absorption of UV-B by urocanic acid which is one of the most abundant substances present in the outer layers of the skin. Whatever the exact cause, immunologists have known for a long time that if a tumour in a white mouse is transplanted into a healthy mouse, the tumour will normally be rejected. But if the unfortunate rodent is then exposed to increased levels of UV-B and the transplant is repeated, the tumour is more likely to be taken up.

What of humans? There is increasing evidence that immunosuppression is reduced by UV-B exposure in all the human population, irrespective of skin type and ethnic origins. Testifying before the US Congress in November 1991, Margaret Kripke of the University of Texas outlined this possible effect upon the population at large. 'There were no differences in the immunosuppressive effect of UV radiation in persons with white, brown or black skin, indicating that pigmentation is not protective against this form of UV-induced immunosuppression. Thus, the population at risk of immunological damage from UV radiation is much larger than that at risk of developing skin cancer.'

And most worrying of all is a possible linkage with the spread of the AIDS virus. Some immunologists have suggested that the relatively high incidence of skin cancer in patients with HIV is significant. Most epidemiologists will not be drawn publicly on this matter, but off the record, many believe that it cannot be discounted as a possibility. This is by far the most insidious long-term effect that may follow from ozone depletion, and urgent research into this aspect is needed.

In early 1992, Bedrich Magas unexpectedly received an instrument with which to measure the amount of UV-B reaching the ground in Punta Arenas. It arrived in the post as a consequence of the international publicity about the bug-eyed bunnies and his obvious interest in the subject. The Solar Light Company of Philadelphia had sent him a radiometer with which he could measure that part of the ultraviolet which causes sunburn during the following austral spring. After the instrument had spent an inordinate time in customs, Bedrich was indeed able to use it and, at face value, his measurements indicated something serious. On certain days, *twice* as much UV-B was recorded when the ozone hole passed above. Yet this observation is not as portentous as it may at first seem, for UV-B measurements are notoriously misleading.

Scientists who monitor UV-B at ground level are hampered by many constraints. They have to take into consideration many factors which cannot be represented with any degree of accuracy: for example, the natural variability of stratospheric ozone, the effects of clouds and atmospheric dust and the distinct effect of localised pollution. The characteristics of the instrument used have to be well known and it needs to be continually calibrated if any meaningful comparisons can be made. For these many reasons, it is not an easy matter to interpret most UV-B data.

Although UV-B levels have unambiguously increased owing to ozone depletion over Antarctica, there are parts of the world where they seem to be on the decrease. Ground-based ultraviolet measurements are infinitely more complex and contentious than even the skin-cancer statistics quoted previously. A prime example is the 'Scotto affair', which resulted from publication in the late 1980s of ground-level UV-B measurements from the United States. After analysing the levels of UV-B measured at a variety of sites on the ground between 1974 and 1985, Joseph Scotto of the National Cancer Institute and his colleagues reported that they were actually *decreasing* across sizeable portions of the country. However, most of the instruments used in the Scotto survey were located at military or commercial airports, where increased air-traffic pollution is likely to have masked the underlying picture of how much UV-B was being let in through the ozone layer.

Subsequent work by Shaw Liu of the NOAA Aeronomy Lab revealed that increases in sulphate emissions due to pollution have reduced UV-B incidence anywhere between 6 to 18 per cent across the United States. Other observations show that cleaner air transmits ultraviolet radiation more effectively: results published in 1992 showed that twice as much UV-B was reaching the ground in New Zealand as in Germany (at 45°S and 49°N respectively). This had nothing to do with stratospheric ozone depletion, as some environmentalists claimed; rather, German clean-air laws had yet to show any effect in the atmosphere. To some extent, tropospheric ozone produced by the photochemical effect of sunlight upon pollution protects the surface. Interestingly, measurements made by Mario Blumthaler at the University of Innsbruck have shown that UV-B levels have been increasing by roughly 1 per cent per annum since 1984 in the Austrian Alps above much of the pollution. Taken all together, these observations reveal that the underlying picture of increased ultraviolet is extremely complex.

Instruments used to measure UV-B incidence are, in some sense, the big brothers of the Dobson instruments. But where the latter measure discrete wavelengths to compare absorption by stratospheric ozone, ultraviolet radiometers measure rather greater ranges of wavelength. Accordingly, these instruments are more complicated than the venerable Dobsons, but thanks to the electronics revolution of the 1980s, they are much less cumbersome. The instrument which Bedrich received is of the Robertson–Berger type and makes one measurement in the space of a few seconds across a small part of the ultraviolet spectrum centred on the region where the effects of erythema, or sunburn, are most marked. In fact, the express purpose of the Robertson–Berger instrument is to produce 'a climatology of sunburn', and they are limited in their suitability for the determination of trends in ambient UV-B (as the Scotto survey found in the late 1980s). And, as we will see, the Robertson–Berger instruments are not ideally suited for the task since they measure just one small portion of the ultraviolet spectrum and are not well calibrated.

The effects of different wavelengths of ultraviolet radiation are not equal. Different biological systems respond differently to the UV-B spectrum and each is most susceptible to damage at a specific wavelength. The spectral response of sunburn is different from that of, say, melanoma and is as different again from that which will damage phytoplankton or plants. Each phenomenon has its own 'action spectrum' to which instruments have to be tuned to gain an understanding of how they will respond to increases in UV-B. The Robertson–Berger instruments mimic, so far as possible, the response of sunburn at the shortest wavelengths in the ultraviolet spectrum.

Nevertheless, during the austral spring of 1992, Bedrich Magas attempted to make systematic, scientifically valid measurements of the erythemal UV-B reaching the ground in Punta Arenas. 'Data without a baseline make no sense,' he said. 'So we compare the data record obtained at the same latitude north as we are south.' He thus obtained systematic observations correlated by the instrument's co-inventor, Daniel Berger, in 1981. They are a statistical average of measurements throughout the daylight hours at different times of the year.

Throughout the course of a day, the ultraviolet spectrum observed at the ground will change: when the Sun is low in the sky, radiation has to pass through a thicker portion of the stratosphere than if the Sun were directly overhead. This will reduce the amount of biologically active ultraviolet at the shortest wavelengths reaching the ground. The data to which Bedrich compared his own observations were obtained at specific times of the day in a variety of weather conditions at a comparable point in the season. He chose data from Belsk-Duzy in Poland, which at 51.8°N nearly matches the latitude of Punta Arenas at 53°S. Both places are also virtually at sea level. 'It's not a perfect equivalent,' he said, 'but it is near enough.'

The initial results showed that during September and October 1992, there were roughly two to four times more UV-B reaching the ground in Chile than in Poland. These observations correlate well with ozone column measurements obtained from the ground. On 4 October, the Nimbus 7 TOMS observed a column abundance of 185 Dobson units above Punta Arenas. 'On that date there was 3.7 times more UV-B causing biologically active erythema,' Bedrich said. One rule of thumb is that a 1 per cent change in the ozone column would result in an increase of 2–3 per cent UV-B. 'Our data agrees very well with that,' Bedrich said. 'There was on average a 2.2 per cent correlation between ozone loss and UV-B increase. On 4 October it was 5.5 per cent, which is more than a little worrying.'

As a result, Bedrich has recorded his observations on a specially dedicated phone line so that concerned citizens can be informed of the UV-B flux. After making measurements throughout the day, he will estimate the UV-B which has reached the ground in Magallanes. In this he is not alone, for weather services around the world have started to make UV-B forecasts to remind people of the dangers of sunburn. But some scientists question the usefulness of such forecasts, as the experience on two separate days in October 1992 in Punta Arenas show. Tuesday 20th was cloudy and windy in southern Chile; the next day was bright and clear. One would have thought that there would be more UV-B reaching the surface on the Wednesday, with less in the way of meteorological obstacles. Yet Bedrich's Robertson–Berger observations clearly show the opposite: 3.48 times more UV-B was received on the cloudy day than on the clear day.

Part of the problem in releasing data from Robertson–Berger instruments is that they do not give any indication of the total ultraviolet flux reaching the surface at any one time. To make these rather more complicated measurements requires much more sophisticated spectroradiometers. Over the space of a few minutes they will scan the whole of the ultraviolet spectrum, usually at steps of a nanometre or so. Unlike the Robertson–Bergers, these 'spectral' instruments return an absolute measurement of the UV flux. They also have the added advantage of an internal calibration lamp which allows their operators to check on the instrument's performance right across the UV spectrum. The main reason why many authorities have not used these instruments thus far is their prohibitive cost: they are in the $20,000 to $50,000 range, whereas a simple R–B radiometer costs around the $500 mark.

Many scientists around the world are now involved in setting up networks of these spectroradiometers of determine how the total UV flux is changing as a result of ozone depletion. Brian Gardiner of the British Antarctic Survey (joint author with Joe Farman and

Jonathan Shanklin of the original ozone-hole paper in *Nature*) is now helping to establish a dedicated network of instruments across Europe, with some as far north as Norway and others as far south as Greece. 'The way forward, as I see it,' he said, 'is to establish a network of these spectral instruments.' He believes that having absolute measurements will enable intercomparisons of data from around the world to become more accurate and valid. Once these comparisons can be made, it will be possible to quantify the processes that can affect the transmission of ultraviolet light throughout the atmosphere. To some extent it would then be possible to calculate exactly what the total UV-B flux at any time would be if factors such as ozone column abundance, Sun angle and cloud cover were known.

'We would like to show convincingly that we can account for the passage of radiation through the atmosphere and how it relates to our measurements at the ground,' Gardiner said. 'The variations are very different at each wavelength. There will be bigger variations at shorter wavelengths because that is where ozone absorbs most strongly.' And, thus, spectral instruments will give the best indication yet of increases across the whole of the ultraviolet spectrum. In June 1993, for example, John Frederick of the University of Chicago released observations that as much as 50 per cent more ultraviolet radiation at 305 nanometres had been measured at Ushuaia on Argentinian side of Tierra del Fuego after the Antarctic vortex had broken up. 'These are the first measurements to show a large increase in UV radiation over a populated area of the world,' Frederick told reporters. What was perhaps more worrying was that these measurements were obtained in December 1990, after the most intense period of ozone depletion had occurred, and with less chlorine loading of the stratosphere than now present.

'Our quantitive understanding of the biological effects of the UV radiation is far behind our understanding of ozone depletion,' Frederick added. 'Maybe this report has biological relevance and maybe not. There is still a lot to learn about biological effects.'

The only place in the world where it has been firmly established that there is a marked increase in UV-B is the last great unpolluted continent. Given that each spring nearly two-thirds of the ozone effectively disappears above Antarctica, it hardly comes as a surprise to learn that surface UV-B levels there have increased dramatically. John Frederick of the University of Chicago used a narrow-band spectroradiometer to make measurements from McMurdo Sound. In October 1990, his measurements showed that there was three times as much UV-B at the wavelength of 305 nanometres reaching the ground as would normally be expected. In 1990, Dobson observations showed that stratospheric ozone levels were 145 Dobson units, some 100 units less than they had been in October 1988. Frederick and his colleagues concluded that the resultant UV flux 'may be the largest experienced in this region of the world since the development of an ozone layer on Earth'.

How far does this obvious encroachment in enhanced ultraviolet levels stretch northward? This question is being addressed by a project funded by the Chilean national government and is known as Fondecyt 1143 after the principal funding agency (an acronym of the Spanish for National Foundation for Scientific Technological Development in Chile). It employs a number of instruments – spectroradiometers and others with narrow windows at UV-B wavelengths – to monitor ultraviolet flux from the north of the country (18°S) down to Chilean Antarctic stations on the Palmer Peninsula at 64°S.

'We have the whole spectrum covered across the whole of the country,' said Professor Sergio Cabrera, the project leader, who is a biologist at the University of Chile in Santiago. As the north of Chile is rather more mountainous, the emplacement of instruments at different altitudes in close proximity will ensure that measurements can account for obvious altitude effects which would otherwise swamp latitudinal ones. 'Before we can assess what the effects of the ozone hole are, we need data,' said Dr Cabrera. 'We're now getting it.'

Although the network of instruments was still being tested and calibrated in October 1992, the northern extent of the ozone hole was clearly discernible. Dr Cabrera himself had spent time in Antarctica using portable instruments to check on TOMS measurements. 'All our measurements correlate well with the ozone hole,' Cabrera said. These observations will help assess the extent of change in UV-B flux across the whole of the country as a result of ozone depletion. 'Here in Chile we have the responsibility to show politicians around the world what is happening so they can prepare policy more effectively,' Cabrera said.

The Fondecyt 1143 project was set up in March 1991 with funding for three years, although measurements did not begin until early 1992 because of unexpected delays. Of the $154,000 allocated to purchasing spectroradiometers from the United States, nearly $40,000 has gone on sales tax and customs charges. Sadly, Chilean scientists are subject to the same funding pressures as their colleagues elsewhere in the world.

The most pressing impact of Antarctic ozone depletion concerns organisms which are, for the most part, unknown and unseen by the general populace. Phytoplankton are minute marine life forms which derive their name from the Greek word for wandering: collectively, they have been called the 'pastures of the oceans'. The food chain – the biological hierachy of feeding – ultimately rests upon them. In Antarctica, in particular, the local populations of whales, krill and fish depend on the annual production of phytoplankton for food. Once again, stories emanating from the southernmost continent about the role of these tiny creatures hint at a global disaster in the making.

One central issue in discussions of Antarctic phytoplankton involves the contentious issue of how ultraviolet light is absorbed by water. Not all wavelengths are absorbed equally, and those radiations at the shortest wavelengths will penetrate the furthest because they are that much more 'energetic'. Normally, most solar radiation will penetrate only a few metres because most sea water is not clear and will absorb much of the transmitted radiation. In any case, many aquatic life forms can naturally 'handle' levels of UV-B radiation.

Phytoplankton, for example, have photoreceptors which determine the incident UV-A radiation by sensing in the blue part of the visible spectrum. If there is an increase in ultraviolet flux, the phytoplankton will descend to a greater depth of water (if prevailing currents allow), but the phytoplankton will still need to access less energetic ultraviolet radiation (at longer wavelengths), which they require for photosynthesis. One point which tends to be overlooked in press reports is that water which contains phytoplankton tends to be turbid, and this turbidity serves to scatter incident ultraviolet radiation more effectively. So, to some extent, phytoplankton are protected against UV-B flux by their own scattering of light underwater.

Phytoplankton have an inbuilt 'sunscreen' which consists of amino acids, and laboratory experiments have shown that different species produce slightly different sorts of sunscreens of varying degrees of resistance to UV-B. In other words, not all phytoplankton are sensitive to the same extent to increased ultraviolet flux. Other laboratory experiments to date have not been able to reproduce the exact 'dosages' across the UV spectrum which will result from ozone depletion in Antarctica. Similarly, the cumulative effects of the biologically active ultraviolet have not been measured with any degree of accuracy on different aspects of the cell integrity within the phytoplankton themselves. Nutrient uptake, protein content, sunscreen-pigment production and so on each have their own different 'action spectrum' which requires investigation. A great deal remains to be learned about the way in which phytoplankton will respond to increased UV-B flux in laboratory experiments.

In the field, however, there is now a well-established correlation with phytoplankton destruction by the ozone hole. Over six weeks in the austral spring of 1991, two scientists from the University of California at Santa Barbara quantified phytoplankton exposure to ultraviolet from the research vessel *Polar Duke* in the Bellingshausen Sea off the Antarctic continent. Ray Smith and Barbara Prézelin suspended a variety of phytoplankton species in polyethylene bags at different depths underwater to test their response to the increases in UV-B as the ozone hole passed above. Some of the bags contained filters to act as shields for specific wavelengths of UV-B, while others did not. The former thus served as a control against which the effects of ultraviolet could be gauged.

Overall, the results showed that anything between 6 and 12 per cent of phytoplankton production was inhibited by increased UV-B which resulted from the ozone hole. The *Polar Duke* made regular trips to coincide with the passage of the ozone hole directly above, and by use of a special underwater spectroradiometer, Smith and Prézelin observed that UV-B penetrated as far as 70 metres below the surface. Normally, springtime in Antarctica brings with it a 'bloom' of biological activity as marginal sea ice melts and phytoplankton reproduce in the relatively stable water which remains.

With truly consummate bad timing, the ozone hole opens at the worst time of year 'when most of the population is in the nursery', as Barbara Prézelin told reporters when her findings were published in *Science* in February 1992. Privately the UC Santa Barbara researchers acknowledge that the destruction of phytoplankton could be much higher than the observed 6–12 per cent. They are continuing with a series of experiments in Antarctica to investigate more extensively this very issue.

Many press reports of the ramifications of this observed reduction in phytoplankton reproduction veered towards environmental apocalypse. The disruption of the food chain which may result has been likened to the collapse of a pack of cards, with consequent pronoucements that fishery stocks around the world are at risk. Yet many photobiologists exercise caution in extrapolating these findings because there is much to be learned about the sensitivity all marine species will exhibit to increased ultraviolet flux.

So it is a little premature to make predictions about the effect of Antarctic phytoplankton on the rest of the world's marine populations. One US scientist interviewed for this book went so far as to say:

> This may sound a little cavalier, but in one sense, the 'best' place you could have an ozone hole is over Antarctica. Despite the fact it contains high productivity areas – the southern ocean has wonderful productivity – no one in particular depends on it for their food. If all the animals and plants above 60°S died, it would not have a dramatic effect on the food supply of the Earth. But it would be an unmitigated disaster for the Antarctic ecosystem.

This last point was reiterated by Dr Sergio Cabrera of the University of Chile, leader of the Fondecyt 1143 UV-B measuring project. 'This increase in UV-B is a very serious threat because the Antarctic food chain is very simple,' he told me in Punta Arenas shortly after a field trip to the Palmer Peninsula in October 1992. 'We expect that several species cannot synthesise protective sunscreen molecules and they will be killed for a long time.' And others point out that phytoplankton themselves have another important role in the terrestrial biosphere: they also serve to 'fix' carbon dioxide levels in the atmosphere to offset its warming influence on the Earth. Extrapolating how this may affect the rest of the world cannot be attempted with any degree of accuracy at the moment, but it nevertheless represents an additional cause for concern.

Phytoplankton destruction is obviously not the only concern for the Antarctic ecosystem. To understand how life forms may be affected by continuing ozone depletion – and to predict its effects with a greater degree of accuracy – requires more information obtained over many years. Only then will the underlying trends become apparent. Writing in the journal *Antarctic Science*, Deneb Karentz of the University of California at San Francisco summarised biologists' difficulties with crisp simplicity: 'Analyzing the genetic and ecological effects that increased UV exposure may have had on Antarctic organisms is hindered by the lack of UV photobiological research conducted before the development of the ozone hole.'

At the start of 1992, the US National Science Foundation began a six-year programme to look at Antarctic wildlife. In particular, the Long Term Environmental Research Program will observe the migration and breeding patterns of penguins. Penguins, like humans, are near the apex of the food chain and any marine food shortages below them in the feeding hierachy would be very serious indeed. One researcher has likened the role of penguins to that played years ago by canaries in mines – to warn of danger. By the time the survey is complete in 1998, the cumulative damage of ozone depletion on penguins will perhaps be readily apparent to the species at the top of the food chain – us. By then, human beings may have a better idea of how ozone depletion worldwide will affect crops and plants upon which our own food supplies ultimately depend.

The fax from Greenpeace in Santiago described him as a botanist. But the ruddy-complexioned man named Ricardo Borrquez Schultz is actually a market gardener who has lived and worked in Punta Arenas all his life at the family home on the Avenue Senoret. It was clear that the environmental pressure group wanted to exaggerate the importance of his noticing strange changes in the plants, flowers and shrubs which he tends, imbuing them with a spurious expertise and non-existent objectivity. Like most of the evidence for ozone-related damage, Schultz's findings have little basis in scientific reality.

Jardin Schultz is, by European standards, a small enterprise, more like a municipal allotment than a thriving commercial garden. Nevertheless trade is brisk in this part of the world. Plants and daffodils are grown in small rows at the rear of the house and there are two large greenhouses within which more exotic species may be found. As always, the blustery winds in the Magallanes region often crack window panes unexpectedly and many of them are covered in nylon as temporary replacements.

As someone who is well accustomed to the peculiarities of the local climate, Schultz noticed during the mid-1980s that some plants would very quickly turn yellow and die unexpectedly. From 1987 onwards, Schultz, and another commercial gardener, Lothar Blunck, were certain that there were discernible changes during the springtime. 'This is the period when the plants blossom and they can easily be damaged,' Schultz explained. The large greenhouses contain limited UV protection in the form of those temporary nylon sheets, and increasingly, Schultz noticed that plants beneath the broken panes seemed to suffer more damage. The nylon sheeting also started to crack before its guarantee expired.

Could it be caused by the ozone hole which opened each springtime? Or was it a problem with the soil? As a matter of routine, Schultz has to send soil samples each year to the laboratories of the local Agriculture and Farming Service to renew his trading licence. 'If there are any diseases, then I would be stopped from selling plants,' he said. But each spring, the answer was the same: the soil was normal. Yet Schultz was insistent that the plants' flowering patterns had changed, as had weather patterns in this part of the world. When I interviewed him in October 1992, he was concerned that crocuses and alyssa had bloomed six weeks earlier that spring than they had ever done previously. 'I have no explanation for why this should have happened,' he said. It may seem churlish, but to my untrained eyes, many of the plants exhibiting damage looked as though they needed to be watered a little bit more often. It is fairly obvious that the damage seen had little to do with increased ultraviolet flux.

How Soon Is Now?

But what of hard, scientific information on the effect of ozone depletion upon plants elsewhere in the world? The results obtained thus far are often ambiguous and contradictory. Laboratory experiments are limited in their applicability to the real world because of the complicated inter-relatedness of factors which cannot yet be modelled. This much is amply demonstrated by laboratory work carried out at the same latitude north as Punta Arenas is south.

Attaching a 'BEWARE RADIATION!' sticker to a plant experiment which simulates increased levels of UV-B may seem like overstating the case. But for Geoff Holmes of the Botany Department at Cambridge University, it is born out of expediency: the warning serves to discourage vandals from damaging his experiments which are carried out in the grounds of the department building. Some of the delicate electronics which he employs to perform control measurements have been vandalised. It is just another vexation for scientists who are working in a notoriously underfunded area of research where much remains to be learned. Holmes is just one of a handful of photobiologists around the world who are investigating the effects of increased UV-B on plants. In many ways, the native flora and fauna of Cambridge are roughly similar to those found in southern Chile. 'In terms of knowledge of the effects of UV-B on plants it really is early days yet,' Holmes said. 'We've had some interesting results so far, but I don't think we're in a position to extrapolate forward.'

In the same way that the human populace has developed skin pigmentation to handle the normal amount of UV-B they would reasonably experience, plants too have adapted to natural levels of ultraviolet. The mechanisms by which they do this remain largely mysterious, though Holmes himself remarked: 'Plants won't waste energy to develop the necessary pigmentation when they are affected by small changes in incident UV-B.' Work at Cambridge, as elsewhere, has shown that most plants are susceptible to irreparable damage by increases in UV-B incidence of around 15 per cent. Such an increase has not been measured on the ground anywhere in the northern hemisphere. 'If there is an increase in UV-B at the surface it will definitely have an effect,' Holmes said. 'The question of how much of an effect remains open.'

Deadly Springtime

At the Botany Department in Cambridge, Holmes has thus far performed three sorts of experiments. Controlled cabinets inside the laboratory can simulate levels of ultraviolet radiation above and beyond that which is now being received at the ground and revealed the 15 per cent limit of sensitivity. Glasshouse and outdoor experiments have also been carried out using supplementary UV-B to see whether the effects of this increased radiation has a noticeable effect compared to natural sunlight. In all cases, a computer-controlled dosimeter automatically records the real-time ambient ultraviolet flux: the computer then instructs lamps to increase UV-B. In the early 1990s, Holmes performed experiments with 20 per cent more UV-B and began repeating them with 25 per cent more. So far as possible, these experiments are triggered by solar output and the electronics will also switch the UV light source on half an hour before dawn and off within thirty minutes of dusk. Where other experimenters have only been able to increase biologically active ultraviolet in 'jumps' over long periods, Holmes can mimic the amount of radiation reaching the ground (which may vary with cloud cover etc.) to within five minutes.

Holmes's first set of experiments focused on a range of tree species and were, to say the least, surprising. A handful of species common to northerly latitudes were grown with 20 per cent more UV-B than normal. Comparing the dry weight of the samples at the start of the experiment with those at the end, the results were as follows after three seasons:

sycamore (*Acer pseudoplatanus*)	−18 per cent
birch (*Betula pendula*)	−22 per cent
beech (*Fagus sylvatica*)	+14 per cent
oak (*Quercus robor*)	+16 per cent

The following species decreased by the following percentages after just one growing season and these results are thus very preliminary:

ash (*Fraxinus excelsior*)	−10 per cent
Corsican pine (*Pinus nigra*)	−12 per cent
Scots pine (*Pinus sylvestris*)	−6 per cent
lime (*Tilia cordata*)	−12 per cent

From these observations it might be concluded that beech and oak will actually flourish under increased levels of UV-B. But these findings may have more to do with the parasites that feed on the leaves having been killed off. Interestingly, in both cases, the leaves of these species appear greener because the enhanced UV-B has allowed them to photosynthesise for longer. It would plainly be premature to suggest that if there are marked increases in UV-B at the ground, the British landscape will soon be peppered by beech and oak and all the other trees will be killed off. Over three growing seasons, both these species grew leaves for six weeks longer than they normally would. These observations have to be weighed against the distinct latitude dependence which all trees exhibit (particularly towards temperature). 'If these tree species start growing three weeks earlier in Britain,' Holmes suggested, 'then they may suffer from frost damage.'

Similar complications are also apparent in work which Holmes has performed outdoors with grassland species in a seminatural habitat. By observing growth rates in 'grassland perennials' which are normally found in the chalky downs of southern England, for example, he quantified their apparent sensitivity to increases in UV-B, thus:

perennial thistle (*Cirsium acaule*)	resistant
rough hawkbit (*Leontodon hispidus*)	intermediate
ribwort plantain (*Plantago lanceolata*)	intermediate
small scabious (*Scabiosa columbaria*)	sensitive

In all cases, increased UV-B causes the growth of these species to slow down appreciably and a simple extrapolation shows that thistle will eventually dominate. But these plants have not been grown in competition with each other nor in proximity to other plant species. 'We don't know what will happen to their natural parasites in the wild,' Holmes said. 'It's a bit premature to say that the English downlands will be overrun with thistles.'

The global implications of these results are hardly clear. In time, Holmes hopes to repeat these sorts of experiments with ever more complicated ecosystems to determine the relative importance of other factors affecting plants in the wild. In much the same way that UV-B measurements are now starting to be funded (Holmes is biological adviser to the European UV-B Monitoring Network), its effects on plants are now receiving necessary and appropriate levels of finance. In Britain, NERC and the Agricultural and Food Research Council have funded several biologists (Holmes included) to the tune of nearly £1 million to investigate the effects of UV-B on other species.

One hope is that when it has been determined which genes are responsible within plant cell DNA for tolerance towards UV-B, it will be possible genetically to engineer more resistant species of tree. This would probably be near impossible for plant species in their natural habitats, and many people might find the inducement of such artificial changes unethical. Much work remains to be done, and if new research in Germany is anything to go by, food manufacturers may become interested in providing funding for this sort of research. Dr Manfred Tevini in Karlsruhe is investigating how increased UV-B may affect the taste of herbs out of the largesse of Knorr, a leading soup manufacturer in Europe.

The damage to crop productivity from increased UV-B flux is another area of research where little information is at present available. In the United States, for example, work undertaken by the Environmental Protection Agency has shown that over half of nearly 300 crop species examined appear to be sensitive to increases in UV-B radiation. Among the most sensitive are peas, beans, squash, melons and cabbage. Some species produced fewer flowers, which affect their fertility, and soya beans, for example, showed a reduction in their oil and protein content. Although many of these experiments were performed in the laboratory, it was concluded from the EPA results that ozone depletion 'has the potential to exert very substantive effects on the environment'.

Only a dozen studies have been performed in field experiments with about twenty species of crop. In nearly half these studies, crop yield was reduced when increases in UV-B were simulated. In the early 1990s, biologists at the Australian National University in Canberra found reduced yields in soya beans, rice and peas when repeatedly exposed to low-level bursts of UV-B. The plants were irradiated for six to eight weeks with 5–10 per cent more UV-B than normal, and some species showed that growth was reduced by up to 70 per cent. 'What we're seeing is damage to the reproductive apparatus of the plants,' said Dr Malcolm Whitecross, a botanist on the team. 'I don't think there's any doubt ... it is damage at the molecular level.'

On the other hand, research by biologists at the Brookhaven National Laboratory on Long Island reveals little discernible damage when increased UV-B was incident upon alfalfa seedlings. These species seemed to be extremely resistant to increased radiation right across the ultraviolet spectrum. 'Some of the previous work was really very scary,' Dr John Sutherland of BNL told reporters after their findings were published in *Nature* in August 1992. 'The extent of damage has enormous economic implications and we must ensure our figures are accurate.' Some authorities exercise caution because other climatological factors may have to be taken into consideration when trying to determine those implications. Geoff Holmes at Cambridge, for example, is now looking at oil-seed rape, which is one of the few plants that will actually start growing in the springtime when the most pronounced ozone losses are apparent in northern mid-latitudes.

So, although it is clear that increased levels of UV-B will affect crop yields, not enough information is yet available upon which to base sound economic forecasts around the world. Many strains of crop are, to some extent, designed for the latitudes at which they are grown and those found nearer to the equator will be more resistant. Rice is a prime example, for its various strains exhibit wide degrees of sensitivity and resistance to UV-B increases. Some seem better suited to increases in UV-B because they are naturally grown at latitudes where there is a greater ultraviolet flux. Alan Teramura of the University of Maryland has been working with Indonesian authorities to quantify which strains of rice would be better suited to increased UV-B. Some biologists suggest that genetically engineered crops with greater resistance to increased UV-B may help offset some of the economic problems which may result.

Another economic problem raised by ozone depletion is the cumulative damage to building materials such as wood and plastic. Increase in ultraviolet radiation at the ground will cause more rapid light-induced degradation of such materials and mean that they will have shorter overall lifetimes. This will obviously add to the cost of materials which are used in the building trade around the world. Though 'light stabilisers' are often currently employed in plastics, their usefulness at shorter wavelengths into the ultraviolet is not well known.

The same could also be said of materials like rubber, wood, textiles and paints, whose response to UV-B is also not well documented. While a handful of experiments have been carried out in this area, the plastics examined have been those which tend to be used indoors, and, to quote a most recent UNEP study on materials damage, 'realistic assessment of the increased damage to materials under different scenarios of partial ozone layer depletion remains a difficult exercise'.

The economic consequences of ozone depletion cannot yet be assessed with any degree of accuracy. Much more work is needed before the true cost of crop and building-material damage will become apparent.

The evidence, then, for present damage to life on Earth as a result of ozone depletion is at best sketchy and often exaggerated. Virtually all the reports about health effects in southern Chile can be dismissed as hearsay, psychosomosis or wrong-headed overstatements. The only quantifiable damage that has been established beyond any doubt is to the phytoplankton in Antarctica. And yet it is unclear what these observations actually mean for the global ecosystem as a whole, a comment which could be made of most of the observed biological effects from ozone depletion. As chlorine loading of the stratosphere increases, it will become clearer what, if any, long-term damage will ultimately result.

If there are any health effects from ozone depletion, they are still most likely to be found in South America, so it would be foolish to ignore what may happen there in the future. Bedrich Magas used the analogy of a tall building on fire. 'If the first floor is burning, here in Punta Arenas we are on the second,' he said. 'The smoke is giving us reason to give the alert to the third-floor occupants – the rest of the world.' Further scientific enquiry and regular monitoring of the environmental effects of ozone depletion will let us know whether there truly can be smoke without fire.

9

Dr Watson Goes to Washington

> The success at Montreal can be attributed to no single prime cause. Rather, a combination of planning and chance, of key factors and events made an agreement possible. [First] and foremost was the indispensable *role of science* in the ozone negotiations. Scientific theories and discoveries alone, however, were not sufficient to influence policy. The best scientists and the most advanced technological resources had to be brought together in a cooperative effort to build an international scientific consensus. Close collaboration between scientists and government officials was also crucial. Scientists were drawn out of their laboratories and into the negotiating process, and they had to assume an unaccustomed and occasionally uncomfortable shared responsibility for the policy implications of their findings. For their part, political and economic decision makers needed to understand the scientists, to fund the necessary research, and to be prepared to undertake internationally coordinated actions based on realistic and responsible assessments of risk.
>
> Richard Elliot Benedick, *Ozone Diplomacy: New Directions in Safeguarding the Planet*

WHEN Robert Tony Watson arrived in Washington in late 1980 to manage NASA's upper-atmosphere research, few could have predicted the central role he would soon assume in the unfolding saga of ozone depletion in the decade which followed. He would become central to the drawing-together of a worldwide scientific consensus on the policy implications presented by stratospheric research, and, as a result, lead politicians through its intricacies to forge international legislation to ban CFC production. Without Bob Watson's participation, it is quite possible that there would have been no protocols to banish chlorofluorocarbons at all.

It may have helped Bob Watson's success that he doesn't exactly look the part of a diplomatic mover and shaker. A short, dark-haired man with the type of curly beard more familiarly sported by wild people who live in caves, he looks more like a biblical prophet than a scientific one. His obvious integrity and intelligence were to impress many diplomats and politicians who had learned nothing more about science since they had left high school. His powers of persuasion and his loquaciousness are legendary: whenever he appears in footage from international meetings, he can be seen in the background somewhere – talking. Bob Watson has a remarkable gift of the gab, redolent of the street traders from his native Romford. And even after his twenty years' residence in the United States, his accent has not been diluted to the mid-Atlantic drawl of most expatriate Brits.

Watson's appointment to the normally conservative civil service enclave of NASA Headquarters in 1980 was an unusual choice. Bob Watson had loved his research work in Pasadena at the Jet Propulsion Laboratory and had developed a decidedly West Coast way of life. He rarely wore a tie, affected the sorts of shirts preferred by undercover detectives in badly made TV series, and hadn't even considered life as a programme manager when the previous incumbent left. Then in his early thirties, he faced a classic dilemma in the career path of the research scientist: each promotion multiplied his administrative duties and restricted the time available to do research. The idea of becoming an administrator full time did not, on the face of it, seem particularly fulfilling or attractive. But after joking that 'you couldn't do much worse than get me in there' with his immediate NASA bosses, he became intrigued by the possibilities. Did he really want to spend the rest of his life measuring rate constants or modelling chemical reactions,

Californian sunshine or no? So he joined NASA Headquarters on attachment, half resigned to the possibility of returning to Pasadena and JPL. if it did not work out. Somewhat to his surprise – and that of his NASA superiors – he loved his new life in the capital.

Characteristically, Watson is modest about the importance of his role, and keen to emphasise that Dr Daniel Albritton, the director of the NOAA Aeronomy Lab, was equally – if not more – influential within the United States, certainly during the crucial period of the mid-1980s. One colleague who knows them both well says they worked together brilliantly in tandem as the 'Mutt and Jeff' of scientific assessments. 'If you find Bob's excitement a little too hard to swallow,' this necessarily anonymous associate said, 'then you'll certainly be inclined to believe Dan. He has the rectitude of a methodist minister.'

Not all all the people he came into contract with were impressed with Bob Watson. One Chemical Manufacturers' Association official vividly recalled her first meeting with him at a State Department meeting in the mid-1980s. He was late, had burst into the room, apologised by making light of it, and later removed his tie and even put his feet on the table. She was the least amused of the several people present and he seemed to delight in knowing it by repeatedly smiling at her, despite the frosty glances which she shot back. 'I cannot begin to tell you how rude I thought he was,' she said later. What makes this story amusing is that the person who told it, the former Elizabeth Gormley, married Bob Watson in May 1992.

The process of bringing about an international agreement to limit the production of CFCs would remind those who tried to effect it of the mythological trials of Sisyphus. No sooner was a compromise or consensus in sight than the intransigence of one or more of the participants would threaten to roll back the whole procedure to the starting point. One of the first international meetings to discuss possible controls to CFC production took place in Munich in 1978. After hours of ever more pointless argument, the conversation took a particularly bizarre turn. The French delegation included a representative from Revlon who said, in essence, that no controls would be possible in his country because French woman would not accept inferior, CFC-free beauty products unlike their pasty-faced, cosmetically challenged American counterparts.

It is easy to forget that the current wisdom was then that the ozone layer would not exhibit signs of damage until well into the next century. One UK official interviewed for this book recalled that until the mid-1980s ozone depletion 'was always at the bottom of my in-tray.' Vacillation and procrastination were thus the order of the day and discussions of possible controls became polarised because there was little sense that they would amount to anything. In essence, Europe wanted a cap on production whereas the United States wanted a worldwide ban on aerosols. Neither side emerges well from close scrutiny of the fine print contained within their proposals.

By the end of the 1970s, the European Economic Community accounted for about 40 per cent of the worldwide production of chlorofluorocarbons. EEC commissioners suggested in 1980 that there should be a 30 per cent reduction in CFC production by the end of 1981, and that no new CFC-11 and -12 manufacturing plants should be built. Though it was announced with no small fanfare that this represented a cutback to 1976 production levels, in reality, it would sanction an expansion of production capacity, for the 30 per cent cutback referred to continuous 24-hour production, which had not been treated as the basis before. The commissioners in Brussels and Strasbourg had actually allowed CFC production capacity to increase by 60 per cent in real terms. The question of introducing replacements for CFCs – particularly in refrigeration and fire-fighting equipment – was also raised in Europe, but most governments expressed concern about the issues of flammability and toxicity, and there, by default, the matter was allowed to rest.

The US government was capable of equally venal intransigence. The incoming Reagan administration let it be known that it would not consider any international legislation until the rest of the world had adopted what was then the US national strategy. In other words, there would have to be an international ban on aerosols before the United States would take any additional steps forward. While US officials made this out to be their moral high ground, the reality was that the American industry did not have to do anything drastic in the meantime. The domestic market for CFC-propelled aerosols was effectively finished but the same could not be said for other uses of chlorofluorocarbons. Domestic controls on any sort of free enterprise were anathema to the Reagan White House, so its call for a pandemic aerosol ban would thus blithely allow US CFC-manufacturers to continue to increase their production much as before.

These positions polarised in the early 1980s. There was a form of diplomatic gridlock between the United States, Canada and Scandinavian countries who wanted an international ban on aerosols, while Japan and the rest of Europe were pressing for a cap on worldwide production. Eventually there would be room for manoeuvre, in no small part due to the efforts of Robert Tony Watson, now firmly ensconced in the corridors of power in the US capital.

The implementation of Public Act 9595, as the US Clean Air Act of 1976 was more formally known, does not, on the face of it, sound particularly portentous for the protection of the ozone layer. It was just one of the many hundreds of bills which Congress routinely had to enact and was certainly one of the least remarkable in the canon of federal legislature. Yet without this otherwise obscure act, it is quite possible that there would have been no international decision to phase out CFCs by the end of the twentieth century. For almost inadvertently, it gave Bob Watson the wherewithal to forge an international consensus on the science of ozone depletion upon which those policy decisions would ultimately be made.

The Clean Air Act mandated the National Aeronautics and Space Administration to report to Congress every other year on the state of the stratosphere. NASA also had to outline how its research efforts had improved our understanding of the complexities of the ozone layer in the interim. As head of stratospheric research, Bob Watson inherited the considerable task of overseeing the work which went into compiling these reports. But when he began his work in earnest, he noticed that there was a pointless and ridiculous duplication of effort. In the two years before he arrived at NASA Headquarters, no fewer than six reports on the state of the stratosphere had been published worldwide. Two had appeared in the United States (NASA's to Congress and the second National Academy study): one was British (under the aegis of the Department of Environment), another came from the European Commission, and there were two further reports from the United Nations Environment Programme and the World Meteorological Organisation respectively.

'This didn't make any kind of sense to me,' Bob Watson said, and not just because of his healthy aversion to the sort of bureaucracy upon which most other civil servants thrive. Not only did many of the same scientists contribute to each of these different reports, they were subsequently further sidetracked from their research by having to write assessments on the differences between them. Most serious of all, these reports were supposed to be used by governments to formulate their approach to the ozone problem. 'Who the hell was a policy-maker going to listen to?' Watson asked rhetorically. At least, he reflected, he was now in a position to do something about this prolixity of paperwork.

So for the next NASA report to Congress, due in 1981, Watson decided to 'internationalise' it as far as possible. Almost straight away he stumbled across another problem. When he called Adrian Tuck, still working within the UK Meteorological Office, to enquire about his possible participation, Watson was surprised to learn that his good friend and colleague could not contribute to any publication sanctioned by the US Congress. As a subject of Her Brittanic Majesty, Tuck could not be seen to be in agreement with the conclusions of a US governmental report. Though the 'special relationship' between Margaret Thatcher and Ronald Reagan was scaling new heights of cloying mutual self-congratulation, their administrations remained implacably opposed in their policies to the CFC problem. Scientists from much less friendly nations would also find themselves isolated by this same division.

Watson soon realised a way out of the impasse. Foreign nationals could quite easily participate if reports were published under the administrative umbrella of international organisations. Accordingly, the 1981 report to Congress by NASA was sponsored by the World Meteorological Organisation, the international body based in Geneva whose remit is to coordinate and plan the world's weather observations. It was a masterstroke on his part and went far beyond any token involvement of scientists from around the world. Watson decided to enlarge the scope of subsequent assessments to draw together as complete a scientific consensus as possible on a problem which still seemed remote. They would involve the participation of an organisation which was then largely unknown outside most governments, although it had been one of the first to address the ozone question in earnest.

The United Nations had set up its Environment Programme in 1972 mainly as a forum to provide information to developing countries on how to avoid damaging their local environment as and when they became more industrialised. Pointedly, UNEP was the first United Nations organisation to be headquartered outside the developed world, in Nairobi, Kenya. Its officials were keen to address the ozone issue, for much of the Third World wished to embrace the technologies of the First, including chlorofluorocarbons. UNEP had been issuing bulletins on the state of stratospheric science and the development of computer models at regular intervals since 1977. It had also set up a Coordinating Committee on the Ozone Layer under the direction of Moustafa Tolba, an Egyptian-born scientist who would become a key player in enacting international legislation to protect the ozone layer.

In 1984, UNEP gladly joined the WMO as co-sponsors with NASA of the most intense international review of the subject ever assembled. Under Bob Watson's direction, over 300 scientists were gainfully employed over eighteen months to produce three weighty volumes known as *Atmospheric Ozone* which, in the end, ran to over 1,500 pages. The 'blue books', as they were soon christened (because of their appropriately azure-coloured covers), exhaustively considered just about everything to do with the science of the issue. They included subtle points raised by inadequate understanding of the atmosphere, the failures and successes of computer modelling, the importance of laboratory work and the need for ever vigilant monitoring – both from the ground and from space – to establish worldwide trends in ozone depletion.

The blue books were a scientific *tour de force*, and the assessments that followed involved a continuing process of peer review at meetings all over the world. For Bob Watson, constantly in motion, forever cajoling, encouraging and talking on the telephone, their rationale became increasingly clear as he chaired meetings in which a multitude of conflicting evidence was moulded into a coherent whole. 'Policy-making is hard enough at the best of times, especially on global environmental issues,' he said. 'International assessments give policy-makers an idea of what we know, what we don't know, which ideas are "solid" and which are conjecture. As a result they can start looking at policy issues.' Never again would politicians be allowed to shelter behind lack of scientific knowledge on so important an issue.

The blue books showed that if CFC production were held at 1980 levels, ozone depletion of 4.9–9.4 per cent could result by 2050. This prognosis was deliberately conservative, for by the middle of 1986 (after the blue books were finally published) production rates would have soared much higher. This reinforced the seriousness of the situation, and the weighty tomes did not pull any punches in their conclusion: 'Given what we know about the ozone and trace gas chemistry climate problems, we should recognize that we are conducting one giant experiment on a global scale by increasing the concentrations of trace gases in the atmosphere without knowing the environmental consequences.'

To reinforce these conclusions, the blue books managed to include Joe Farman's observations – 'by the skin of our teeth', Watson said, as they were about to go to press when the British Antarctic Survey results appeared in *Nature* in May 1985. The Antarctic ozone hole obviously indicated 'that some mechanism is at work in the cold southern polar night or polar twilight that is not generally included in the models. This clearly warrants further investigation.' In due course, it was, but even with this cursory nod in the direction of future research efforts, the blue books formed the basis for an international agreement on the ozone issue which many legislators had said would be impossible to achieve.

Towards the end of his life, Otto von Bismarck is said to have remarked that only three people had ever understood the Schleswig–Holstein question of 1863 – one was dead, another had gone mad and even he, the German chancellor, had forgotten what it was all about. Future generations may come to view the Vienna Convention of 1985 in a similar light, particularly as it was a product of the deliberations of the Ad Hoc Working Group of Legal and Technical Experts for the Preparation of a

Global Framework Convention for the Protection of the Ozone Layer. Yet, when the convention has long since been forgotten or limited to esoteric discussions within doctoral theses as an example of the fire and verve by which international negotiations routinely came about in the twentieth century, its legacy will remain. For some time towards the end of the next century chlorine loading of the stratosphere will have fallen back to levels before CFCs were released into the atmosphere.

In parallel with its international scientific assessments, the United Nations Environment Programme had also convened a committee of legislators and governmental representatives to address the drafting of legislation to protect the ozone layer. The negotiations which resulted were as tortuous and labyrinthine as the full title of the working group itself, even allowing for the usual snail's pace of international diplomacy. Nevertheless, despite uncertainties in the science and polar opposition to the measures required, a consensus emerged at a meeting in Vienna in March 1985. The significant accomplishment of the convention, which was duly signed, was that the ozone issue was serious enough to warrant 'precautionary measures'.

The Ad Hoc Working Group had first been convened with representatives from twenty-four countries in May 1981 to look at the legal ramifications of ozone depletion. After this preliminary meeting, its work started in earnest in Stockholm in January of the following year, planned to coincide with UNEP's tenth anniversary. Unlike the activities a decade later – when you couldn't pick up a newspaper or switch on a television set without being bombarded by references to the Earth Summit in Rio – the Stockholm meeting was purposely low key. There were no photo opportunities, snappy sound bites or politicians. The civil servants met quietly to establish a plan of action. A 'framework convention' would be drafted, around which a legally binding protocol would subsequently be signed to limit worldwide production of ozone-damaging chemicals.

Just getting this Global Framework Convention for the Protection of the Ozone Layer would take three years, endless meetings around the world, and five substantive rewrites before the ink dried on the signatures of representatives from forty-three countries. Strange as it seems, there was no consensus on which substances should be controlled within its terms of reference. Though CFCs were the most obvious ozone-depleting substances, there were others to think about: the halons used in fire-fighting equipment; carbon tetrachloride, found in pharmaceuticals and pesticides; and trichlorethane, a versatile all-purpose solvent, mainly used in cleaning metal parts during the manufacture of industrial equipment.

Although there seemed to be little impetus to effect change at the start of the committee's deliberations, the ever-widening scientific consensus which had emerged under Bob Watson's singular guidance was painting an increasingly disturbing picture of the seriousness of the issue. And so, when representatives of those forty-three nations met in Vienna in March 1985, they finally agreed in principle to a global plan of action. What was later termed 'the first worldwide legal instrument to protect the Earth's atmospheric resources' formally acknowledged the need for additional scientific assessments and also obliged signatories to provide CFC production data. This does not sound particularly Earth-shattering – or, indeed, Earth-saving – but for the legal minds who drafted it, this was assuredly a significant step and an unprecedented departure. Governments had finally woken up to the fact the ozone issue demanded being safe rather than sorry. 'That we are taking the anticipatory approach is a sign, I think, of a political maturity that has developed over the years,' said UNEP's secretary general Moustafa Tolba, when the agreement was signed. '[It] recognises how vital it is that we act to prevent environmental degradation or disaster with wisdom and foresight.'

After the convention was signed, Tolba added: 'Who could forgive us if we reacted too late?' But this concern for the future had yet to be translated into a legally binding protocol. The countries were still gridlocked along the same lines as before about whether to improve an international aerosol ban or a production cap.

It was two totally unexpected developments that would effect a reconciliation. The first came out of the Vienna Convention's declaration that information should be exchanged in a series of ongoing informal workshops. Whereas previously the two sides had snarled like two truculent grizzly bears, an informal workshop held in the United States would see them sit down, metaphorically speaking, and eat honey from a pot. The second was rather more immediate and portentous. Just two months after the ink had dried on the Vienna Convention, Joe Farman's observations from Halley Bay were published in *Nature*. There was no escape from the fact that ozone depletion was 'for real'. The spectre of the Antarctic ozone hole would underpin the urgency in needing to banish CFCs for ever.

With hindsight, the events in the United States in the two years following the discovery of the ozone hole were crucially important. The swirling, psychedelic banshee that graced television screens soon ingrained itself in the public imagination, so that ozone depletion was no longer an abstract concept. During this time, the death knell for the American CFC industry – which accounted for nearly half worldwide production – would be sounded. And thanks to behind-the-scenes efforts of an inner circle of federal government officials, a path was forged through a minefield of right-wing adversity within the Reagan White House. The negotiators included Bob Watson, Dan Albritton and the statesmanlike and somewhat imperious figure of Richard Elliot Benedick.

As head of the State Department's negotiating team on ozone matters, Benedick would excite praise and opprobrium in almost equal measure. It was his peculiar fate to be fêted by environmental groups in the United States for getting the ozone issue to be treated seriously and condemned by others for exactly the same reason. He was not, it must be said, universally popular within the Federal government, and had a way of raising the hackles of colleagues who tell stories of how he would routinely stay in hotels apart from them and often take the credit for breakthroughs which had been achieved collectively. Yet even his detractors say that without him at the helm, the US position would not have been reached so efficiently.

For his part, Benedick later wrote a record of his contributions in *Ozone Diplomacy* which many criticised as self-serving and biased. One reviewer, Fiona McConnell (who had headed the UK delegation), begged to differ from Benedick's view that the United States was the hero and the EC the villain of the piece. If a chronicle had been kept by EC officials, she wrote in the journal *Environmental*

Affairs, then 'it might well have demonstrated that the EC was the true architect of the Montreal Protocol and that the United States nearly wrecked agreement because of its obsession with the use of CFCs in aerosols, rather than CFC emissions from all sources'. However, McConnell did have the good grace to congratulate Benedick on the scope of his work which, curiously, reads at times as though he had been an observer and not a participant. Discretion was perhaps the better part of valour, for Benedick had suffered an indignity which in Soviet government would have resulted in his peremptory removal and airbrushing out of official portraits.

Part of Richard Elliot Benedick's patrician reputation stemmed from the fact that he was formally addressed as Ambassador even though he had never taken charge of a US embassy anywhere in the world. It is probably fair to say that this honorary title contributed to his downfall in the eyes of the Reagan administration. In the late 1970s, Benedick had represented the United States at international population conferences in his capacity as coordinator for population affairs within the State Department. He vigorously upheld the notion that draconian population controls were needed if the Third World was to avoid demographic disaster. Although an appointee of the Carter administration, Benedick found himself siding with George Bush on the issue of planned pregnancies during the run-up to the Republican Party nomination in 1980. After Reagan won, Bush conveniently forgot his pro-choice stance, whereas Benedick was never allowed to. For American 'pro-life' supporters he became as cherished and well-loved a figure as Salman Rushdie is to Muslim fundamentalists. He was subsequently demoted to the position of Deputy Assistant Secretary for the Environment within the State Department. It was the American equivalent to a Soviet official becoming the Deputy Undersecretary for Mining Supplies in Siberia.

When the chief US negotiator on the ozone issue resigned in 1984, Benedick became her replacement until somebody more suitable could be found. More than anything else, it was the White House's lack of interest in the issue which left him there during the crucial 1986–87 period. So by default and oversight, Benedick would become one of the key players, along with Bob Watson and Dan Albritton, in choreographing what the Ambassador was to term 'the interagency minuet' between the State Department, NASA, NOAA and the Environmental Protection Agency.

At the EPA, a young, energetic official called John Hoffman was directing its efforts into the ozone issue with the unbeatable combination of a technical background and a legislative frame of mind. Like Bob Watson, Hoffman was struck by a brainwave which would help unblock the negotiating deadlock between Europe and the United States. In May 1986, Hoffman had experienced first-hand the frustrations of international diplomacy. The first workshop sanctioned directly by the Vienna Convention was held in Rome under the patronage of the European Commission. Both sides continued in their bone-headed intransigence and, according to most participants, the meeting was openly hostile and recriminatory. The only success was an informal dinner, where diplomats could temporarily stop playing at diplomats and relax a little. It dawned on Hoffman that informality would be of the essence at the next workshop – to be hosted by the United States – if significant progress were to be made.

Following publication of the WMO–UNEP 'blue books' in January 1986, it was becoming clear that an immediate phaseout of 90 per cent of all the CFC production would be required to merely stabilise chlorine loading of the stratosphere worldwide. Domestically, the EPA had not effected any CFC-reducing legislation since 1978, and now on the international stage, it wanted to call for a complete phase-out of CFCs. It did not take someone with pointed ears astride the deck of the *Starship Enterprise* to see the illogicality of it all. Accordingly, the EPA announced a Stratospheric Ozone Protection Plan, whose express aim was to 'avoid letting today's "risk" become tomorrow's "crisis"'. Higher officials within the EPA agreed with Hoffman's call that the official US position both domestically and internationally should be a near total phase-out of chlorofluorocarbons.

This significant change was to be announced at the next international workshop, which, directly at John Hoffman's behest, was a more relaxed affair and was held at a spacious, luxurious conference centre in the town of Leesburg in rural Virginia. Informal banquets and square dancing were provided for the delegates. Such events, certainly seemed to help soften the positions, and one UK delegate interviewed for this book said: 'We got to know our opposite numbers as people. It turned out many of them were quite nice, really.' Bit by bit, concessions started to accrue as a result – even the pre-*glasnost* Soviet Union released its CFC production figures for the first time in a decade. 'At that meeting I knew there was going to be a protocol,' John Hoffman later remarked. 'It was only a question then of what percentage reductions were and exactly what structure the protocol had.'

The call for a total phase-out of CFCs within the United States ensured that the writing was on the wall for the American chemical industry. Hitherto it had been playing the same well-worn record that replacements for CFCs would not be feasible for many years. The previous February, at a domestic meeting organised by the EPA, the Chemical Manufacturers' Association said there was little prospect for introducing substitutes. 'Their standard line was they had spent $15 million on alternatives which had been found to be faulty or difficult to use,' said Alan Miller, an environmental lawyer who had been deeply involved with the ozone issue at the Natural Resources Defense Council. But then there was a dramatic turn-around as a result of the Leesburg meeting. The chemical industry's Alliance for Responsible CFC Policy, for example, had been wont to issue statements very much in the style of the Queen of Hearts. Then suddenly in September 1986, it declared that there should be limits to chlorofluorocarbon production. Within days, the world's largest CFC manufacturer, Du Pont, also announced that it would develop substitutes within a decade if appropriate regulations were mandated to make the market economically viable. So although money-making was still the bottom line, Thomas Edison Midgley's legacy was finally at an end. 'Those statements effectively made all that followed possible,' said Alan Miller.

There was, however, one final hurdle to be overcome in the form of a last-ditch anti-regulatory backlash inside the Reagan administration. The new approach to international discussions was formally legislated in something known as 'Circular 175'. It informed federal agencies that there should be a near-term freeze in emissions and a long-term phase-out. But early in 1987, hardline right-wingers within the White House attempted a slamdance with the intent of stamping on the feet of Richard Elliot Benedick. It was also motivated by a sense of pique that officials like Hoffman and Watson had gone too far out on a limb. For his part, Benedick mildly recorded that its critics 'simply had not been paying attention when circular 175 had been developed and approved a few months earlier'. A series of meetings were carried out at the White House, where scientists discussed at length the possible economic and health effects of ozone depletion. Some of the ideologues did not believe Bob Watson's summary presentations and Dan Albritton was later brought in. He presented the same findings with slightly fancier graphics.

Some attendees were resolute in wanting to shoot both the messengers and ignore the message. Their number included the secretaries of the Interior and Energy plus the former head of NASA, William R. Graham, who had subsequently been appointed presidential science adviser. The year before, Graham's first task as NASA administrator had been to issue a memo informing its 10,000 personnel across the country how to wear their security badges. At Senate hearings following the publication of the blue books a few days later, Graham was equally pedantic. 'Projections for the future have a large uncertainty to them and have to be reduced before we take actions for the future,' he declared. A year later, Graham was in a far more powerful position to undermine the scientific constituency which had emerged.

On the subject of the environment, the Reagan years provided some remarkable utterances, the very stuff of which opening monologues on television talk shows are made. The most famous was President Reagan's assertion that 80 per cent of our pollution is caused by flowers and trees. During the White House meetings, the Secretary of the Interior, Donald Hodel, listened to presentations about skin cancers and later publicly declared that they were largely 'a self-inflicted disease'. Later he was to go the whole hog and suggest that Americans should be encouraged to 'wear hats, sunglasses and sunscreen lotions' rather than find CFC substitutes. There was an outcry in the media at the ridiculousness of this 'personal protection plan' and soon a rash of cartoons appeared with fish and animals wearing sunglasses.

Scientists who had spoken at the White House meetings were understandably the most incensed. Dr Margaret Kripke, a leading dermatologist at the University of Texas, would declare in a letter to the Office of Management and Budget that she had not made herself clear. Skin cancers were not the preserve of the idle rich: most of her patients, she noted, were farmers, ranchers and oil workers who had headed south 'because of economic necessity, rather than because they want to play golf year round'. These comments, in particular, would have an impact upon somebody who was generally perceived as spending too much time in the sun with his kitchen cabinet cronies. The president himself, increasingly beleaguered as a result of the Iran–Contra scandal that was enveloping him, realised he could reach for the moral high ground. Political motivations apart, the skin cancer issue was of obvious concern to somebody who had had polyps removed from his colon and growths from his nose.

Reagan overruled those who clamoured for a change in the US bargaining position at a cabinet meeting in June 1987. With this seal of presidential approval, effectively all American roadblocks en route to a protocol had been lifted. It was only a matter of time before Europe would play ball.

Though the road to a protocol limiting international CFC production had been made considerably easier by the 'spirit of Leesburg', its exact resolution was still nowhere in sight. Seventeen months after the Vienna Convention was signed, negotiators met in Geneva to hammer out a draft protocol. Although the US position was now that a 95 per cent cutback in emissions was needed, Europe and Japan still wanted a freeze in production. Subsequent meetings in the first quarter of 1987 gradually began to whittle down the differences between these two bargaining stances. The United States moved towards a ten-year phase-out and the EC became amenable to 20 per cent reductions. The gap became less starkly defined after a negotiating meeting in Geneva where Moustafa Tolba addressed the participants en masse for the first time.

Under his direction, the negotiations made great strides by the expediency of splitting up the larger numbers of participants into smaller groups in closed bargaining sessions. Publicly, Tolba announced that 'no longer can those who oppose action to regulate CFCs hide behind scientific dissent'. A curious yet fragile consensus emerged: most countries agreed that CFC and halon production should be cut in two stages by the end of the century – but only if enough countries agreed to sign. Emerging nations – whose dependence on CFCs was increasing – such as India, South Korea and China did not even attend. The exact wording of a draft protocol was delayed because the UK government wanted only a single, later cut rather than sooner, staged cutbacks in the production of chlorofluorocarbons.

In the spring of 1987, there followed a series of mean and vitriolic exchanges across the Atlantic, which ultimately could be reduced to a kind of Boston Tea Party of the satellite age. Opening salvoes were fired by Richard Benedick, who pointedly told reporters after one negotiating meeting that some countries were obviously 'more interested in short-term profits than in the protection of the environment for future generations'. He was pointing the finger in the general direction of the British, and indeed, throughout his memoir, he portrays the British as the *bête noire* of international diplomatic efforts via their presidency of the EC, through whom negotiations were conducted. Benedick says that he received a cable from the UK government asking the State Department to do something about vociferous American environmental campaigners who were using the media in Britain to great effect. Several UK delegates interviewed for this book say the shoe was on the other foot: that the US ambassador in London was asked in no uncertain terms to tell State Department negotiators to stop sounding off in the press and to lay off the Brits.

So these transatlantic squabbles might well have continued had it not been for the British general election in June 1987. The Thatcher government's recalcitrant record on environmental issues became a key plank in its opponents' attack. To show otherwise, one of the first legislative acts of Margaret Thatcher's third term was a turnaround on the ozone issue and a general agreement with other nations on how best to solve the problem. The enduring consensus that emerged would lead to the signing of protocol in a Canadian city that September.

The timing of the Montreal Protocol was propitious. As last-minute negotiations were being hurriedly carried out, the hundred or more scientists of the Airborne Antarctic Ozone Experiment had descended on Punta Arenas to determine the exact cause of the ozone hole. It is interesting to note that the science which had prompted the protocol had been analysed long before the AAOE had started (from the blue book in fact). Some environmentalists had suggested that the signing should be delayed until the results from Punta Arenas were announced. Wiser heads prevailed. 'The protocol was literally signed without an explanation of what the ozone hole was all about,' said Dan Albritton, who was one of the science advisers to the US delegation. He was, by his own admission, itching to fly south to take part in the AAOE (to join Bob Watson and much of his Aeronomy Lab staff), but he realised that he might be needed in Montreal to help elucidate some of the more complex scientific problems presented by the issue. 'The science debate had been done and it had moved on to economic funds, phase-outs and freezes,' Albritton recalled. 'The results from AAOE influenced the subsequent amendments.'

When representatives from twenty-four nations came to Montreal to ratify a full and legally binding protocol that September, the path of diplomacy still did not run smoothly. Indeed, there were times when they nearly broke down completely. The worst sticking point was when the United States declared that it wouldn't sign unless all the other nations signed first. This sort of posturing may be common in primary school playgrounds, but it dismayed many of the onlookers in Montreal. Many of the scientists felt that they were witnessing a game of political poker with the ozone layer at stake. Discussions between the US negotiating team and that from the European Community continued well into the early hours after the first day of negotiations. Eventually a compromise was struck and it became enshrined in the legalese of 'The Montreal Protocol on Substances That Deplete the Ozone Layer'.

The agreement sanctioned gradual reductions in CFC and halon use: by 1990, consumption was to be frozen at 1986 levels; by 1994, it would be reduced by 20 per cent and then trimmed to 30 per cent by the turn of the century. Critics pointed out that there were no binding controls on trichloroethane or carbon tetrachloride and detected the worst sort of semantic trickery in the word 'consumption'. CFC and halon production could continue as before so long as industrialised nations didn't 'consume' them beyond levels set out in the protocol. This would allow for lucrative exports to developing countries who could continue to use them in increasing numbers. To environmentalists, it was altogether a classically cynical piece of international legislature. But however flawed, it was at least a start, and after signature on 26 September 1987, Moustafa Tolba himself publicly alluded to the words of Benjamin Franklin: 'I consent, sir, to this Constitution because I expect no better.'

In retrospect, the Montreal Protocol was not just 'a rich man's club' merely concerned with ensuring the longevity of well-established CFC manufacturers. The Third World was becoming increasingly important in the industrial production and consumption of ozone-depleting substances. As Richard Elliot Benedick summarised in *Ozone Diplomacy*: 'Industrialized nations with less than 25 percent of the world's population were consuming an estimated 85 percent of CFCs: their per capita consumption was more than 20 times higher than that of the developing nations.' There was a strong sense that the industrialised nations had caused a problem for the Third World: not only would the cost of replacements for CFCs be higher, the new technologies using them would be more expensive. Many countries feared that the comforts available to the First World would never be available to them and wanted financial assistance to make up for it.

The United States, for example, strongly resisted this approach as it set an expensive precedent for each and every other environmental problem. But it was clear that CFC consumption was ballooning in the Third World. China, for example, had increased its CFC consumption by 20 per cent in the 1980s. In 1988, its refrigerator use expanded by some 80 per cent, and the Beijing government had let it be known that it wanted to have a fridge in every Chinese home by the year 2000. If all these fridges employed CFCs, they would account for a far greater amount than the annual EC production. India, too, had even suggested that it might stem the shortfall in Third World consumption by producing CFCs itself. Pointedly, both China and India had refused to sign the protocol because there was no explicit provision for financial assistance.

A way out of this impasse would come from one of the more woolly-worded articles of the Protocol: 'The Parties undertake to facilitate bilaterally or multilaterally the provision of subsidies, aid, credits, guarantees or insurance programmes to Parties that are developing countries for the use of alternative technology and for substitute products.' This flowery declaration would come to have many ramifications for Third World countries with later amendments to the protocol. As one participant in the exhaustive legislative procedures involved dryly noted, 'the beatification of the protocol didn't occur for another three years'. And perhaps most significantly, the protocol showed something else which Moustafa Tolba identified when it was signed: '[We] can act when our scientists tell us that we are facing a distant threat; it proves that we can move before the full magnitude of the disaster is upon us.'

Bob Watson's sense of timing was impeccable. The day after the US Congress ratified the Montreal Protocol, he chaired a press conference which underlined the seriousness of the ozone problem. Discernible ozone losses over the northern hemisphere would have far greater immediacy and seriousness than the springtime losses over Antarctica. And now there was finally irrefutable evidence for it: analysis of the trends in ozone depletion had shown that its loss in the northern hemisphere was twice that which had been expected. For those foolish enough to be reassured that the Antarctic ozone hole was a remote, abstract 'science experiment' phenomenon, there was no escape from the reality of ozone depletion – albeit markedly less severe than over the South Pole – now taking place above their heads. It was the final nail in the coffin for the US CFC industry which had always wanted 'proof' of damage before it took precipitate action. Now, in March 1988, it had evidence in spades.

As we have seen in previous chapters, the question of trends in ozone depletion had proved to be a vexed one. After removing the various underlying factors – dynamics in the form of planetary waves and quasi-biennial oscillations, plus the natural modulation due to the solar cycle – it was not clear in the 1980s that there had been a statistically significant ozone loss worldwide. By 1986, however, scientists at the NASA Goddard Space Flight Center had discerned an appreciable downward trend from both the SBUV and TOMS instruments. Goddard's Don Heath had tentatively reported a 4 per cent loss in seven years of analysed data from

Nimbus 7. Others were decidedly sceptical because ground-based observations from the Dobson network should have picked such a loss up. Bob Watson, for one, was inclined to disbelieve the results and was in a considerable quandary as the head of the space agency's stratospheric research. 'We were in the unenviable position where we couldn't really say if the data were really right or wrong,' he remembered. 'The point is that you can't base international policy on data which is questionable.' As if to emphasise the point, early in 1987 Watson and Heath had found themselves contradicting each other at a Congressional hearing on this very issue.

Once again, Bob Watson came up with a solution to the problem by convening another large, international assessment on the specific issue of trends. This time it was funded by the National Aeronautics and Space Administration as it had a vested interest in the space hardware which seemed to be confusing rather than clarifying matters. Watson would once again chair what came to be known as the International Ozone Trends Panel, which employed the combined talents of 150 scientists who went through satellite and Dobson data with a statistical fine tooth-comb. Significantly, its steering committee included Sherry Rowland, who had shown a way forward in the basic approach of a seasonal analysis of the data from the Dobson instrument at Arosa in the Swiss Alps. In the same way that Don Quixote had charged at the blades of the windmill, previous analyses had tried to determine a signal from globally averaged ozone observations. The Trends Panel would re-analyse the data from 1970 to 1987 latitude band by latitude band and from season to season.

This new approach showed something immediately obvious across the northern hemisphere during the winter season. 'What shot out of the page was a progressively larger signal the further north you headed,' Bob Watson recalled. 'Whereas there was no real trend in the summertime and small trends at around 30°N, there was a statistically meaningful loss between 40° to 60°N in winter.' These worrisome observations were released at a press conference on 15 March 1988. In summary, they were as follows:

Latitude band	Average ozone loss	Winter average	Summer average
53–64°N	−2.3%	−6.2%	+0.4%
40–52°N	−3.0%	−4.7%	−2.1%
30–39°N	−1.7%	−2.3%	−1.9%

The Ozone Trends Panel was keen to emphasise that the key to their analysis was the hundred or so stations in the northern hemisphere which had finally produced statistically meaningful data. It is interesting to note that the panel decided not to use TOMS as a primary source because a totally foolproof method of internally calibrating its data had yet to emerge. Diplomatically, the panel said of the satellite observations that 'the data archived as of 1987 cannot be used alone to derive reliable trends in global ozone'. The panel also pointed out that although there were fewer stations in the south – and thus a much less detailed picture of the underlying trends – it was already clear that the Antarctic ozone hole had distorted the ozone layer across the whole hemisphere. South of 60°S, stratospheric ozone levels had declined by 5 per cent, which showed the pervasiveness of chlorine chemistry within the vortex.

For Sherry Rowland, there was the satisfaction that his original predictions had been vindicated by the Ozone Trends Panel. 'This is the first report to say there are losses of ozone that cannot be ruled out by natural causes,' he said at the press conference. These results unequivocally sent a message to the US chemical industry which it could no longer ignore. Within days, Du Pont announced it would now finally stop production of CFCs as there was now enough evidence to show chlorine chemistry destroyed the ozone layer. Soon enough, industrialised nations around the world followed suit, although not without the usual public kicking and screaming from both the chemical industries and environmental groups in individual countries. More than any other scientific results which had been published before, the trend observations revealed the full potential of the ozone problem.

In the immediate aftermath of the 15 March press conference, however, NASA faced a barrage of criticism for the strangest of reasons. Though Bob Watson's presentations were accompanied by an 'executive summary' of the panel's findings, the report did not appear in its three-volume, 2,000-page entirety for over a year. In keeping with federal practice, the contract for publishing the report went to the lowest bidder, and though well-intentioned, the small company which won was overwhelmed with the work. The trends report did not appear until the spring of 1989. 'We made a mistake,' Bob Watson admitted. 'As a result, we were criticised for "hiding" the data, but that really wasn't the case.'

In the meantime, Allied Industries Inc. decided to perform its own analysis of the same observations from the northern hemisphere ground stations. Two statisticians pored over the raw Dobson data and used an even more complex statistical analysis. 'There were small differences in their findings,' Bob Watson said. 'But they agreed with our basic conclusions.' A similar analysis by Soviet scientists of their own Dobson instruments also confirmed that there was a discernible ozone loss over the northern hemisphere. But what of the rest of the world? Within two years, as a method of self-calibrating space-based instruments became available, the answer would dramatically become apparent.

As a piece of international legislation, the Montreal Protocol did not come into effect until 1 January 1989. Its signatories agreed to annual reviews of the state of knowledge about the ozone layer and planned a further major ratification for the summer of 1990. As scientists watched the ozone hole grow bigger each spring over Antarctica and observed the accelerated stripping away of the ozone layer across the globe, it became obvious that more drastic legislation would be needed. With its newfound willingness to care for the environment, the UK government showed a way forward. Its Stratospheric Ozone Review Group – formed to assess the national implications of ozone depletion in 1987 – suggested that the emission of CFCs into the atmosphere should be peremptorily halted to stabilise the chlorine and bromine loading of the upper atmosphere. Though it would not instantaneously and miraculously stop ozone depletion, there would be a perceptible slowing down as chlorine loading reached a maximum early in the twenty-first century.

This position was soon echoed around the world and the bitter infighting of previous years was, slowly but surely, replaced by a sense of comity and cooperation. On 2 March 1989, the European

Council of Ministers announced that CFCs should be cut by 85 per cent as soon as possible and eventually completely. In effect, this was the death knell for European production of CFCs. A few days later, the newly elected President Bush also declared that CFCs should be phased out completely by the year 2000. Though cynics pointed out that words without commitment were a typical headline-grabbing ploy of a newly incumbent politician, it was an improvement on the stance of the Reagan administration.

The British government's formidable leader was now in her tenth year of office. Margaret Hilda Thatcher had long since taken on the aspect of Queen Boadicea, and, up to this point, had generally viewed environmental issues with that particular disdain she especially reserved for trade unions, European Commissioners, the IRA and objective reporting within the BBC. But early in 1989, she was magically transformed – in the wry phrase of one of her officials – into 'the fairy godmother of the ozone issue'. Suddenly, and totally unexpectedly, Thatcher was gripped by a Damascene conversion to environmental sensitivity. Quite how pragmatic this decision was may be discerned from the fact that the Green Party won 15 per cent of the vote in elections for the European Parliament in May of that year.

At Thatcher's behest, the UK Department of Environment was charged with coming up with a prestigious, headline-grabbing event. The result was a conference entitled 'Saving the Ozone Layer' in March 1989, where representatives from foreign countries could meet and discuss informally. Eighty countries sent their relevant environment ministers to the meeting at which some 120 nations agreed in principle to phase out CFCs completely. 'There was a conscious effort not to pre-empt the first formal review of the Montreal Protocol,' said one DOE official. 'There were no conclusions or declarations. It was an opportunity simply to exchange views.' A few weeks later, Thatcher held a seminar on the problems presented by the environment, assembled all her cabinet and told them to pay attention. More than one participant says that Downing Street had let it be known that any apologies for non-attendance would be 'unacceptable'.

Two of the most famous vignettes from this meeting include Thatcher's continual interruption of the oleaginous Kenneth Baker for repeatedly mispronouncing acronyms and the fact that Sir Geoffrey Howe fell asleep. One of the few non-UK citizens to attend was Bob Watson. He recalled being seated between two ministers at lunch. The one to his immediate left was convinced by the seriousness of the threat to the environment, while his colleague to Watson's right declared that it was an issue which would not have to be tackled for many years ahead. Watson simply moved his chair back slightly and allowed the two ministers to talk to each other about the policy implications. Watson provided scientific information whenever it was needed, and later realised that such interpretation was a vitally important role which scientists could play.

The London conference helped smooth the way for the first full review of the Montreal Protocol held in Helsinki that May. The result was that eighty countries (not just the twenty-four who had originally signed it) did indeed formally agree to phasing out CFCs completely by the year 2000. This meant both production and consumption of CFCs, which was seen as a major breakthrough by environmentalists. The work which had gone into previous scientific assessments was repeated and expanded, to produce some 1,800 pages of text which had involved the work of 500 scientists. Once again, the report was coordinated by Bob Watson and it included the results of both the first Antarctic aircraft campaign and the Ozone Trends Panel which he had also chaired. There was now no escaping the seriousness of the situation worldwide.

Known as *The Synthesis Report*, it convened separate panels on issues raised by the science itself, the long-term impact of ozone depletion on the environment, as well as economic and technological implications for the Third World. By now the process of drawing together these reports was well established and had matured under the remit of both UNEP and the WMO. 'The trends work was a new, important addition,' said Bob Watson. 'By this point, we had the results of both the subsequent industrial and Russian analyses. The results had greater than the 1 per cent accuracy of the satellite instruments, which was important.' With the fact that ozone depletion was unequivocally taking place across the northern hemisphere, the tacit agreement to phase out CFCs completely seemed warranted by the underlying science.

At the next major ratification of the Montreal Protocol, held in London in June 1990, it was agreed that production and consumption of CFCs in developed countries would be halved by 1995, reduced by 85 per cent from the start of 1997 and completely stopped by the year 2000. For developing countries, the timetable for total phasing out was extended by ten years to 2010. The 1990 ratification also introduced controls for carbon tetrachloride and trichloroethane – the former to be phased out by 2000, the latter by 2005. Yet any sense of triumph had to be outweighed against an immutable fact in the ozone issue. Though tightened, the Montreal Protocol would still allow for a further 12 million tonnes of CFCs to find their way into the stratosphere by the turn of the century. Of this amount, some 3 million tonnes would remain there until the year 2010. To the scientist who had discovered the ozone hole, this state of affairs provoked bewilderment when I spoke with him a month later. 'We're monkeying around with something we don't fully understand. We must slow production until we do,' said Joe Farman.

The London ratification was indeed severely lacking in many other respects. It avoided any mention of the 'halon bank', which has quite justifiably been called a timebomb waiting to explode. The role of halons in ozone depletion is less marked than that of CFCs, yet halons are possibly more destructive because of their bromine content. An atom of bromine is believed to cause thirty to fifty times the amount of ozone loss due to a chlorine atom. Although CFC depletion is a hundred times more prevalent, some analyses suggest that halons may cause more than a quarter of the ozone depletion seen throughout the world. Of the halons released in the 1990s, one third will still remain in the stratosphere in the year 2100.

Chlorine usually eats at ozone at very low temperatures in sunlight and most noticeably in the polar stratosphere in the springtime. But bromine is more likely to eat at ozone at any time and at any latitude. Halons are stored for long periods in fire-fighting equipment so are effectively 'banked' within them. Part of the difficulty in estimating what their cumulative, long-term impact will be lies in the irregular, unpredictable use of fire-extinguishers, both in the home and when existing ones are replaced by halon-free models. Nobody knows whether the halons within them will be reclaimed or released into the atmosphere. This so-called 'halon bank' represents a significant danger because the release of halons into the atmosphere cannot be predicted with any degree of accuracy.

More significantly, perhaps, the London ratification avoided any mention of hydrochlorofluorocarbons, or HCFCs, which had already been manufactured as 'safe' replacements for CFCs. By adding a hydrogen atom to a chain of chlorine, the chemical industry had helped circumvent some of the problems associated with CFCs. Some HCFCs – most notably HCFC-22 – had been used within industrialised coolers for many years and were hardly overnight sensations, as many people seemed to think. HCFCs break down much more quickly in the lower atmosphere so little of their chlorine reaches the stratosphere. At face value, the statistics seemed to indicate all was well: a tonne of HCFCs produces 5 per cent of the damage of a tonne of CFCs and it was estimated that only 15 per cent of the market for CFCs would be replaced by HCFCs over the decade of the 1990s.

But, as we will see, the models used to analyse the impact of HCFCs were two-dimensional and employed only gas-phase chemistry. By the turn of the 1990s, it had become clear from more advanced models that HCFCs could contribute to far greater ozone depletion in the short term. 'Anything other than modest substitution of HCFCs for CFCs could both increase the peak chlorine loading and sustain for decades unprecedented levels of stratospheric chlorine,' a UK Stratospheric Ozone Review Group report concluded in the late summer of 1991. As one of the authors of that report, Joe Farman continued to rail against the stupidity of phasing them in. To him, at least, it made as much sense as somebody putting down a glass of strychnine only to be offered arsenic instead. In a BBC interview broadcast just after the UNEPSORG report was published, he told Claire Harrison of the *Newsnight* programme: 'The sad fact is we haven't got time to use these things. We've got so much [chlorine in the stratosphere] already, we don't want any more at all. We're already past the danger point.'

At the end of 1991, the third UNEP–WMO Scientific Assessment was published. The work of some eighty scientists, coordinated as ever by Bob Watson and Dan Albritton, was released first as an executive summary at the end of October, and a few weeks later in its entirety. The main thrust of its findings was that ozone depletion was no longer confined to Antarctica nor to the northern hemisphere in winter: further trend analyses had shown that it now occurred at middle and high latitudes over both hemispheres in spring and summer as well. 'On a global average, there has already been a total loss of 3 per cent of the ozone layer,' Moustafa Tolba told reporters at a specially convened press conference held at the United Nations in New York. 'Another 3 per cent loss is expected to take place within the next ten to twenty years as a result of the chemicals already in the atmosphere.'

By far the most significant development ushered in by the 1991 report was the first ever reliable trends observations from the global perspective of satellites. Using the 'wavelength pair justification' technique (see Chapter Five), the SBUV and TOMS data had independently confirmed observations from the ground-based Dobson stations, most of which are located in the northern hemisphere. And, more importantly, this confirmation had resulted in the first reliable trends measurements across the southern hemisphere. The degree to which the ozone layer worldwide had depleted became most apparent latitude band by latitude band, and from season to season. 'Suddenly it became clear that we weren't just seeing ozone depletion in the northern winter,' Bob Watson said. 'We were seeing it in the spring and the summer as well. And we were seeing at virtually all latitudes, apart from the tropics.' The percentages were as follows:

	Dec–Mar	May–Aug	Sep–Nov
TOMS: 1979–91			
45°S	−5.2 ± 1.5	−6.2 ± 3.0	−4.4 ± 3.2
Equator	+0.3 ± 4.5	+0.1 ± 5.2	+0.3 ± 5.0
45°N	−5.6 ± 3.5	+2.9 ± 2.1	−1.7 ± 1.9
Ground-Based: 26°N–64°N			
1979–91	−4.7 ± 0.9	−3.3 ± 1.2	−1.2 ± 1.6
1970–91	−2.7 ± 0.7	−1.3 ± 0.4	−1.2 ± 0.6

To Bob Watson, this ubiquitous ozone loss heightened the reality of an oft-stated concern: 'If the consequences of ozone depletion are an increase in ultraviolet at ground level, there's more concern if it will happen in the spring and summer,' he says. 'That's when crops start to grow, and, to a lesser extent, when people spend more time outdoors and sunbathe.' And while he acknowledges the 'red herring' nature of skin cancer predictions, it is nevertheless a worrying development, particularly if the populace at large does not change its lifestyle as the seriousness of the situation undoubtedly warrants. Most press coverage of this latest assessment zeroed in on the health implications, and so we may hope that the message was not lost on sizeable portions of the world's population.

In its concluding chapter, the 1991 assessment also considered the policy implications of the enhanced ozone depletion which it had unequivocally confirmed. The report reiterated the basic message that the release of chlorine and bromine in whatever forms should be curtailed as soon as possible. Coming at the start of both the European Arctic Stratosphere Ozone Experiment and the NASA Airborne Arctic Stratosphere Experiment II, the UNEP–WMO findings reinforced the need for these extensive new campaigns. The Executive Summary of the assessment had concluded with two findings which, though they received little press coverage, more than anything else illuminated some of the fundamental scientific mysteries which remained. 'Present models containing only gas-phase processes cannot simulate the observed seasonal ozone depletions at middle and high latitudes,' the assessment summarised. '[Heterogeneous] models simulate most of the observed trend of column ozone in middle latitudes in summer, but only about half of that in winter.' As if to underscore this point, the assessment also pointedly added that 'there is not a full accounting for the observed downward trends in global ozone'. In other words, though there was evidence for worldwide ozone loss, the exact mechanisms behind it remained as tantalisingly elusive as they always had been.

For Bob Watson, the significance of these findings reinforced the need to maintain a continuous ozone-monitoring system as chlorine loading increased during the 1990s. 'Our experience shows that we should not rely on one system alone,' he said. 'We need both satellite and ground-based stations together.' And although self-calibration is now readily available with newer satellites (such as that aboard the Meteor TOMS instrument), experience has shown that each individual instrument has its own idiosyncrasies which need time before they are understood completely. Watson believes that there should be a 'dovetailing' of coverage from each successive instrument launched in the future, although he acknowledges that with the current funding pressures on NASA and its planned Earth Observing System, there may be unavoidable gaps. 'The philosophy is there,' he said of the desire to maintain coverage. 'Whether we can achieve it, unfortunately, is another matter.'

The London Amendment to the Montreal Protocol came into effect on 1 January 1992. Most of the European countries had agreed to adopt legislation agreed within the European Community to speed up the complete phasing-out of chlorofluorocarbons by the end of June 1997. Germany, with its strong history of environmental politics, has been the best of the west European bunch and completely banned the sale of products containing CFCs in the last week of January 1992. German industry had also agreed to stop using CFCs by August 1991, and across Europe, many CFC production plants were being phased out much earlier than envisaged as a result of the worldwide economic recession.

The situation for developing nations was not nearly so rosy. By the start of 1992, only two Third World countries had actually ratified the London amendment – China and India – and both had made it abundantly clear that they would act on it only if developed nations would increase financial and technical aid. India, in particular, wanted to be reassured that it would get the $1.2 billion it needed to switch to ozone-friendly technology as part of its burgeoning industrial development. At a press conference in October 1991 to announce the European Arctic Stratospheric Ozone Experiment, then UK environment minister David Trippier declared that developing countries would not even attend the next amendment to the Montreal Protocol planned for late 1992 'unless there's aid and technology transfer from the developed countries'.

At the start of 1992, the omens weren't exactly good. A new clause had been put into the revised protocol in London to provide for a three-year 'Interim Financial Mechanism' to help developing nations convert to ozone-friendly technology. A glossy brochure produced by Trippier's Department of the Environment painted a reassuring picture of the benefits: 'The Interim Financial Mechanism provides for an Interim Multilateral Fund of between $160–240 million which will be financed by contributions from developed countries. As well as paying the incremental costs the developing countries will face in complying with the revised Protocol, the fund will help with the cost of feasibility studies and technical assistance.'

By January 1992, however, only $9.3 million of the $240 million pledged had actually been paid. Environmentalists pointed out that this sum of money was much less than would be needed to minimise the pollution from just one average-sized power station. Behind the scenes, a rather more complex picture emerges of the desire of developed nations to avoid acquiescing to outlandish plans from Third World countries (like China's fridge-in-every-home declaration). In the same way that Vietnam peace talks floundered over the shape of the negotiating table, funding plans for the Interim Mechanism were stuck over which projects should be funded and how. One UK official involved with the negotiations said quite bluntly: 'We aren't going to put money in until there are firm plans as to how it's going to be spent. We're not going to bankroll Third World countries so they can do more or less do as they please without telling us.'

It was obvious from case studies on a country-by-country basis that different Third World nations had different needs, which obviously depended on their size and state of development. For example, Mexico would need around $10 million to convert to newer technologies, whereas Malaysia might need ten times as much as that. During 1992, estimates of how much money would reasonably be needed to convert to new technology and to underwrite subsidies for HCFCs, for example, came under review. By the time of the third amendment to the Montreal Protocol in Copenhagen came to be signed, studies had been completed for those Third World countries that accounted for some 75 per cent of existing CFC usage. Specific plans to eliminate some 25,000 tonnes of ozone-destroying chemicals had been approved for fourteen countries. It was only a matter of time before that funding was forthcoming after detailed plans were submitted, and then the financial mechanism itself would be permanently established.

One of the most controversial aspects of the most recent scientific assessments has involved the calculations of how much damage will be done in the short term by the so-called 'transitional substances' such as HCFCs. Touted by the chemical industry as 'safe' replacements for CFCs, they are, as we have seen, more damaging to the ozone layer in the short term than was originally realised. Partly this was due to the dependence on a concept known as Ozone Depleting Potentials (ODPs) which were developed in earlier assessments to show politicians how much relative damage would be done by each individual substance as it accumulates in the stratosphere. Calculations for ODPs, however, were based on two-dimensional models which employed only gas-phase chemistry as it was understood in the mid-1980s. As those models could not reproduce the workings of today's atmosphere with any degree of accuracy – nor indeed the gestation of the Antarctic ozone hole – it hardly seemed like a valid basis upon which to formulate policies concerning the future of the ozone layer.

'That was what worried me about this whole issue,' said Susan Solomon of the NOAA Aeronomy Laboratory. 'It prompted my interest in ensuring that we could properly account for ozone depletion by these transitional substances.' She was not alone in these worries, for other scientists and many environmentalists realised that the concept as originally envisaged was entirely inappropriate. The 1991 Stratospheric Ozone Review Group (UNEPSORG) report says quite pointedly that 'although ODPs purport to give a true basis for comparison between halocarbons, the models of atmospheric chemistry by which they are calculated cannot reproduce measured ozone depletions. Furthermore ODPs are calculated with reference to CFC-11 (a long-lived compound) and are seriously misleading when applied to short-lived compounds in the short-term.' Because long-term ODPs are compared to species which remain in the stratosphere for centuries, it had appeared that HCFCs would do little damage. But it had become clear that HCFCs release their chlorine far sooner than previously believed. Hitherto, it had been thought that the destruction would only reach a 'steady state'

– when ozone would be destroyed as fast as it was being created – until centuries hence.

The steady state ODP of HCFC-22 (in use as a refrigerant since the 1960s), for example, suggested that it would only cause 5 per cent the cumulative damage of CFC-11 over five centuries. Calculations by Susan Solomon and Dan Albritton (the director of the Aeronomy Lab) showed, for example, that HCFC-22 will cause 19 per cent of damage that CFC-11 would cause over five years, 17 per cent over ten years and 13 per cent over twenty-five years. The 1991 UNEP–WMO assessment squared up to the fact that ODPs based upon theory are limited in their usefulness and it was clear that a new basis for calculations was sorely needed to clarify the issue. The subsequent work by Solomon and Albritton introduced a 'semi-empirical basis' for calculations. What they defined as 'time-dependent' ODPs used actual measurements of chlorine and bromine species in and around the Arctic stratosphere combined with updated theory of atmospheric circulation.

They produced what is essentially an *à la carte* menu of the damage which may be done by replacements, which was published in *Nature* in May 1992. They noted: 'To the extent that they are needed, substitute compounds can [help] to protect the ozone layer. But the next 20 years are likely to encompass the highest atmospheric halocarbon abundances that the Earth will (presumably) experience.' Their calculations showed quite clearly that not all HCFCs are created equally and will damage the ozone layer significantly as chlorine loading continues unabated. A comparison of the most 'popular' HCFCs in use today shows just how damage is done in the short term. Note the wide variations between their relative damage over the crucial first twenty years after release:

Time in years since release (compared to CFC–11

	5	10	15	20	25	40	100	500
HCFC-141b	54%	45%	38%	33%	30%	22%	13%	11%
HCFC-22	19%	17%	15%	14%	13%	10%	7%	5%
HCFC-123	51%	19%	11%	8%	7%	4%	3%	2%
HCFC-124	17%	12%	10%	8%	7%	5%	3%	2%

As ever, Albritton and Solomon avoided advocating any particular policy apart from the self-evidently sensible observation that the use of chlorine should be stopped as quickly as possible. They added that such a course of action would 'minimise the risks of further surprises produced by the non-linear chemistry of unprecedently large halocarbon abundances and its coupling to unpredictable influences such as volcanic eruptions.'

The work of Solomon and Albritton thus acted to help policymakers prepare for the next ratification of the Montreal Protocol. In particular, Albritton saw this research as an ideal opportunity for scientists to explain the subtleties of stratospheric science, in keeping with the work which he coordinated with Bob Watson which went into the UNEP–WMO assessments. 'Industry will not want to invest huge amounts in substitutes and then drop them after five to ten years,' he said. 'We need to help advise what sort of balance needs to be struck. The main focus for policymakers is to get back to normal chlorine levels. What we can say, as scientists, is if you use these HCFCs in this amount, here are the changes you will get in ozone depletion. As a scientist I can say: "if you choose to do *this*, then the atmosphere will respond like *that*." What we have done is to present a spectrum of choices.

Increasingly, the question of substitutes will involve an interplay with the role of tropospheric chemistry, particularly the OH radical whose presence will dramatically alter the balance of how compounds are removed on a global basis. Work by A. Ravishankara and G. Vaghjiani also at the NOAA Aeronomy Lab has shown that the OH radical removes methane much less efficiently than originally thought. This suggests that the lifetime of HCFCs will be shortened, although more work needs to be performed in this notoriously difficult area of research. 'We're relying on the lower atmosphere to remove these substitutes before they get to the stratosphere,' Dan Albritton elaborated. 'Much of the stratospheric protection in the future is going to be focused on understanding tropospheric chemistry. Once you mention the troposphere, you can't view it as isolated chemistry. The dynamics and transport come in very strongly.' Albritton sees that over the next few years there will be a new emphasis in research involving laboratory chemists, modellers and experimentalists to track down the relative importance of the myriad reactions in tropospheric chemistry. 'The more you drink coffee between those groups, the better the science,' Albritton added.

By the time of the signing of the third amendment to the Montreal Protocol in Copenhagen in November 1992, there was a sense of maturity and, if not finality, then growing consensus on the ozone issue. While nobody was deluded enough to suggest that the problem had reached anything that remotely approached a solution, there was much less adversity and recrimination than had accompanied earlier negotiating meetings. 'Recession or no recession,' said Moustafa Tolba at the final meeting, 'the next generation to live on this planet deserves an ozone layer as good as nature can provide.' And so far as possible, that is what legislators representing ninety-three countries attempted to enforce in Copenhagen.

The agreement is outlined below. In essence, the findings of the 1991 assessment buttressed the need to bring forward the phase-out of CFCs and the sanctioning of legislation on HCFCs and methyl bromide for the first time. The schedule for phase-outs can be tabulated as follows, and dates refer to the first of January in each case:

	CFCs	other CFCs	halons	carbon tetrachloride	1,1,1 trichloroethane
Freeze	in place	–	in place	–	1993
20%	–	1993	–	–	–
50%	–	–	–	–	1994
75%	1994	1994	–	–	–
85%	–	–	–	1995	–
100%	1996	1996	1994	1996	1996

There are actually two groups of chlorofluorocarbons accounted for in the Montreal Protocol and subsequent amendments. The 'CFCs' listed above include the most commonly used, numbered 11, 12, 113, 114 and 115 in the family. The 'other CFCs' account for the rest, including the fully halogenated ones.

Production of HCFCs will be capped to production levels as they stood at 1989, and will be phased out as follows:

Freeze	1996
35%	2004
65%	2010
90%	2015
99.5%	2020
100%	2030

It had been hoped by some countries that HCFCs could have been completely phased out by the year 2005. The United States pressed for a 'basket' of uses, whereas Europe predominantly wanted to limit the production. Some countries, however, were vociferous in wishing to keep HCFC production going for as long as possible, because some industries wanted time to recoup their investment costs. In the end a compromise was struck: although 99.5 per cent of HCFC production will be finished by the end 2020, that extra half a per cent will allow for continued air-conditioning uses within the United States.

Perhaps more encouragingly, bromine-containing compounds have also been more severely cut back, as warranted by their marked propensity to destroy ozone. There is a freeze in the production of methyl bromide at the start of 1995 and a 100 per cent cutback on hydrobromofluorocarbons (HBFCs). All in all, as Moustafa Tolba concluded: 'The measures agreed to in Copenhagen are the strongest package of global environmental law ever enacted.'

The only opposition to the third amendment was exhibited by some environmental groups who decided to pull their usual kindergarten theatricals in the meeting halls by sounding off klaxons. Quite what this was supposed to achieve remains obscure, but beneath these antics and the shrillness of their perennial complaints that legislation is too little, too late, there is a growing recognition that agit-prop antagonism is limited in its efficacy for persuading politicians or business executives to change their attitudes to issues affecting the environment. Greenpeace UK, for example, has produced a brochure entitled 'Making the Right Choices' which is designed for industries to find out about alternatives to CFCs and HCFCs. Quite a few small-scale companies worldwide may not have registered the fact that at the end of 1995, CFCs will no longer be available.

The Greenpeace brochure advocates the use of alternative refrigerants such as ammonia, hydrocarbons and water, which can be used 'safely', and lists addresses where companies can buy or, at the very least, learn more about newer technologies, that are generally available. 'Choosing HCFCs now may make another expensive change necessary soon,' the introduction quite sensibly points out. Many companies are – of their own volition – 'ecologically friendly' and by far the most notable are those industries which have not had to base their work on traditional chlorine-based chemistry. The computing and electronics industries which previously would have used CFCs to clean circuit boards, for example, have now switched over to water-based cleaning solvents. Some authorities have pointed out that lemon juice can even be used in circuit-board applications, and throughout the industry generally, the use of transitional substances has been entirely avoided.

Things are not so cut and dried for other industries and the dilemmas raised by refrigeration, for example, are particularly thorny. In Britain, CFC recycling from refrigerators has largely been unsuccessful: a government-led campaign which started in 1990 has reclaimed at most about 1 per cent of the CFCs in general refrigerant use today. Similarly, there is concern in the United States that some car air-conditioning companies have been stockpiling CFCs, because they have offered warranties with automobile purchases beyond the deadline after which they will be phased out. ICI, Britain's leading chemical manufacturer, has been marketing a hydrofluorocarbon for refrigeration known as HFC-134a which contains no chlorine at all. However, it has a global warming potential some 3,200 times greater than that of carbon dioxide, and so its inadvertent release would aggravate the greenhouse effect quite markedly (HFC-134a is also starting to be used in air conditioning by many US car manufacturers). On the other hand, Greenpeace has made a song and dance about a fridge made in the former East Germany by DKK Schjerfenstein, which uses propane and butane for coolants, both of which have virtually no global warming impact. However, it is really a larder, as it can only be set to one temperature, and to keep such a larder working requires far greater consumption of electricity than that of most domestic refrigerators. If sufficient homes adopted them, their cumulative effect would be to add to the carbon dioxide released into the atmosphere from power stations.

Clearly the conflicting demands of the ozone layer and the greenhouse effect will not be easy to reconcile. The global warming impact of replacements for CFCs involves both a direct effect (their own capacity to absorb heat) and an indirect effect (how they chemically alter abundances of other trace gases). A concept known as the Total Equivalent Warming Impact, which takes into account both these effects, has been developed to account for these factors. A report carried out for the Alternative Fluorocarbons Environmental Acceptability Study in December 1991 concluded: 'Replacement of CFCs with suitable HCFCs or HFCs yields a dramatic benefit in reducing UNEPTEWI for all CFC end-use applications considered; reductions in UNEPTEWI range from 10 per cent to 98 per cent, depending on application. The UNEPTEWI improvements are most striking in relatively high-loss applications such as solvent cleaning, automotive air conditioning, and commercial building roof insulation.' And given the time-honoured cliché that as industry caused the problem it should be responsible for sorting it out, there is some hope from the most recent UNEP–WMO assessment, whose technical and economic panel noted: 'Technological optimism is amply justified by the historical record. Innovation to replace CFCs has been rapid, effective and economical.'

By far the most contention is over methyl bromide, which has been used since the early 1960s to sterilise soil and help in the fumigation of crops so that they can be grown all year round. Because of its bromine content, however, methyl bromide could do more damage in the short term. A re-evaluated short-term, semi-empirical assessment of its ozone-depleting potential shows that methyl bromide may do twenty times the damage of CFC-11 over an initial five-year period. In fact, it is so toxic that when it is sprayed on crops, farmers have to wear respirators. It leaked into water supplies in the early 1980s in the Netherlands and has been banned by the Dutch government ever since.

The industrial uses of methyl bromide have mushroomed in recent years: between 1984 and 1990 its production increased by 50 per cent and its consumption is increasing at about 6 per cent per annum. Of an estimated 66,000 tonnes used worldwide in 1991, it is estimated that some 80 per cent is used for sterilisation purposes so that crops can be grown under glass all year round and provide the Western world with out-of-season fruit. Most developed countries require the fumigation of fruit and timber products before they are imported from elsewhere. The remainder of the consumption is taken up by fumigation and quarantine for things as diverse as ship containers, art collections and beverages such as coffee and tea. The picture is, however, complicated because methyl bromide is also produced naturally, mainly by algae in the oceans. Estimates for this naturally produced methyl bromide may account for anything between 30 and 70 per cent of the total amount which accumulates in the atmosphere.

While there is little that can be done to mitigate against natural sources, at Copenhagen it was agreed that methyl bromide production should be frozen at 1991 levels by the start of 1995. The government was amenable to a complete phase-out by the year 2000, but vociferous opposition from the Israelis, in particular, saw only a freeze eventually being legislated. However, a non-binding resolution suggested that there should be a 25 per cent cutback by the turn of the century and, as one participant in the negotiations reflected: 'The key thing was getting it in.' Encouragingly, a report by an industrial users group convened by UNEP in Washington in June 1992 reported that in 30–90 per cent of cases when methyl bromide is used for sterilising purposes, substitutes could be used by the end of the century.

At Copenhagen, the issue of funding for Third World countries to develop alternative technologies received a considerable fillip as the Financial Mechanism was permanently established, its budget increased to $500 million, and provision made for internal review and a wider-ranging examination of its future. Whereas before, promissory notes had been issued by national governments to Third World countries, money has been transferred into the mechanism's account directly so that it can be issued accordingly. Since the Copenhagen signing, government representatives have also met to discuss the potential minefield of 'essential uses' – such as medical inhalers and explosion suppressants – which currently still require the use of CFCs. They account for a small percentage of worldwide production.

The next amendment to the Montreal Protocol is scheduled for the end of 1994 and the meeting is to be held in Bangkok. The issues of methyl bromide, essential uses and financial mechanism will doubtless continue to be sticking points, as will the need for tightening the uses of transitional substances. The sad fact remains that during the 1990s, ozone depletion around the world will continue and gather pace. In the manner of firemen alert for backdraughts within a smouldering building, scientists will be on the lookout for the latest surprises and endeavour to explain their environmental impact to governments. The scientific and technical assessments will doubtless continue to reflect a fragile consensus on the issues, though, sadly, their execution will lack one of their key participants.

A few weeks after the Copenhagen signing, Moustafa Tolba retired from his position as executive director of the United Nations Environment Programme, and formally did so on the twentieth anniversary of its creation. For seventeen of those twenty years, Tolba had directed the programme with the skill of a high-wire artist, and can be forgiven an element of self-congratulation in his retirement speech, for he could take a great deal of credit for UNEP's success. At a time when the agency's funding was still under threat from economic recession, Tolba forcefully concluded: 'We have given the world very good value for money. Let UNEP take credit where credit is due. We fathered the ozone agreements – two and half million people who would have got cancer will be spared.'

In the week before the US presidential election of 1992, George Bush made the last – and perhaps most revealing – of a series of political miscalculations. Trying to ridicule the Tennessee Senator who was standing as Bill Clinton's running mate, Bush openly referred to him as 'the Ozone Man', 'Mr Ozone' or plain 'Ozone'. 'If I want foreign-policy advice,' Bush declared at one rally, 'I'd go to Millie before I'd go to Ozone and Governor Clinton.' In Gore's own state, just four days before the election, Bush tried to poke fun at the

Senator's interest in protecting both spotted owls and the ozone layer. 'You know why I call him "Ozone Man"?' Bush asked. 'This guy is so far out in the environmental extreme we'll be up to our neck in owls and outta work for every American.'

Collectively, the American electorate responded by metaphorically sinking the Bush administration up to its neck in owl droppings. The patrician Texas-based politician who had once seriously claimed that he was the 'environmental President' lost the election. In the place of the accident-prone J. Danforth Quayle came a Vice-President whose sensibilities reflect the maturity of environmental awareness in recent years. Twenty years before, concern about threats to the ozone layer was often perceived as way out and decidedly Californian in nature; that Bush had decided to characterise them in the same way helped contribute to an overall impression that he was out of touch. For politicians of the baby-boom generation, an interest in the environment is not based in political expediency alone.

In many ways, Albert Gore has come to personify much of the new consensus about the environment and has eloquently expressed his viewpoints in the best-selling *Earth in the Balance*. Not only is it intelligent and articulate (two adjectives which are notably absent from descriptions of most political tracts), the book seems to have been written by the Vice President himself. He originally heard about the greenhouse effect as an undergraduate at Harvard in the late 1960s. Employing what many reviewers called 'Kennedyesque' passion, Gore points out that ozone depletion represents just one of the many complex issues which now face the inhabitants of the planet, and that '[the] struggle isn't just about CFCs. Ultimately it's about the entire relationship between human civilisation and the global environment.'

So Bill Clinton and Al Gore swept to a landslide victory with the incongruous soundtrack of a fifteen-year old Fleetwood Mac single. Die-hard Republicans braced themselves for the worst, while Washington – and, for that matter, the rest of the world – prepared for the prospect of a presidency no less enthralling than that of John F. Kennedy. It was fifteen years since a Democrat had last been in the White House, and over thirty since the most glamorous incumbent had announced his new frontier. The 'New Camelot' was fast becoming a banal argot of the times. It was clear that Bush's recalcitrance on questions of global warming and population control would be swept away. What else could follow?

The answer turned out to be very little. Within days of taking over the reins of power, the Clinton White House had reversed its position on Haitian refugees, became embroiled in arguments over homosexuals serving in the armed forces and found some of its nominated officials rejected because they had knowingly employed illegal aliens or exhibited amnesia towards the payment of federal taxes. In other words, it was business as usual in Washington politics. It was clear within the first few months of the Clinton administration that environmental legislation would take a back-seat to the twin vexations of the federal deficit and the establishment of a health-care programme. Apart from fine-tuning taxation to include a carbon tax, hardly any wide-ranging or sweeping reforms on the environment were proposed, particularly as the administration was forced into hand-to-hand combat with an increasingly truculent Congress on much more visible legislative issues. Al Gore himself seemed largely absent from the fray.

Despite the fact the Clinton administration has not introduced any fundamental legislation on the environment, it has inherited a well-oiled machine to effect change on the ozone issue. The Environmental Protection Agency, once slow to act during the Reagan era, now employs a large group of people to enforce specific legislation ratified by Congress and the Senate. The 1990 Clean Air Act, which resulted from the London amendment to the Montreal Protocol, came into force in 1992, and has raised the phasing-out of CFCs to a new level of importance. CFCs in motor vehicle air conditioning and refrigerants now have to be recycled by certified technicians. Safe disposal of CFCs after they are no longer used is another new development which will ensure that they will not be leaked into the environment inadvertently. Increasingly, this will be unlikely as prominent labelling of products which contain ozone-depleting substances is also now demanded by law. Along with a ban on non-essential products, there is also a well-established programme of avoiding the procurement of CFC-containing substances within federal agencies. This self-evidently sensible approach to the release of CFCs is yet another manifestation of the maturity exhibited towards the ozone issue within governmental circles in the United States.

Further legislation will undoubtedly follow when the Copenhagen ratification comes into effect. The scientific basis for its efficacy will be oiled by one of the key players in the story chronicled in these pages, who has now risen to an unprecedented level of influence. In April 1993, American newspapers carried a small announcement of a new appointee to the then increasingly beleaguered Clinton White House. The President's Office of Science and Technology Policy is charged with coordinating scientific policy under the presidential science adviser. A new candidate had been announced to take charge of all matters concerning the environment. After Congressional hearings which rubber-stamped the appointment, Robert Tony Watson assumed the responsibility for this prestigious and prominent post in the early summer.

In his new capacity within the Clinton White House, Bob Watson is attempting to bring together the same processes to bear on both the questions of global warming and biodiversity that he largely pioneered over the ozone issue. He has already been intimately involved with the former as part of his work for the Intergovernmental Panel on Climate Change, which was also set up in the late 1980s by UNEP and the WMO to tackle issues raised by climate change. Watson chaired a large working group within the IPCC on greenhouse gases and aerosols, and sees it as an outgrowth of the work begun under his aegis with the first UNEP–WMO ozone assessment in the mid-1980s. 'The very last chapter of the blue books from 1986 looked at the ozone chemistry contribution to the climate debate,' he said. And, now, as a matter of course, the IPCC has used the same basic approach of peer-reviewing the work of large groups of scientists culled from around the world.

The future work of both the IPCC and the ozone assessments will reflect the overlap of the science concerning ozone depletion and global warming and lead to a merger of traditionally diverse scientific disciplines. The next major scientific assessment on the ozone issue will be finished by the end of 1994, by which time an interim progress report by the IPCC will appear. A year later, there will be a smaller ozone assessment and a full IPCC report. In both these sets of publications, there will be joint chapters and common workshops on what Watson terms 'chemistry climate issues'. These include the thorny topics of the relative radiative forcing and global warming potential of trace gases involved in ozone depletion. This joint approach will, once again, avoid pointless duplication and with luck reinforce the findings of both.

In one respect, the work which results in the IPCC reports is different to the ozone assessments. Before its findings are published, drafts of the IPCC text are reviewed by governments and can formally be commented upon before they appear in press. As a matter of courtesy, Watson will send the ozone assessments to governments (although not formally mandated to do so) but will not amend any of the findings if any points are raised. 'We must never politicise the technical and scientific assessments,' Watson said. 'No government has ever asked me to change any of our findings for political reasons.' And this clearly defined lack of censorship has, above all else, ensured that the science behind ozone depletion is presented to policy makers with as little ambiguity as possible.

There are other pressures upon the process by which international assessments come about, as Watson is only too well aware. They have tried to accommodate a full range of opinion from as wide a cross-section of the scientific community as possible. Watson's rationale is that the 'best' people should be involved, which includes recruiting scientists from developing countries and not just from the ranks of 'the great and the good'. As Watson sees it, if scientists have relevant technical credentials to take part in the debate, then they will be invited to participate, and that includes qualified people from both industry and environmental pressure groups 'wherever appropriate'.

Joe Farman believes that the assessments often gloss over some of the complex issues by simplifying matters for the consumption of economists and politicians. 'They do not represent scientific assessment as I would like to see it,' he told me. 'They are more of a consensus and weaken the overall message. They should make statements more forcefully.' Ever candid, Watson agreed. 'I think there is room for improvement in what I would call "the substantive minority viewpoint",' he said, 'If one scientist in a hundred was in disagreement, it would be difficult to accommodate his or her ideas;

but if, say, between five and twenty-five dissented, then 'we must put their ideas in, and what's more, explain why it is a minority view'. Watson also believes that such a process of self-correction is vitally important in advising governments, and added quite openly that 'we've got to be the first to admit our mistakes'.

On the issue of scientific advocacy, however, Watson is adamant that scientists who help formulate governmental policies should not prejudge the issues relating to policy. Though he believes that scientists should be free to act as advocates if they so wish, there is no place for advocacy in the forging of scientific assessments. 'If policy-makers thought for one moment that the science was being driven by preconceived policy decisions, that would be an absolute disaster,' he said. 'Portraying the science to policy makers in intellectually understandable language which also honestly reflects the state of science is something we must strive for and continue to improve.'

And in that sense, what is past will undoubtedly be prologue as the issues raised by global warming come into focus. The process of gathering together scientists, peer-reviewing their work and presenting their collective findings to policy makers independently serves as a template for dealing with future environmental problems. And although this approach is now taken for granted, virtually all the scientists interviewed for this book agree that without Bob Watson it might never have happened so quickly or effectively in the first place.

10

Planet in Peril?

> The world faces a growing list of problems: overpopulation, dwindling resources, environmental degradation, industrial pollution, ozone depletion, global warming and the rest. It is already too late to avert them. But we can still mitigate some of their effects and prepare ourselves for a different sort of world. Over the last twenty years there has been some progress towards this goal. A remarkable change in public awareness has led individuals, groups, governments and international bodies in their different ways to take the first steps towards wisdom. These are to recognise that these problems exist and to begin – albeit in piecemeal fashion – to do something about them.
>
> Sir Crispin Tickell, Address to the British Association,
August 1991

> The prophecy given by Apollo ran that if Croesus made war upon Persia, he would destroy a mighty empire. Now in the face of that, if he had been well advised, he should have sent and inquired again, whether it was his own empire or that of Cyrus that was spoken of. But Croesus did not understand what was said, nor did he make question again. And so he has no one to blame but himself.
>
> Herodotus, *History*

High above the populated continents below or at exotic-sounding locations as far as possible from humanity's polluting influence, the good news was, for once, unequivocal. Five years of international controls had had an effect on the stratosphere. And the measurements that showed it had been made at the behest of an industry which had foisted the ozone problem on an unsuspecting world in the first place.

Air samples from 'clean air' stations in Samoa, at the South Pole, Cape Grim in Australia, Alaska, Hawaii, Canada and Colorado provide an unarguable daily record over many years of molecules such as CFCs in the atmosphere. Transported back to the United States, gas chromatography in its most advanced form enabled researchers at the NOAA Climate Monitoring and Diagnostics Lab in Boulder to sniff out one molecule in a trillion and instantly recognise subtle differences in the abundances of each and every member of the chlorofluorocarbon family. These samples represented as wide a spectrum of unpolluted air as could be obtained across virtually the whole of the planet. Analysis of the different air samples in one laboratory circumvents errors in measurement which otherwise would be made at each of these stations. As a result, the Boulder laboratories could easily discern any subtle changes in the handful of molecules which wreak so much damage to the ozone layer above our heads.

By 1992 the message from the air samples was clear: the amounts of CFCs pouring upwards were still increasing but at a far slower rate than had ever been seen before. The stratosphere was responding to international controls sanctioned by the Montreal Protocol: by the end of 1991, global CFC use was 40 per cent less than in 1986, the last uncapped year. Other researchers in networks operated by the World Meteorological Organisation, examaining air samples taken from aircraft at the threshold of the ozone layer, were finding similar results. The NOAA research was the first to be reported in the pages of scientific literature, in September 1992, and in time other similar measurement would corroborate their findings. Yet there was little room for celebration.

A recurrent feature of the protracted history of ozone depletion is that bad news follows good. The fact that the rate of increase of CFC levels had slowed in the autumn of 1992 had to be reconciled with the fact that the Antarctic ozone hole that year was the largest, deepest and quickest-forming ever reported. And though the reports of the decreased emission of CFCs were undoubtedly good news, there was a significant caveat: the danger posed by CFCs was far from over. The NOAA clean-air team concluded in *Nature*: 'If the observed deceleration continues at 1990–91 levels, global atmospheric CFC-11 and -12 concentrations would reach a maximum before the turn of the century and thereafter begin to decline.'

In other words, it was going to get worse before it would get any better, a point which was underlined by the gestation of the ozone hole at the same time as these measurements appeared in print. During September and October 1992, the ozone hole grew to its largest size ever seen: on 5 October, it covered 23 million square kilometres compared to 20 million on that date the year before. Balloon measurements from Antarctica revealed that ozone was totally destroyed at altitudes between 14 and 18 kilometres, whereas in 1991 it had been largely absent in the two kilometres above the 16 kilometre height.

The situation became even worse by the same time the next year. To some extent, the stratosphere over Antarctica in 1992 reflected the presence of the aerosol from Mount Pinatubo, which would provide surfaces upon which ozone-destroying reactions could be enhanced. By the end of 1993, however, they had largely precipitated out and yet signs that ozone was starting to strip away could be discerned in the middle of August, days earlier than ever before. It is clear that despite its rate of increase slowing down, increased chlorine loading of the stratosphere is taking its toll on the Earth's environment. Accordingly, the hole which opened in the stratospheric ozone layer over Antarctica was larger, deeper and quicker-forming than even the previous year.

The year 1992 had established a record for the lowest ever measurement of ozone then seen of 108 Dobson units. On 5 October 1993, ozone depletion at the South Pole claimed a new record of 100 Dobson units. Something like two-thirds of all the ozone above Antarctica had disappeared in just under six weeks. Brian Gardiner, joint author with Joe Farman and Jonathan Shanklin of the original announcement of the existence of the hole, said:'Once you've got down to a hundred units, then there's not an awful lot of ozone left in the lower stratosphere.' As chlorine loading heads towards its maximum by the turn of the century,

How Soon Is Now?

Antarctic ozone will doubtless continue to be lost sooner in the spring and thereafter plunge to record-breaking minima. 'When I look at this year's results, I don't think anybody knows whether it will get any worse,' said an ever-cautious Brian Gardiner. 'The future is going to bring many other surprises and that, I think, is the most worrying thing of all.'

It had become clear by the start of 1994 that atmospheric circulation patterns across much of the northern hemisphere had been subtly altered due to heating and/or cooling by the Pinatubo dust cloud. This had changed the efficacy of the transportation of ozone from equator to pole, thereby thinning the ozone layer at various times and locations around the world. Immediately after the Pinatubo eruption ozone levels over the tropics in the lower stratosphere declined. As the aerosol layer heated up, it rose and drew up air from below which was much poorer in ozone. As circulation patterns changed around the world, so too did the normal spread of ozone, as witnessed in the northern winter which immediately followed (1991–92). The long-term chemical effects ushered in by the Pinatubo aerosol were no less intriguing although by the end of 1993 it was obvious that predictions of losses of 10–20 per cent around the world had been exaggerations.

Yet there should have been some chemical effect from the sheer amount of aerosols the volcano had spewn into the stratosphere. Mount Pinatubo had released an estimated 20 million tonnes of sulphur dioxide, thereby making sulphate aerosols much more readily available throughout the ozone layer worldwide. The surface areas available upon which heterogeneous reactions could take place had increased anywhere from ten to a hundred times. The sulphur content of the aerosol was also relatively high, a factor above all else believed to speed up the efficiency of ozone-destroying reactions.

The acid test, as it were, of the pervasiveness of the Pinatubo chemistry could be gauged from measurements over the northern hemisphere in the winter following the eruption. The key to the chemistry involved concerns the way in which oxides of nitrogen are converted to nitric acid on aerosols or polar stratospheric clouds so that it essentially 'freezes out' of the stratosphere. As a result, the atmosphere's natural buffering towards active chlorine is removed.

As we have seen, both the European Arctic Stratosphere Ozone Experiment and the NASA AASE-II aircraft campaigns observed significantly reduced ozone levels over the winter of 1991–92.

Partly it was due to the ridge of high pressure which had sat over the north Atlantic, but both balloon flights from Sweden and ER-2 sampling from Maine had revealed the presence of the Pinatubo aerosol. Over the Arctic in early 1992, the stratosphere denoxified as much as it could, but it had reached what might be termed saturation point. Jim Anderson used the analogy of teenagers raiding a fridge full of food. 'Three of them could quite easily empty it in an hour,' he said. 'Even if they invite fifteen more of their friends in, there'll still be no food left.' In other words, no matter how much more aerosol was added to the polar regions, it could not affect the ozone depletion there above and beyond a certain point.

Over northern mid-latitudes, however, there was discernibly greater denoxification caused by slightly different reaction from those on polar stratospheric clouds. Over the pole, chlorine nitrate will react with water (within the clouds) to denitrify at cold temperatures: but at warmer temperatures encountered further away from the vortex, a temporary repository of the oxides of nitrogen – nitrogen pentoxide, N_2O_5 – will react on sulphate aerosols. This will remove the naturally occuring oxides which would otherwise buffer the chlorine chemistry and, as we will see, significantly alters the resilience of the ozone layer over northern mid-latitudes. Because it does not require freezing conditions, its effects are rather more pervasive well into the spring season as far as the northern mid-latitudes.

The absence of nitrogen oxides throughout the northern hemisphere because of the Pinatubo aerosol was observed by both the EASOE and AASE-II campaigns. One difference between the American and European campaigns was that although the ER-2 gave a longer, more extensive 'slice' of observations than the balloons, it did so at lower altitudes. Some of the EASOE balloon flights could reach altitudes of 40 kilometres and grab samples as they descended or else make spectroscopic observations remotely.

'There was more aerosol at higher altitudes following the eruption of Pinatubo,' said one of the EASOE coordinators, John Pyle. 'It was very clear that columns of NO_2 were about 30 per cent down.'

How Soon Is Now?

Profiles returned by some of the larger balloon experiments confirmed Jim Anderson's 'hungry teenager' analogy of the reactions triggered by the volcanic aerosols as measured by ER-2 instruments. 'In the wintertime, the aerosols played no role in the chemistry over the Arctic,' agreed another key player in EASOE, Jean-Pierre Pommereau of the French Service d'Aéronomie. 'In the spring, however, they do help remove some of the oxides of nitrogen.'

The conventional wisdom before both campaigns was that if nitrogen dioxide was removed, then activated chlorine would persist for much longer. Yet observations of the presence of NO_2 in the spring outside the vortex showed that the situation was not as clear cut as was previously suspected. There was some evidence from EASOE that its abundances were rather more variable and very much dependent on temperature. There is also a distinct possibility that the nitrogen dioxide – and the stratosphere's ability to recover from ozone destruction – returns far quicker in the spring when the Sun is once again present. 'I think there is more to the story than we know at present,' said John Pyle. 'We know some of the chemical processes involved, but there are plenty of inconsistencies.' The modelling of how reactions take place on aerosols and polar stratospheric clouds is not, he believes, tied down and there is much work to be done in the future.

One effect of Mount Pinatubo was to reveal that so far as future ozone depletion over Europe and the United States is concerned, we are literally living under the volcano. The availability of volcanic aerosols seems to aid ozone depletion at lower latitudes far more significantly than had previously been thought. The inopportune explosion of another volcano may yet exacerbate ozone depletion as we enter the greenhouse world of the future. It is even more disquieting that nobody can predict the likelihood of such an event or how severely it will affect the ozone layer as chlorine loading reaches its peak.

Away from the spotlight of media interest, scientists involved with the second Airborne Arctic Stratosphere Experiment spent more than a year analysing their data, re-evaluating their understanding of conditions in the stratosphere over the winter of 1991–92 and, finally, writing up their results in the formal, rigorous and (invariably) monotonous style required by scientific journals. A handful of papers appeared in print on 27 August 1993 in *Science*, and made barely a ripple in media terms, partly because there was no accompanying NASA press release, but mainly because news editors probably thought that the results were old hat. Reading between the lines of the scientific papers, however, revealed some deeply puzzling and, indeed, disturbing observations.

In an overall summary article, Dr Jose Rodriguez of Atmospheric and Environmental Research, Inc., based in Cambridge, Massachusetts, concluded that 'measurements by the AASE-II campaign have confirmed some aspects of our understanding of stratospheric ozone. However, the ozone bill of health is precarious enough that future checkups are warranted.' By watching the evolution of the Arctic vortex over the winter of 1991–92, it was possible to perform a tally of nearly all the chemical species involved in the processing of air within the vortex. New instruments aboard both the ER-2 and the DC-8 had revealed new perspectives on how the Arctic stratosphere primes itself for ozone destruction. The possibility of an ozone hole had tended to colour all reporting, but in the final analysis, the main results from the campaign showed that conditions outside the polar vortex were the most interesting – and disturbing – of all.

'Most puzzling is the whole question of hydrogen chloride,' said Dr Rodriguez. 'It is normally thought to be a stable reservoir for chlorine within the stratosphere.' As expected, AASE-II experiments revealed that inside the vortex, whenever active chlorine is present as chlorine monoxide, levels of HCl were comparatively low. In other words, the coldest conditions and the polar stratospheric clouds which resulted were responsible for shifting the industrial chlorine from its reservoir to its active, ozone-depleting form. Previous campaigns, however, had been able to measure only column abundances of HCl, but thanks to a new 'tunable' spectrometer known as ALIAS flown aboard the ER-2, it was possible to measure localised concentrations. Measured abundances of HCl outside the vortex were, according to Rodriguez, 'at much lower levels than we expected. This was surprising and unexplained as there were obviously no PSCs present.'

This implies that there are fundamental gaps in our understanding of the fine details of chlorine chemistry in the Arctic stratosphere. The way in which the chlorine shifts back to its inactive forms may be more complex than originally thought. 'This raises many questions about our understanding of the total chlorine budget of the stratosphere,' suggested Rodriguez. 'There may be other chlorine species we haven't accounted for yet.' Like the EASOE observations about the variability of nitrogen oxides, these measurements from AASE-II may help explain why ozone trends at mid-latitudes in the northern hemisphere cannot properly be accounted for at the moment.

Another intriguing finding at mid-latitudes came as a result of the ubiquitous aerosols from Mount Pinatubo. Outside the vortex there is evidence that at very cold temperatures, heterogeneous reactions were nearly as efficient on aerosols as they would be on polar stratospheric clouds. On 19 January 1992, as the DC-8 flew roughly along the Arctic Circle from Iceland to Greenland, it encountered air which was heading towards the northeast and heavily laden with volcanic aerosol. Using a laser-ranging instrument, scientists aboard were able to measure how much hydrogen chloride and chlorine nitrate were present above the plane. Once again, although temperatures were above those at which polar stratospheric clouds formed, it was clear that both chlorine reservoirs were reduced. Chlorine had been activated outside the vortex by the Pinatubo aerosol, and the extent to which this was done showed that it was more efficient than had been previously suspected.

But most significant of all were measurements made at mid-latitudes at the start of the AASE-II campaign in October 1991, well before the Pinatubo aerosols had arrived and been injected into the polar stratosphere. 'The main story isn't just that this immune system is repressed because of Pinatubo,' explained the AASE-II project scientist, Jim Anderson of Harvard University. 'It was never there in the first place to defend us.' This worrying statement requires a brief recap on the complex merry-go-round of reactions involving the oxides of nitrogen in the stratosphere.

Most of the oxidised nitrogen in our atmosphere reaches the stratosphere in the form of nitrous dioxide, N_2O, which is then immediately broken down by ultraviolet light. In the complex melange of radicals which result, they react and recombine into various 'reactive nitrogen species' of one sort or another. In the form of NO_2 or NO, they present a buffer for ozone depletion by tying up any active chlorine which happens to be around. Collectively they are called NOx. In all their other forms, including nitric acid and nitrogen pentoxide, they are known as NOy. The ratio of NOx to NOy is thus an index of the amount of 'buffering' available at any one time.

The NOx/NOy ratio will, of course, change most dramatically due to the formation of polar stratospheric clouds at around −78°C, when nitric acid freezes out. 'Once it is tied up in PSCs it can no longer be photolysed,' Anderson said. 'This means that NOx cannot be made available to help off high levels of ClO.' On previous aircraft missions, both NOx and NOy could only be measured separately, so the value for the ratio was largely unknown. On AASE-II, however, an instrument developed by David Fahey of the NOAA Aeronomy Lab could measure both species simultaneously. On the first flight of the ER-2 in the AASE-II campaign, there were between three and four times less NOx than expected. This had nothing to do with the Pinatubo aerosol, which was only later observed to arrive and saturate the lower stratosphere.

These measurements were, to quote Anderson, 'supremely unsettling' because they showed that our understanding of how the stratosphere had supposedly protected itself from ozone destruction was completely off beam. Small wonder that mid-latitude trends cannot be accounted for, as the NOx/NOy measurements reveal that a piece has been missing from the chemical jigsaw puzzle. 'It's a superb example, along with Joe Farman's discovery of the ozone hole,' Jim Anderson said, 'that we are sadly lacking in our understanding of the ozone layer.' As chlorine loading continues, it is clear that the stratosphere is becoming increasingly sensitive to ozone depletion.

It was early 1994 before the final results from the European Arctic Stratosphere Ozone Experiment appeared in the hallowed pages of a scientific journal. Well before then, however, it was clear that its key findings echoed those from the NASA aircraft campaign during the winter of 1991–92. Perhaps the most obvious result was that the Arctic vortex had filled with chlorine in January 1992 owing to the cold conditions. 'When we saw that,' said John Pyle, 'it was a surprise to us and very newsworthy. Now, I think we could infer that from temperature maps of the area.' Once temperatures drop below the threshold of stratospheric polar cloud formation, it is known that the chlorine would be activated as quickly as it could. 'We have a good picture of the chlorine activation,' Pyle added. 'There's a lot more to be done on chlorine deactivation and how it goes back into reservoir forms. We got a lot of interesting clues from EASOE.'

Whereas AASE-II revealed unsuspected results about HCl, the European campaign noted something intriguing about the other reservoir species, chlorine nitrate. At the start of the winter season in November and December, there was an appreciable ozone loss even when it wasn't cold enough for polar stratospheric clouds to have formed. Though it amounted to a few per cent, it was still significant when the underlying dynamic changes were accounted for. There is some evidence that this loss may be due to a cycle of reactions involving chlorine nitrate, for when it is photolysed, it briefly produces active chlorine. It had been thought that this relatively slow cycle would not be important, but in the early winter when sunlight is still plentiful, it could cause a loss of ozone of perhaps a few per cent.

Another overall aim of the EASOE campaign was to see how ozone depletion within the polar regions was affecting mid-latitudes, and Pyle was candid enough to say that 'the connection was not really made between the two'. This was partly because of the limited geographical extent of observations, and partly because of the meteorological conditions. During the decade of the 1980s, TOMS aboard Nimbus 7 saw that, on average, the ozone layer over the northern hemisphere mid-latitudes had thinned by around 8 per cent. There are three possible mechanisms that could account for this discernible loss: dilution by polar air containing less ozone mixing in at lower latitudes; air being chemically processed through the vortex; or chlorine activation on aerosols or polar stratospheric clouds outside the vortex.

Virtually every scientist interviewed for this book seemed to have his or her own pet theories, and insufficient evidence has yet been amassed to settle the dilemma at present. As both EASOE and

AASE-II have shown, the situation is even more complex than it first appeared, with or without the influence of Mount Pinatubo to contend with. 'EASOE gave us a lot of clues,' John Pyle reflected. 'Our follow-on mission will specifically look at many of these questions.'

The conditions observed over the Arctic in the winter of 1991–92 acted in many ways as a harbinger of the ozone losses that were actually seen a year later. An Arctic ozone hole first became a distinct possibility in the late spring of 1993. As we have seen, in early February 1992, the Arctic stratosphere warmed and conditions that might otherwise have led to severe, prolonged ozone depletion were averted. If it had been colder for longer, then a marked loss of ozone might well have resulted. Over the winter of 1993 however, conditions did conspire to remain cold well into the sunlit period in March, and researchers were able to learn more about the mysteries of the chemical processing within the Arctic stratosphere. Though there were no large-scale experiments being undertaken, evidence from smaller balloon campaigns, ground-based stations and satellites helped clarify the situation.

Quite categorically, there was no Arctic ozone hole in the spring of 1993 – although, it must be said, some British newspapers gave the impression that there had been one at the end of February 1993. It all started with some deft media manipulation by environmental groups, for, at the time, John Major's hapless government was suffering handicaps enough without the vexation of being accused of hiding an ozone hole from the public.

A number of MPs began to ask questions, with Greenpeace cheering from the sidelines. In a written Parliamentary answer, David MacLean, who was then junior environment minister, asserted that levels of ozone were 'within previously observed ranges for the time of year'. Quite whether he was referring to the previous year's EASOE results or earlier statistical data was not clear, but even he did not deny that there was significant ozone loss over the northern hemisphere.

Indeed, ozone measurements worldwide had shown that ozone columns across northern mid-latitudes were well down – between 9 and 20 per cent. But an ozone hole? The title of a press release from the World Meteorological Organisation denied it: 'NO OZONE HOLE OVER EUROPE AND SIBERIA BUT CHEMICAL DESTRUCTION IS POSSIBLE.' And observations revealed that ozone columns were still well within the 320–40 Dobson-unit range which might be reasonably expected. What seems to have triggered the scare story was the passage of the polar vortex over the northerly extremities of the UK on certain days in late February. This resulted in the ozone column being reduced by up to 25 per cent. 'One low value doth not an ozone hole make' quipped one exasperated British researcher to whom I spoke in early March.

It was clear, however, that in February and March 1993, the vortex had persisted in the cold stratosphere which had wobbled as far south as the Mediterranean on certain days. On a glorious mid-March afternoon I visited Jean-Pierre Pommereau of the Service d'Áeronomie of France's national scientific research organisation, CNRS, in the unlikely setting of a château in a wood close to Verrières-le-Buisson, south of Paris. The walls of his office were plastered with the latest satellite print-outs and balloon measurements from around Europe. The southward extent of the vortex had been self-evident to Pommereau when he launched a number of small balloons from Aire-sur-l'Adour outside Toulouse. On 10 March, for example, a balloon launch from there had found signs of chemical processing in the stratosphere as the polar vortex seemed to be relatively intact. 'We saw the same conditions in the Arctic in March that we find in the Antarctic in September,' Pommereau added.

Throughout the early spring of 1993 there seemed to be confirming evidence from ground-based observations and other balloon flights in Scandinavia that mini-holes had opened up in the Arctic vortex. Pommereau was at pains to point out that mini-holes were not the same as the 'mother of all holes' which opens up over Antarctica each austral spring. 'Mini-holes show you that there has been chemical processing, but you are downwind of it,' he explained. When I interviewed him in March, Pommereau believed that the cold temperatures which had persisted may well have had something to do with the circulation changes resulting from the climatic effect of the Pinatubo aerosol. 'A key issue is to understand the evolution of stratospheric temperatures,' he added. 'The future will very strongly depend on the climatology.'

How much had the lingering presence of the Pinatubo aerosol contributed to the observed ozone loss? Many authorities believed that it was significant, for the coldness of the winter of 1992–93 could not fully account for the observed ozone losses. '1990 was, statistically speaking, a cold winter and the stratospheric warming came late on,' said Susan Solomon of the NOAA Aeronomy Lab. 'But we didn't see anything like this in 1990.' As for a full-size hole in the future, the climatology will probably ensure that we are, to some extent, protected. But that may be missing the point with regard to the measured ozone loss in the springtime in the north. The transport of ozone from equator to pole is at its strongest at that time and, normally, ozone levels would be on the increase.

To Susan Solomon, the situation is akin to watching a fire hydrant burst – so much is happening, both dynamically and chemically, that it is difficult to make out the details of conditions within the Arctic ozone layer. 'Even if total ozone remains fairly constant over the North Pole,' Solomon said, 'it is probably being destroyed as it should be on the increase at that time.' The problem is, to continue her metaphor, we have no idea how long the hydrant will stay on in the future as chlorine loading reaches a peak after the turn of the century. Satellite results published a few weeks later showed just how precarious the flow of ozone from equator to pole really is.

On 15 April 1993, the front cover of *Nature* carried an eyecatching cover illustration. Two polar projections of both hemispheres of the Earth were displayed with doughnut-shaped, psychedelically hued blobs around their respective poles. Closer inspection revealed that they represented levels of chlorine monoxide in the stratosphere some six months apart. What was most surprising, perhaps, was that the blob over the Arctic in February 1993 was more pronounced than that over the South Pole in August of the previous year. Heralded as 'New angles on ozone depletion', a paper by Joe Waters and colleagues chronicled the first eighteen months' findings from the Microwave Limb Sounder.

As we saw in Chapter Five, the MLS was least disturbed by the Pinatubo aerosol in the immediate aftermath of the explosion and, thus, it had produced scientifically interesting observations of polar ozone before any of the other UARS instruments. Waters's paper chronicled the initial findings over the South Pole but of rather more immediate interest were its most recent observations, barely a few weeks old. At first glance, the MLS measurements during the late winter of 1992–93 showed that ozone loss had resulted from the prolonged cold and was rather more severe than that observed in 1991–92. At a comparable altitude in the stratosphere, Waters et al. declared that 'vortex-integrated ozone decreased by 0.7 per cent per day on average between 10 February and 1 March, whereas during the same period in 1992, it increased by 0.3 per cent per day on average'.

As late as 4 March, chlorine monoxide stayed almost at 1 part per billion but it was not possible to see the exact distribution of the ozone loss towards the mid-latitudes in the absence of tracer measurements. Of the two UARS instruments capable of observing tracers, CLAES had by this time virtually stopped working and the HALOE instrument was looking in the opposite direction from Waters's instrument. Nevertheless, it was clear from MLS observations that the delay in the stratospheric warming meant that ozone levels across the northern hemisphere were on average 10 per cent lower and, towards the poles, some 20 per cent lower than normal. These findings agreed with the ozone trends measurements obtained by the various TOMS and SBUV instruments, which were published a week later in *Science* magazine.

Curiously, the TOMS observations made a lot of newspaper headlines, whereas the MLS ones did not. Joe Waters and his colleagues had observed something of fundamental importance which TOMS could not. It was clear that ozone supplied from the equator had kept the spectacle of an ozone hole from occurring. But, as Joe Waters made clear when he discussed these initial results from 1993 with me a few days later, it may not always be thus.

As the rise of carbon dioxide continues unabated, surface temperatures will increase throughout the globe. While climatologists can argue until they're blue in the face about the exact amount of that rise, it is clear that the temperature difference between the equator and the pole will decrease. As we saw in Chapter Six some authorities believe that mid-latitude storm systems (which provide Britain with its eternally bad weather) will get weaker and the stratosphere will be cooled as wave activity from below will not deliver warmer air. Others believe that storm systems will become more active as evaporation from the oceans will increase and serve to warm the stratosphere. Whichever situation results, however, the distribution of ozone from the equator to the pole will be fundamentally changed in ways which remain unknown. 'In the future, all that is saving us from an Arctic hole is the transport of ozone by atmospheric circulation,' Joe Waters said. 'I think that's something which we're only just starting to realise and I really do believe that it's a real cause for concern.'

In at least one respect, the ozone story has turned full circle. For the issue that catalysed the first concerns about human influence on the stratosphere two decades ago is now once more starting to preoccupy atmospheric scientists around the world: supersonic air travel. Its effects on ozone depletion are being considered in the light of findings from the most recent experiment campaigns and their observations of the role of sulphate aerosols.

Models of stratospheric chemistry which do not include heterogeneous processes predict that the emission of nitrogen oxides from the next generation of supersonic aircraft will significantly reduce ozone levels worldwide. However, when heterogeneous reactions – particularly that of nitrogen pentoxide on sulphate aerosols – are included, little damage results. Clearly it is a contentious area of research which will require greater clarification, for the real atmosphere is infinitely more complex than even the best mathematical computer models that attempt to represent its workings.

As part of its High Speed Research Program, NASA will spend nearly $40 million until 1995 to look at questions raised by technological issues such as engine design and pollution suppression. Concurrently, another group of scientists has been meeting since 1990 under the aegis of NASA's Atmospheric Effects of Stratospheric Aircraft (AESA), a working group which has focused on the scientific questions posed. 'What we are trying to do is to understand the background conditions in the atmosphere in which future aircraft emissions will be injected,' said Richard Stolarski of NASA's Goddard Space Flight Center. 'To some extent, the altitude at which they will fly is crucial, for the role of NOx changes quite distinctly at the tropopause.' As most aircraft will fly in the lower stratosphere where the ozone layer is at its most sensitive to change, there is concern that the situation will have its greatest potential for harm.

Although the AESA Working Group will not report its complete findings until 1995, preliminary calculations have received a great deal of press interest. At the Paris Air Show in June 1993, reporters were excited when a NASA official told them that a fleet of 600 supersonic aircraft would deplete perhaps as much as 1 per cent of the ozone layer worldwide. But the models used did not have the benefit of heterogeneous chemistry within them, and, as Stolarski noted, 'the keys to the kingdom are held in the atmosphere. We're not yet there in terms of the modelling as we are in a data-poor situation.' Through the funding made available to AESA, two aircraft campaigns have been sanctioned which will help remedy this situation.

The need to move ever increasing numbers of passengers ever more quickly comes from the rise of economies in the Far East and along the Pacific rim. As Asia and Australia industrialise, supersonic aircraft will play a part, for little opposition is going to be expressed about sonic booms across vast unhabited stretches of the Pacific. Some estimates suggest that by the year 2005 – when, coincidentally, chlorine loading of the stratosphere will have reached a peak – more than 300,000 passengers will be travelling 5,000 kilometres every day. The aircraft used will probably be behemoths having twice the range of Concorde and three times as many passengers, yet equipped with lean-burn engines to minimise the emission of nitrogen dioxide which would help catalyse the destruction of ozone. 'Aircraft manufacturers won't commit to a design unless they know what level of emissions are going to be acceptable,' said Howard Wesoky, manager of NASA's High Speed Research Program. At an estimated cost of $15 billion to develop, the aircraft will need to meet environmental impact standards.

Scientists involved with the AESA will look at the input of NOx while technologists with the HSRP will consider the engines which are going to be used. It is already clear that there may be 'corridor' effects caused by repeated emissions along the most popular routes. To try to quantify such small-scale perturbations, there will have to be considerable improvements in the reliability of atmospheric models. 'If the results look promising, then we'll continue with the development of the technology based on emission levels which are appropriate,' said Wesoky. Once the UNEP–WMO assessments have considered the HSRP findings, the International Civil Aviation Organisation will set a standard for engines, no earlier than 1998.

Of the two aircraft campaigns sanctioned by AESA, one has already finished. In October 1992 and May 1993, a NASA ER-2 was fitted out with a new suite of experiments as part of a campaign called SPADE – the Stratospheric, Photochemistry, Aerosols and Dynamics Expedition. From flights out of NASA Ames at both sunrise and sunset, scientists have been able to learn more about the heterogeneous processes which seem to continue when gas phase photochemistry shuts off when the Sun is gone. 'On SPADE we had instruments to measure each of the major families of free radicals, plus a number of reservoirs,' said Stolarski. For the first time it has been possible to measure all the chlorine, bromine, nitrogen and hydrogen radicals – ClO_x, BrO_x, NO_x and HO_x – involved in catalytic destruction of ozone. The last group includes the ever elusive OH radical, which is known to drive a lot of the photochemical reactions, and an instrument designed by Jim Anderson's Harvard team was capable of the sensitivity required to sniff them out. 'We've got some useful numbers,' he told me after the most recent phase of the SPADE campaign had finished.

The next aircraft campaign will use the know-how gained by SPADE to make far more sensitive measurements. Known as the Measurements of the Atmospheric Effects of Stratospheric Aircraft – MAESA – it started in March 1994 and employed a Perseus drone flown out of Darwin, Australia. The campaign's aim is to quantify the balance between creation and loss of ozone high above the tropical regions over a six-month period. Since the tropics serve as the production line for stratospheric ozone, fleets of high-speed aircraft which traverse the central Pacific may adversely affect how that ozone will subsequently be transported around the globe. Once again, before it is possible to predict future changes with any degree of accuracy, it will be first necessary to quantify the state of the tropical stratosphere there today. At present, nobody really knows whether heterogeneous chemistry will play an important role there, or, indeed, the exact details of how the tropics exchange ozone to higher latitudes. At the same time as the Perseus flights within MAESA, another NASA campaign using the ER-2 will help provide answers to these and other questions by flying out of Christchurch, New Zealand.

The United States is not alone in funding research into future aircraft emissions. A smaller, but no less diverse, effort is being coordinated via the European Commission and is termed AeroNOx. After the media hoopla in the early 1970s, there is a sense this time around that scientific agencies will bend over backwards to avoid the controversies in which they found themselves embroiled last time. Of all the interviews carried out for this book, my interviews with NASA's Wesoky was the only one attended by a public information 'minder'. 'In the 1970s, the scientists and the representatives from the aircraft industry worked separately,' he explained when asked about this point. 'Now we're working together and we hope to avoid some of the contentiousness.'

While American scientists have headed south, their European colleagues have once more headed for Kiruna, this time determined to avoid the pitfalls of the EASOE campaign two years earlier. This latest campaign will last for over eighteen months and involve smaller, intensive bursts of activity, mostly concentrated in the springs of 1994 and 1995. Its aim is to focus European research on the chemistry of the stratosphere across the whole of the northern hemisphere. 'We want to have a new approach,' said Jean-Pierre Pommereau. 'We want to be able to measure what we really want to measure and not just those things that we can. It is definitely the 'son of "EASOE".'

In 1992, the shorter, more intense EASOE campaign exacted a toll which few scientists wished to experience again. As well as having workloads dictated by the vicissitudes of Arctic weather, many scientists had either to commute to Kiruna repeatedly or else endure separation from families and loved ones for weeks (and suffer the gastronomic limitations of the canteen at the launch base). And the larger payloads were often unstable as they ascended leading to their loss. It is a measure of the antipathy the French balloon teams felt towards the follow-on that they suggested it be called the 'French Acronym Campaign' to make an acronym whose pronunciation would allow them to vent their frustrations. In the end, they settled on the SEcond Stratospheric Arctic and Mid-latitude Experiment (whose acronym is easier on the Gallic tongue than EASOE ever was).

Perhaps the greatest problem during any intensive campaign is the detailed analysis of data returned as the experiment progresses. Each successive balloon flight will hopefully reveal answers to certain questions, pose further ones and then help piece together a clearer picture of how conditions evolve as winter passes into spring. Ideally, there should be sufficient time to prepare for the next flight, allowing scientists to amend where the instruments will need to look or where samples will be taken. Experimenters may also need time to debate whether it would be a better idea to fly different combinations of instrument packages, if possible. During EASOE, however, planning time was at too high a premium and the scientists involved felt they were drowning in a deluge of data. The scientists coordinating the SESAME campaign hope that by extending the time over which the flights are carried out, it will be possible to obtain an optimum scientific return.

The modelling developed during the previous Kiruna campaign will be expanded so that near real-time data from the SESAME campaign may be plugged into the British universities stratospheric model to act as an operational tool. 'We've made a number of improvements to our models in the two years since EASOE,' said John Pyle, who has been aided in the task by colleagues in the Chemistry Department at Cambridge University. 'I think we're in a much stronger position for this longer, follow-on campaign.' Once every week, Pyle hopes that by accessing his model, which runs on the Cray Y-MP at the Rutherford Appleton Lab, he will be able to make a forecast of the state of the Arctic stratosphere over the following few days. Actual meteorological data will be used to initialise the model and predict where the most interesting chemistry will be found. As a result, balloon and aircraft flights can be coordinated to investigate these particular regions. 'My hope is that we'll be able to say, for example, "Next Friday afternoon would be a good time to measure chlorine nitrate within the vortex," Pyle suggested. The models are also meant to enable SESAME scientists to be wise after the event, as it were, by assimilating the data to help diagnose the way in which the stratosphere actually behaved.

The longer SESAME campaign will allow for a more sedate pace of activity so that data can be sufficiently analysed in between flights. As well as having greater numbers of experimental packages, SESAME will cast its geographical net much wider. Once again, the overall campaign will comprise ground-based observations, balloons and aircraft measurements but they will not just be concentrated on Kiruna and Scandinavia. SESAME will try to extend coverage longitudinally so that experimenters will observe what happens to air that is processed within the Arctic vortex and how it subsequently circulates outside it. For example, it is planned that ozone sondes will be launched as far east as Zhigansk in Siberia. 'We'll be seeing the same ozone that has passed on to a different area,' Pommereau said. 'If there are any chemical effects, they should be amplified.'

There will also be more measurements obtained at lower latitudes and at all points in between. Previously, measurements of ozone in and around the vortex made at ground-level missed a whole swathe of Europe centred on 60°N. This meant there were no observations from this crucial region over which the vortex was still holding together in the late winter when the Sun returned. 'We've observed the coldest areas when there is no sunlight,' Pommereau explained. 'But when the sunlight returns, we only have observations of the warmer regions near to the vortex edge.' This means that there has been little ground-based coverage of the area where ozone is most likely to be depleted. So whereas EASOE had involved painting broad strokes across the canvas that is knowledge of the stratosphere, it is hoped that SESAME will result in finer portraiture.

When SESAME began in earnest in January 1994, Jean-Pierre Pommereau was excited by some of the wilder proposals that had been suggested, including his own contribution in the form of a balloon he calls Le Montgolfiér infrarouge, or Infrared Montgolfier. Photographs strewn around his office at Verrières-le-Buisson revealed that it bears a resemblance to a giant bag made of kitchen foil, for it is open at the bottom and fills up with air as it is heated up by the Sun. As a result, it rises during the day and slowly descends in the evening. By having one side made of aluminium and the other of plastic, the heat can be trapped to keep the balloon afloat for many weeks at a time. As there is no pressurised gas within the balloon envelope, it will not lose buoyancy through the escape of gas. 'The Infrared Montgolfier should enable us to follow the vortex for as long as we like,' he said.

SESAME may also involve other 'crazy ideas' as Pommereau described them. Italian scientists, for example, want to fly experiments on a Russian plane known as *Geophysika*. The French space agency, CNES, hopes to fly a Perseus drone out of Kiruna in 1995 whenever polar stratospheric clouds appear, so that instruments aboard could sample them directly for the first time ever. But undoubtedly the craziest proposal of all is the most richly ironic. Thirty years after it was touted as the very incarnation of environmental evil, Pommereau hopes to be able to borrow an Air France Concorde to fly in and out of the polar vortex. The hope is that its exhaust plumes will interact with the chemistry there to produce stratospheric clouds. If it were flown in a zigzag pattern off the Norwegian coast, ground-based instruments could quantify the effects of Concorde's emissions, which would go some way to help unravel how stratospheric chemistry may be changed by future supersonic aircraft. 'It's going to be an interesting few years ahead,' Pommereau said with a smile. Though SESAME will not be the last word on the subject of the Arctic stratosphere, it will result in the most detailed picture yet obtained.

For many years now, the US Navy has run Operation Deepfreeze from Christchurch International Airport on the sheltered eastern seaboard of New Zealand's South Island. What sounds suspiciously like the name of something from an old episode of *The Man from Uncle* turns out to be the collective description for the logistical operations which supply American bases in Antarctica. As a result, the local inhabitants have become used to the sight of large Hercules transporters heading south during the summer months. In the austral autumn of 1994, the South Islanders will become inured to the spectacle of regular flights of the ER-2 in and around Christchurch. The Airborne Southern Hemisphere Ozone Experiment (ASHOE) will be based in the same hangars as Operation Deepfreeze from the middle of March to the end of October 1994. It will herald the most detailed examination of ozone depletion over the whole of the southern hemisphere ever attempted.

'We want to cover the whole hemisphere and follow the evolution of the polar vortex in its entirety,' said Adrian Tuck of the NOAA Aeronomy Lab who will, once again, serve as the all-important mission scientist. 'We want to get into the region around the edge of the vortex where air is exchanged with the rest of the hemisphere.' Like SESAME, the NASA campaign will involve two-week bursts of activity, but over a shorter, six-month time period. ER-2 flights will be centred on early April, early June, early August and throughout October, the all-important time when the ozone hole forms and opens over Antarctica. A specially dedicated ER-2 will fly chemistry and tracer experiments, occasionally interspersed with dynamics and radiation packages which will be flown both north and south. The experiments involving chemistry will profile the same regions in and around the vortex on each successive flight to gauge how much processing has occurred and where the air parcels have moved in the interim.

The surface winds at Christchurch are comparable to those at Stavanger or Punta Arenas. Climatically, there is a greater statistical chance of being able to fly, and logistically, the airport already handles all US Navy flights south. 'We will be able to fly when we want and deep into the vortex,' Adrian Tuck said. The ER-2 will be able to penetrate as far south as 67°S, well within the boundaries of the vortex, whose boundaries wobble around 65°S. One major difference compared to earlier campaigns will be the absence of the DC-8 which, in any case, is booked for another NASA research campaign. The main reason is that with the Upper Atmosphere Research Satellite up and running for over two and a half years, its observations will provide a greater 'hemispheric' view into which the ER-2 observations can be fitted. 'With UARS up there, we have a much more powerful technique to get a global picture, 'Adrian Tuck added. 'UARS has transformed how we look at the stratosphere.'

Because of the interests in looking at the hemispheric implications of future supersonic aircraft fleets, the production and loss of ozone at tropical latitudes will also be investigated during ASHOE to complement the MAESA campaign which will be carried out more or less simultaneously. En route to Christchurch, the ER-2 will perform measurements from Hawaii down to the tropics and back. 'The exchange of air between the tropics and higher latitudes is a very interesting area of research,' Tuck explained. Perhaps the most important question to be answered by the ASHOE campaign overall concerns the extent to which chemical processing within the vortex affects the rest of the southern hemisphere stratosphere. Unlike the north, where climatological distortions generally disturb the atmospheric circulation around the pole, the flow of air around the South Pole is much easier to understand. As a result, it may be possible to check the validity of a theory for which Adrian Tuck has become a champion: that the vortex behaves like a 'flow processor'.

Conventional wisdom has it that the Antarctic vortex is completely isolated from the rest of the atmosphere, and Tuck acceded that 'to the simplest approximation that is very much true'. Yet there is a growing body of evidence which hints that the vortex is rather leakier than had originally been supposed. Various UARS observations during previous austral springs have shown distinct blobs of processed material effectively 'thrown off' the vortex and extending towards lower latitudes. Tuck believes that air descends into the vortex from above, dries out as it descends and, after being chemically processed, is ejected sidewards around the tropopause. One of the strongest pieces of evidence for such a scenario comes from the UARS Halogen Occultation Experiment which scanned the Antarctic stratosphere over two complete spring seasons (1992 and 1993) and part of the 1991 spring after it was launched. 'It is clear that there is dehydrated air which stretches from Antarctica to the tropics,' says the HALOE principal investigator, Jim Russell III of NASA's Langley Research Center. 'There is some mixing in with other air masses and it seems likely that it originates at the pole.'

In the complex chemistry of the stratosphere, the total amount of hydrogen present should never change. Hydrogen from both water vapour and methane (which is oxidised to produce water vapour) account for some 6.7 parts per billion by volume as seen by HALOE in the upper stratosphere. In the lower stratosphere, however, there is at most 4 ppm, which – to quote Russell directly – 'means you have removed some hydrogen, probably due to dehydration which takes place on PSCs'. In September and October, there are losses of around 8–10 per cent of total hydrogen across much of the southern hemisphere mid-latitudes.

Other HALOE observations show that there is intense descent within the centre of the vortex, which means that photochemically enriched air is being brought down from above. Russell, like many other atmospheric scientists, is not as certain as his Aeronomy Lab collaborator of the unequivocal existence of a flow processor. 'I think it's largely a question of semantics,' Russell explained. 'There is no doubt there is some processing which stretches out beyond the vortex. We need more extensive observations to show how and where air is exchanged.' The ASHOE campaign, buttressed by HALOE observations of chemical tracers, will go some way to returning new data to confirm the possibility of the flow processor, and Tuck himself adds that 'there are lots of pieces which we need to fit together'.

Again, at first glance, such discussions about the extent to which air is being processed may seem like esoterica of interest only to scientists intimately involved with the nuances of complex research. However, the notion of a flow processor may portend something more troublesome. Adrian Tuck, for one, firmly believes that no matter what the degree to which flow processing affects the rest of the southern hemisphere, its cumulative effect may not result in a slow encroachment. 'What concerns me most is that the ozone hole, for example, suddenly appeared,' Tuck explained. 'It seems more than possible to me that the cumulative effects of flow processing across much of the southern hemisphere may also become apparent in a similarly dramatic fashion.'

Only time and intense scrutiny by aircraft and satellites will tell us how close to reality this most worrisome of possibilities may become as we approach the new millennium.

Twenty years after ozone destruction by CFCs was first postulated, the science of the stratosphere in all its paradoxical and contradictory manifestations continues to perplex researchers around the world. The nearest to a last word on the subject should surely go to the scientist whose most enduring contribution remains its most spectacular and frightening discovery. It was Joe Farman's characteristic persistence and single-mindedness to what was dimly perceived as an obscure backwater of science that led him and his British Antarctic Survey colleagues to stumble upon the ozone hole in the first place. Nearly a decade later, it still remains an abhorrently fascinating reminder of how shocking developments in ozone research can truly be. The same quiet, pipe-smoking charcter he always was, Joe Farman is called by dint of what he terms his 'horrible discovery' to officiate at any number of meetings, conferences or colloquia. This he does with equanimity, if not pleasure, for he will gladly discuss the problems which ozone depletion could bring about.

In many ways, human experience provides a better marker than dry scientific facts. Joe Farman's career spans the full breadth of the modern phase of the ozone story, for he was born into a world where CFCs had just been discovered and the existence of the ozone layer would shortly be unequivocally confirmed. Starting with his work on the ice in the International Geophysical Year, his rigorous and methodical observations provided the basis upon which the certainty for ozone loss could be gauged. While the debate intensified, particularly in the United States in the early 1970s, Farman kept his head down, continuing with his measurements, and stayed largely incognito. After the announcement of the ozone hole, he has become a key player in the story, and is today more likely to be attending conferences on clinical dermatology or environmental effects on ecosystems than colloquia on the iniquities of stratospheric chemistry.

For Farman, the lesson from the last decade is that education is the key to the future and the surprises which will doubtless appear. 'We must train people to ask the right questions,' he says. For him the scientific method remains the only hope for the future as research becomes ever more specialised. A good grounding in science generally will help to understand the complexities presented by ozone depletion in all its forms. Many people will be attracted to the subject for they feel they will be able to help 'solve' the problem, and Joe Farman himself says quite candidly that 'it's very difficult to keep your emotions apart from your scientific work'. As he has become older and the ozone problem self-evidently more serious, his own concerns have intensified. Where once he had the quintessential English reticence toward anything vaguely troublesome, he is today unavowedly green in his views. 'I've had ministers say to me: 'Joe, you don't have to go on like that. I get it every morning at the breakfast table from my sons,'" he recalled with glee.

As to his own part in the story, he views himself as the right person at the right time who happened to make a series of observations which could not be ignored. For him the full significance of his findings can be evoked from one of his strongest memories of the largely unknown continent at the bottom of the world. In 1972, he visited the geographical South Pole, where the United States has a scientific base, on a courtesy trip. His most abiding memory of the trip provides a telling encapsulation on the whole ozone saga: 'I can remember standing there and looking at the signpost to all the different capital cities. There I was at the bottom of the world just taking in the beauty and remoteness of it all. And I can remember thinking: Here's the last place on Earth that we're going to damage. Just evoking that memory brings the horror of ozone depletion all the more clearly to me.'

The story chronicled in this book may, at times, have read like a particularly obscure parable from the annals of science fiction. It is filled with heroes, villains, odd events, strange coincidences and peculiar twists and turns. To me, at least, it is reminiscent of that seminal film of the science fiction genre from the 1960s, *The Day The Earth Caught Fire*. The film's slightly preposterous dénouement has the Earth heading towards the Sun and an untimely – not to say fiery – demise. Judicious triggering of nuclear explosions, it is hoped, will veer the Earth away from certain destruction. In the final reel, the camera pans across two newspaper headlines prepared before the fate of the Earth is known. One reads WORLD DOOMED!, the other WORLD SAVED!

In many ways, that is exactly how the problems raised by ozone depletion have been reported. If in the 1980s each and every development in ozone science was taken as a portent of doom, they now seem to reinforce the impression that it might all have been wishful thinking by environmentalists. Some elements of the media have latched on to the Ozone Backlash and two of its seemingly credible shibboleths. The fact that the rate of CFCs heading into the stratosphere are decreasing and observations that ultraviolet light reaching the ground has not palpably increased are both taken as evidence that the world is now somehow miraculously saved from danger.

This is a palpably false conclusion.

Despite the decrease in the rise of CFCs, they are still accumulating in the stratosphere and will remain there for another century. The complications seen thus far from just a decade's existence of an ozone hole in Antarctica are a taste of what may yet come. The atmosphere does not behave quite as even the most perspicacious scientist can predict. The most worrying aspect of the story is that totally unexpected developments have occurred about which scientists have been as open-mouthed as the rest of us. The Antarctic ozone hole, the continuing inexplicability of ozone loss at northern mid-latitudes and the total absence of part of the atmosphere's natural buffering to ozone depletion are just a handful of examples. There is precious little comfort from the fact that, thus far at least, the tattered ozone layer has not gone into irreversible decline.

Doubtless the future will see the ozone issue continue to be reported like those final headlines in *The Day The Earth Caught Fire*. As this book has shown, it would be ridiculous to say both that the problem has somehow sorted itself out or that the very fate of the Earth is threatened. It is simply prudent for scientists to strive to understand the complexities of the stratosphere.

It is abundantly clear that nature will always be one step ahead. If there is a lesson to be learned from the ozone issue thus far, it is surely that it is relentless and unremitting in its capacity to spring surprises. Perhaps the most worrying aspect of the ozone issue for the remainder of the 1990s is that so much remains to be learned. Scientists must keep a constant vigil and be alert for any surprises upon which they will doubtless continue to unwittingly stumble. We should, at the very least, be thankful for that.

Acronyms

AAAS	American Association for the Advancement of Science
AAOE	Airborne Antarctic Ozone Experiment (1987, out of Punta Arenas)
AASE	Airborne Arctic Stratosphere Expedition (1989, out of Stavanger)
AASE II	Airborne Arctic Stratosphere Experiment II (1991/1992, out of Bangor, Maine)
AEC	US Atomic Energy Commission
AESA	Atmospheric Effects of Stratospheric Aircraft (NASA Working Group, Convened 1990)
ASHOE	Airborne Southern Hemisphere Ozone Experiment (1994, out of Christchurch, New Zealand)
BAS	British Antarctic Survey
CFCs	Chlorofluorocarbons
CHEOPS	Chemistry of the Polar Stratosphere (French balloon campaigns, 1987–1990)
CLAES	Cryogenic Limb Array Étalon Spectrometer (instrument aboard UARS satellite)
CMA	Chemical Manufacturers' Association
CNES	Centre National D'Études Spatiales (French national space agency)
CNRS	Centre National pour la Recherche Scientifique (French national research body)
EASOE	European Arctic Stratosphere Ozone Experiment (1991/92, out of Kiruna)
ECMWF	European Centre for Medium-range Weather Forecasting (based in Reading)
EOS	Earth Observing System

ACRONYMS

EPA	US Environmental Protection Agency
ESRANGE	Space Operations Centre, operated by Swedish Space Corporation, Kiruna
GISS	Goddard Institute for Space Sciences (NASA establishment in New York City)
HALOE	Halogen Occultation Experiment (instrument aboard NASA UARS satellite)
HCFCs	Hydrofluorocarbons
IDDEA	Environmental Protest Group, Punta Arenas
IGY	International Geophysical Year (1957/58)
IMOS	Presidential Task Force on the Inadvertant Modification Of The Stratosphere (1975)
IPCC	Intergovernmental Panel on Climate Change
ISAMS	Infra-red Stratospheric And Mesospheric Sounder (instrument aboard UARS satellite)
JPL	Jet Propulsion Laboratory
LPMA	Limb Profile Monitor Of Atmosphere
MAESA	Measurements Of Atmospheric Effects of Stratospheric Aircraft (Aircraft campaign out of Darwen, 1994)
MLS	Microwave Limb Sounder (instrument aboard UARS satellite)
MRF	Met Research Flight
NAS	US National Academy of Sciences
NASA	US National Aeronautics & Space Administration
NOAA	US National Oceanic & Atmospheric Administration
NOZE	National Ozone Expedition (1986, based in Antarctica)
NSF	US National Science Foundation
OCTA	Oxidising Capacity of Tropospheric Atmosphere (European Aircraft Experiment, 1992/93)
POCC	Payload Operations Control Center ('Mission Control' at NASA Goddard Space Flight Center, Greenbelt, Maryland)
QBO	Quasi-Biennial Oscillation
SBUV	Solar Backscatter Ultraviolet (instrument aboard Nimbus 7 satellite)
SESAME	Second Stratospheric Arctic And Mid-latitude Experiment (1994/95, around Europe)
SORG	UK Stratospheric Ozone Review Group
SPADE	Stratospheric Photochemistry & Dynamics Experiment (US Aircraft Campaign, 1992)
SSBUV	Shuttle Solar Backscatter Ultraviolet (SBUV flown on Space Shuttle)
SST	Supersonic Transport aircraft
TECHNOPS	French Balloon Campaign, 1989
TEWI	Total Environmental Warming Impact

TIROS	Television & Infra-Red Observation Satellite (first US weather satellite, 1960)
TOMS	Total Ozone Mapping Spectrometer
TRACE-A	Transport & Chemistry Near The Equator in the Atlantic (aircraft campaign, October 1992)
UARS	Upper Atmosphere Research Satellite (launched September 1991)
UGAMP	Universities Global Atmosphere Modelling Project
UNEP	United Nations Environment Programme
UV-B	Biologically-active ultraviolet radiation
WHO	World Health Organisation
WMO	World Meteorological Organisation

Glossary

Aerosols – Small particles suspended in the atmosphere upon which ozone-destroying chemical reactions can take place. Most aerosols in the stratosphere contain sulphur and result from volcanic eruptions or the burning of fossil fuels.

Air Mass – An arbitrarily-defined 'packet' of air whose motion over the space of a few days can easily be tracked, thereby allowing scientists to determine how much ozone depletion has taken place within it.

Catalytic Cycle – A set of chemical reactions the net result of which is for one chemical species (the catalyst) to be regenerated. In the catalytic destruction of ozone, that fate befalls the chlorine monoxide radical.

Chemical Species – the generic name given to compounds involved with chemical reactions.

Chlorine Monoxide (ClO) – The smoking gun in ozone depletion – one chlorine monoxide molecule can destroy several hundred thousand ozone molecules. When CFCs are broken down by solar radiation, chlorine monoxide results thereby, catalysing the destruction of ozone but itself remaining unscathed.

Chlorine Monoxide Dimer – A special form of chlorine monoxide formed when two chlorine monoxide molecules combine. Its stability at low temperatures allows it to persist in the polar stratosphere in wintertime.

Chlorofluorocarbons – Compounds consisting of chlorine, fluorine and carbon with many industrial uses such as refrigeration, insulation and the cleaning of electronic components. Invented by Thomas Midgley in 1928, they are extremely stable in the lower atmosphere. Their only known loss is slow upward mixing where they are broken down in the stratosphere by solar radiation. Here they break down and release chlorine atoms which destroy ozone molecules.

Column Abundance – A measure of the total number of atmospheric molecules in an arbitrarily-defined area above the Earth. Usually the column abundance in stratospheric research refers to the number of ozone molecules in an imaginary column stretching throughout the atmosphere whose base is a square centimetre.

Free Radicals – The name given to unstable molecular fragments which are involved in catalytic destruction cycles such as ozone depletion.

Gas Phase Chemistry – Chemical reactions which occur between atoms and molecules which are gaseous only.

Heterogeneous Chemistry – Chemical reactions which occur on the surfaces of atmospheric particles, such as aerosols. Chlorine can be shifted into its reactive, ozone-destroying forms due to heterogeneous chemical reactions which take place upon particles within polar stratospheric clouds.

Hydrochlorofluorocarbons (HCFCs) – Compounds which consist of hydrogen, chlorine, fluorine and carbon. Envisaged as short-term replacement compounds for CFCs, now known more damaging than originally thought. They are destroyed much faster in the lower atmosphere so a proportionately smaller fraction of HCFCs enters the stratosphere where they will damage ozone.

Hydrofluorocarbons (HFCs) – Compounds consisting of hydrogen, fluorine and carbon which contain no chlorine so pose no threat to the ozone layer. Like HCFCs, they are destroyed in the lower atmosphere.

Limb – The name given to the curvature of the Earth's atmosphere as seen from space or in the stratosphere.

NOx (Nitrogen Oxides) – These refer to naturally-occuring oxides of nitrogen which can participate in the destruction of ozone. Collectively, the term refers to nitric oxide (NO) and nitrogen dioxide (NO_2).

NOy (Odd Nitrogen) – The collective description for all forms of nitrogen oxides, including those which do not destroy ozone directly e.g. nitric acid, which effectively 'tie up' oxides of nitrogen which would otherwise be available to stop ozone depletion.

NOx/NOy ratio – An indication of the natural buffering to ozone depletion afforded by the oxides of nitrogen in the atmosphere.

Ozone – An unstable molecule made up of three oxygen atoms. Roughly 90 per cent of the ozone in the Earth's atmosphere is found in the stratosphere and is referred to as the ozone layer. Ozone absorbs ultraviolet-B radiation (UV-B) and acts as the primary shield against this dangerous radiation reaching the Earth.

Ozone Depletion – Loss of ozone due to human activity is more accurately referred to as chemical ozone loss, for there are significant natural losses due to dynamical motions within the atmosphere. Ozone is continually being created and destroyed in the stratosphere by a variety of chemical reactions. Because these reactions depend on

GLOSSARY

temperature and the availability of sunlight, there are large variations in the naturally-occuring presence of ozone depending on the season and latitude. Unequivocal ozone loss by chemical means occurs when it can be ascertained that the rate of ozone destruction exceeds that of its natural production.

Ozone Hole – A term used to describe the visual impression of the marked seasonal loss of ozone over the South Pole during the austral spring. Ozone depletion is much less severe in the northern hemisphere in the corresponding Arctic spring season.

Ozone Layer – The region of the stratosphere which contains the bulk of the atmosphere's naturally-occuring ozone, concentrated between 10 and 25 miles (15 and 40 kilometres) from the surface.

Polar Stratospheric Clouds (PSCs) – Microscopic particles of water ice and hydrates of nitric acid which form in the stratosphere above the poles in very cold temperatures. Chemical reactions occur on the surfaces of PSCs to convert chlorine into its reactive forms which destroy ozone.

Polar Vortex – A term used to describe the self-contained body of air above the polar regions which is effectively cut off from the rest of the atmosphere during the wintertime. That over the North Pole is much more 'leaky' than that above the South Pole, and this lesser degree of isolation generally precludes the formation of polar stratospheric clouds and attendant ozone loss.

Stratosphere – The layer of the Earth's atmosphere which lies above 10 kilometres from the surface to around 50 kilometres (roughly 8 to 30 miles). Temperatures in the stratosphere increase with altitude due to the absorption of ultraviolet radiation by the ozone layer.

Tracers – Atmospheric compounds whose abundance in the presence of ozone is well established. They can thus be used to determine whether ozone has been destroyed chemically or dynamically within a given air mass.

Tropopause – The boundary between the lowest part of the atmosphere (troposphere) and the stratosphere above it. Its height above the Earth varies due to the motions of storm systems below which serve to distort it.

Troposphere – The lowest layer of the atmosphere extending up to anywhere between 8 and 15 kilometres (about 5 and 10 miles) depending on weather conditions.

Selected Bibliography

The primary source for information in this book came from the interviews with scientists listed later. But some written material was indeed consulted. 'General' material is similar to the level of this book: 'technical' material would require more advanced scientific knowledge.

General Books

Oliver E. Allen & The Editors of Time-Life, *Atmosphere* (Amsterdam, Time-Life, 1983).
Richard Elliot Benedick, *Ozone Diplomacy – New Directions in Safeguarding The Planet* (Cambridge, MA & London, Harvard University Press, 1991).
Nigel Calder, *Spaceship Earth* (London, Viking, 1991).
Lydia Dotto & Harold Schiff, *The Ozone War* (New York, Doubleday, 1978).
John Gribbin, *The Hole In The Sky – Man's Threat To The Ozone Layer* (London, Corgi, 1988).
John J. Nance, *What Goes Up – The Global Assault On Our Atmosphere* (New York, William Morrow, 1991).
Sharon L. Roan, *Ozone Crisis: the 15-Year Evolution Of A Sudden Global Emergency* (Chichester & New York, Wiley Science Editions, 1990).
Stephen H. Schneider, *Global Warming – Are We Entering The Greenhouse Century?* (New York, Vintage Books, 1990).
Jonathan Weiner, *The Next One Hundred Years* (London, Rider Books, 1990).

Selected Bibliography

Technical Books

Robin Russell Jones & Tom Wigley (eds), *Ozone Depletion – Health & Environmental Consequences* (Chichester, John Wiley & Sons, 1989).

J. T. Houghton et al. (eds), *Climate Change: The IPCC Scientific Assessment* (Cambridge, Cambridge University Press, 1990); *Climate Change 1992: The Supplementary Report To The IPCC Scientific Assessment* (Cambridge, CUP, 1992).

J. T. Houghton, *The Physics of Atmospheres* (Cambridge, CUP, 1977).

NASA Upper Atmosphere Research Program, *Research Summaries 1990–1991* (January 1992).

Collections of Papers

The UNEP/WMO assessments are standard references, collecting together the wisdom at the time of publication. The 'blue books' appeared in January 1986, the more detailed 'Synthesis Reports' in 1989 and the most recent assessment in December 1991.

The UK Stratospheric Ozone Review Group reported annually at the behest of the Department of the Environment since 1986. The SORG reports are all available from Her Majesty's Stationery Office in London.

Complete results from the Airborne Antarctic Ozone Expedition were published in the *Journal of Geophysical Research* published by the American Geophysical Union. They are:

J. Geophys. R, Vol 94, No. D9, 30 August 1989, No. D14, 30 November 1989.

Complete Results from the Airborne Arctic Stratospheric Expedition, were published in:

Geophysical Research Letters, 17, Number 4, March 1990, Supplement.

A most recent issue concerned aspects of both Arctic and Antarctic ozone depletion:

Polar Ozone Special Issue:

J. Geophys. Res., 97, No. D8, 30 May 1992.

Complete results from AASE II and EASOE are referred to in Chapter 10 references.

Recent Technical References

Martyn Chipperfield, 'Satellite maps ozone destroyed', *Nature*, 362, pp. 592–593, 15 April 1993.

David W. Fahey et al., '*In Situ* measurements constraining the role of sulphate aerosols in mid-latitude ozone depletion,' *Nature*, 363, pp. 509–514, June 1993.

J. F. Gleason et al., 'Record Low Global Ozone in 1992', *Science*, 260, pp. 523–526, April 1993.

Alan Plumb, 'Mixing And Matching', *Nature*, 365, pp. 489–490, 7 October 1993.

J. A. Pyle et al., 'Ozone loss in Antarctica: the implications for global change', *Phil. Trans. R. Soc. Lond.*, 38, pp. 219–226, 1992.

W. J. Randel et al., 'Stratospheric transport from the tropics to middle latitudes by planetary-wave mixing', *Nature*, 365, pp. 533–537, 7 October 1993.

'The Upper Atmosphere Research Satellite (UARS): Results From the First Year and a Half of Operations', *Geophys. Res. Letts.*, 20, No. 12, pp. 1215–1330, 18 June 1993.

J. W. Waters et al., 'Stratospheric ClO and ozone from the Microwave Limb Sounder on the Upper Atmosphere Research Satellite,' *Nature*, 362, pp. 597–602, 15 April 1993.

Interviewees

My primary source for information came from interviewing the following people. There were more than a dozen people in Britain and the United States who preferred to remain anonymous. Those scientists whose names are asterisked I contacted on further occasions and the interviews cited here are the first of many. The interviewes detailed here are arranged chronologically with an indication of where they took place:

1992

AT ESRANGE, Kiruna during EASOE:
Neil Harris*	March 13,14 & 16
Jean-Pierre Pommereau*	March 15 & 16
Claude Camy-Peret*	March 14
John Starkey	March 14 & 15
Henri Ovarez	March 15 & 16

AT JPL, Pasadena, California:
Dennis Flower	May 19
Joe Waters*	May 26

At U. Cal Irvine:
Sherry Rowland*	May 29

At NASA Ames:
Philip Russell	June 3
John Arvesen	June 4

At Lockheed Missiles & Space Co:
Aidan Roche	June 4

How Soon Is Now?

At NASA Ames:
Jim Barrilleaux — June 5
Estelle Condon — June 5

At NCAR:
Sasha Madronich — June 8

At NOAA CMDL:
Jim Elkins — June 12

At NCAR:
Rolando Garcia — June 15
Alan Fried — June 15
Geoff Tyndall — June 15

At NOAA CMDL:
Sam Oltmans — June 15 & 18

At NOAA AL:
Adrian Tuck* — June 17
David Fahey — June 17

At NCAR:
Andy Weinheimer — June 19
John Gille — June 19
Michael Coffey — June 19

At U. Col. Boulder:
Gary Rottman — June 19

At Chautauqua Park, Boulder:
Stephen Schneider — June 20

At Hotel Boulderado:
Guy Brasseur — June 20

At NOAA AL:
Daniel Albritton* — June 22
Michael Profitt — June 22
Susan Solomon* — June 22

At Denver University:
Frank Murcray — June 22

At NASA GISS:
James Hansen* — June 25

At Royal Commission:
Sir John Houghton — September 4

INTERVIEWEES

At BNSC:
John Harries*   September 14

In Punta Arenas:
Bedrich Magas   October 19–23
Oscar Riquelme   October 19
John Gibbons   October 20
Walter Ulloa   October 20
Jaime Abarca   October 21
Sergio Cabrera   October 22

In London:
James Lovelock   November 13

At St. Thomas's Hospital:
Robin Russell Jones   November 16

Telephone interview:
Desmond Walshaw   November 26

At Met Research Flight:
Danny McKenna   December 18

Telephone interview:
Keith Shine   December 22

1993:

At NASA HQ:
Joe McNeal   January 5
Michael Kurylo   Jan 5
Bob Watson*   Jan 5 & 14
Shelby Tilford   Jan 5

AT NSF:
John Lynch   Jan 6

At Greenpeace, Washington DC:
Damien Durrant   Jan 7

At Harvard:
Jim Anderson*   Jan 8

At NASA Goddard:
Gene Feldman   Jan 11
Arlin Krueger   Jan 11
Rich Stolarski   Jan 11 & 12

How Soon Is Now?

Anne Thompson	Jan 12
Mark Schoerbel	Jan 12
Carl Reber	Jan 13
Ricky Rood	Jan 13
Ernest Hilsenrath	Jan 13
John Speer	Jan 13

At Center For Global Change, College Park:
Alan Miller — Jan 13 & 15

At CMA:
Elizabeth Watson — Jan 14
Bob Orfeo — Jan 14

At AFAES:
Katie Smythe — Jan 14

At NASA HQ:
Howard Wesoky — Jan 15

Telephone Interview:
S. Fred Singer — February 26

At Imperial College, London:
Joanna Haigh — March 11

At Université Curie, Paris:
Claude Camy-Peyret — March 12

At the European Space Agency, Paris:
Philip Goldsmith — March 15

At CNRS Service D'Áeronomie, Verrìeres-le-Buisson:
Jean-Pierre Pommereau — March 15

At the University of Cambridge:
Rod Jones — March 23
Michael McIntyre — March 23
Geoff Holmes — March 23

At the British Antarctic Survey:
Jonathan Shanklin — March 24
Brian Gardiner* — March 24

At Oxford University:
Fredric Taylor* — March 31

At the Rutherford Appleton Lab:
Lesley Gray — March 31

Interviewees

At ICI, Northwich:
Michael Harris — April 13
Archie McCulloch — April 13

At Greenpeace, London:
Corin Milais — April 22
Sue Mayer — April 22

Telephone interviews:
Oliver Schein — May 19
John Austin — August 6
William Brune — August 6
John Langford — August 7
Tony Cox — August 18
Jose Rodriguez — October 7

Index

AAOE *see* Airborne Antarctic Ozone Experiment
AASE *see* Airborne Arctic Stratosphere Experiment
AASE II *see* Airborne Arctic Stratosphere Experiment II
Abarca, Dr Jaime, 2, 3, 287–91
Abbot, Dr Charles Greeley, 11
ABC News, 278
Ad Hoc Working Group of Legal and Technical Experts for the Preparation of a Global Framework Convention for the Protection of the Ozone Layer, 323–4
AEC *see* US Atomic Energy Commission
aeronomy, 45, 86
Aero Nox, 378
aerosols, 5, 62–3, 70, 72–73, 319–20, 327, 358
  see also spray-can propellants
AESA *see* Atmospheric Effects of Stratospheric Aircraft
Africa, annual biomass burning in, 131, 132
ageing process, 285
Agricultural and food Research Council, 313
AIDS virus, 298
air conditioning, 56, 57, 73, 352, 353, 357
Airborne Antarctic Ozone Experiment (AAOE), 93–7, 99, 110, 248, 275, 332–3
Airborne Arctic Stratosphere Experiment (AASE), 99–102, 117, 233, 283, 340
Airborne Arctic Stratosphere Experiment II (AASE II), 28, 104, 106, 110–14, 117, 131, 133, 187, 232–3, 242–5, 247–9, 344, 365–71
Airborne Southern Hemisphere Ozone Experiment (ASHOE), 120, 381–4
Aire-sur-l'Adour, near Toulouse, 17, 31, 372
Alaska, 108, 113, 362
Albritton, Dr Daniel, 318, 326, 327, 330, 333, 343, 348, 349–50
Alconbury, Norfolk (USAF base), 117
ALIAS spectrometer, 367
Allen Woody, 192
Alliance for Responsible CFC Policy, 329
Allied Industries Inc., 338
Alternative Fluorocarbons Environmental Acceptability Study, 353
altitude: air pressure decreases with, 37; DC-8 and, 113; and nitrogen, 100–101; of ozone depletion, 80; temperature decreases with, 37, 38; temperature rises with, 37
Amdahl computer, 221, 222
American Academy of Dermatology, 287
American Association for the Advancement of Science, 250, 257
American Chemical Society, 55, 70, 259
American Geophysical Union, 86, 275
American Meteorological Society, 244
Ames Research Center, California *see* NASA Ames Research Center
amino acids, 4, 306
ammonia (as refrigerant), 55, 352
Anchorage, 112, 113
Anderson, Jim, 233, 245, 249, 365, 366, 368, 377; and AAOE, 97–9; and AAOE II, 112; Bastille Day results

INDEX

(1977), 75–6; and chlorine monoxide, 75, 111, 234; and ER-2s, 137; and Farman's paper, 82; launches new technique, 76; measures OH radical, 74–5, 377; and ozone layer, 369; at press conference, 235–7; and *Wall Street Journal*, 244

Angola, 132

animals: alleged damage to, 2, 272, 277–9, 289, 294; evolution of, 5; research on penguins, 308

Antarctic Science, 308

Antarctica, 47, 88, 140, 381; CFC level measurement, 62, 63; first years of ozone observations, 49–51; formation of chlorine monoxide, 13; and International Geophysical Year, 48; low levels of ozone, 50, 85, 303–4; NOZE mission, 88–91; and NOZE II, 96; phytoplankton of, 305–9, 315; ozone layer starts to change, 78; UV-B levels, 303–7

see also under ozone hole; vortex, polar

antigens, 298

Apollo spacecraft, 144

Arab-Israeli wars (1967 and 1973), 108

Arctic: chemical processes in stratosphere over, 13; CHEOPS campaigns in, 26–7; ISAMS and ozone loss, 168; lowest ozone values ever seen, 27; ozone hole potential, 8, 13, 28, 29, 102, 104, 212, 213, 235, 237, 241, 242, 245, 371, 373, 374, 375; and Pinatubo aerosol, 365; temperature, 14, 26, 28–9;
see also under vortex, polar

Arctic Circle, 12, 368

Argentinian Air Force bases, 94

Arizona, University of, 291

Arosa: Dobson record re-examined in, 85, 175; as a spa, 44; spectrograph measures ozone, 43; spectrometer in, 43–4, 336; and TOMS drift, 174, 175

Arvesen, John, 114, 115, 117, 140

Ascension Island, 126

ASHOE *see* Airborne Southern Hemisphere Ozone Experiment

Asia, 376

Atlantic magazine, 95

Atlantic Ocean, 28; 132, 365; ozone bulge, 131

Atlantis, 177

atmosphere: carbon dioxide and, 308 CFCs in, 62, 63, 68, 69, 324, 362; gases remain in same proportions as at ground level, 37; hydroxyl radicals in, 129–30; lower *see* troposphere; ozone increased in, 127; ozone spread to the poles by circulation of the, 46; temperature measurement, 146; tritium in, 66; warming up, 38

Atmosphere, Oceanic and Planetary Physics building, Clarendon Laboratory, 161

Atmospheric and Environmental Research, Inc. (Cambridge, Mass), 367

atmospheric chemistry, 67, 81

Atmospheric Effects of stratospheric Aircraft (AESA), 375, 376, 377

atmospheric gases, measurement of, 62

atmospheric modelling, 194–208, 376; blueprint of, 199; climate models, 205–7, 210, 221, 222–3; and clouds, 210–11; extrapolation, 198; and oceans, 210, 211; '1-D' models, 196; parameterisation, 210, 211; 3-D models, 197, 198, 212, 224, 225, 226, 227 two-dimensional models, 82, 86, 196, 219, 225, 226; weather forecast, 205, 206, 225, 227, 228

Atmospheric Ozone, 322

atomic power, 57, 65

aurora borealis, 17

Aurora Flight Services, 137, 139

Austin, John, 212, 213

Austral University of Moldavia, 278

Australia, 128, 292, 294–5, 376

Australian National University, Canberra, 313

aviation, and ozone layer, 58–61

Azores, 119

B-52 bomber, 183–4

Backscatter Ultraviolet (BUV) instrument, 148–9

Baeriswyl, Fernando, 282–83

Bahrain, 126

Baker, Kenneth, 339–40

Baker, Dr Richard, 36

Balloon Intercomparison Campaign, 76, 77

balloons, scientific, 10; aircraft and balloon scientists try to coincide observations, 240, 379; and CFCs in stratosphere, 74, 264; and chlorine monoxide, 75; crewed balloons abandoned, 37; first flight (1784), 37; and industrial pollution, 129; hydrogen-filled, 37, 38; Infrared Montgafier, 380–81; and infrared radiometer, 146; North Pole experiments, 12, 13, 15–31; OClO measurement, 240; oxides of nitrogen measurement, 240; and ozone hole, 363; ozone measurements, 58, 89, 127, 371; ozone sondes, 85; and Pinatubo aerosol, 365; SESAME,

403

378–81 and stratosphere, 37, 76–7, 148, 372; and volcanic aerosol, 365–6; and wind conditions, 105, 108
Bangkok 355
Bangor, Maine, 28, 104, 111, 112, 113, 116, 233–6, 240, 243
Barrilleaux, Jim, 103–4, 109, 112, 114, 115, 116, 118–21
BAS *see* British Antarctic Survey
basal-cell cancers, 286, 294
BBC, 339, 342
Bedrich Magas Kusak, 275–7, 283, 284, 287, 289, 299, 300, 301–2, 315
Bellingshausen Sea, 306
Belsk-Duzy, Poland, 301
Benedick, Richard Elliot, 316, 326, 327, 329–30, 332, 334
Benidorm, 296
Berger, Daniel, 301
Berkeley, 194
Bertrand, Abbé, 37
big science, 185, 186, 255
biomass burning, 131, 132, 264, 265
Bishop Tuff, California, 267
Blackbirds, SR-71, 120, 136
Blake, Donald, 133
'blue books' (*Atsmopheric Ozone*), 322–3, 328, 333, 358
Blumthaler, Mario, 300
Blunck, Lothar, 309
Boeing, 59, 60
Bolin, Terry, 135–6
Bondi, Sir Hermann, 254
Bondi Beach, 294, 297
Boots, 296
Boscombe Down, Wiltshire, 122
Boston bombers, 122; first measurements within the stratosphere, 46
Boston Herald, 202–3
Boulder, Colorado, 60–61, 72, 81, 88, 134, 166, 190–91
Boville, Byron, 196, 197, 229
Bracknell, Berkshire, 204
Braniff Airways, 93, 110
Bransfield, 79
Brasilia, 132
Brasseur, Guy, 197, 198, 219
Brazil, 131, 132
Brewer, Alan, 46, 122, 124, 133
British Antarctic Survey (BAS), 6, 50, 78, 83, 253, 254, 302, 323, 385
Brodeur, Paul, 256
bromine, 164, 212, 338, 341, 344, 353, 377
Bromley, D.Allan, 239, 244
Brookhaven National Laboratory, Long Island, 314
Brune, William H., 102, 129, 130–31

Buisson, Henri, 39–43, 45
Burford, Anne M., 73–4
Bush, George, 238, 239, 240, 244, 327, 339, 355–6, 357
butane, 352
Butchart, Neal, 212
butterfly effect, 206, 226
BUV *see* Backscatter Ultraviolet instrument
Buyukmihci, Dr Nev, 279
Byron, George Gordon, 6th Baron, 190, 217

Cabrera, Professor Sergio, 304–5, 307–8
California, University of (Davis), 279
California, University of (Irvine, Orange County), 66, 68, 71, 133
California, University of (San Francisco), 308
California, University of (Santa Barbara), 306, 307
California Institute of Technology 35, 83, 128
Cambridge University, 226, 310, 311, 379
Campo Gibbons, Otway Sound, 279–81
Camy-Peyret, Claude, 13, 18–21, 24, 29–32
Canada, 362; and CFCs ban, 320
Canadian Journal of Physics, 216
Canberras, 124, 125, 126
Cancer Council of New South Wales, 295
Cancer Research Campaign, 296
cancers *see* basal-cell cancers; skin cancers
Cape Canaveral, 141, 154, 177–8, 181
Cape Grim, Australia, 362
Cape Horn, 94, 273
carbon, 55, 133, 357
carbon dioxide, 133, 164, 239, 352, 353; and phytoplankton, 308; and potential Arctic ozone hole, 213; rising levels, 49, 207, 211, 212, 213, 374
carbon monoxide, 134, 162
carbon tetrachloride, 325, 333, 350
carbon-14 radiodating, 65
Cars: air conditioning, 352, 353, 357; carbon monoxide and, 134; and Smog, 35–6
Carson, Rachel, 57–8, 64
Carter, Jimmy, 327
catalytic cycles, 60, 87, 90
cataracts, 2, 277, 278, 279, 281, 282, 283, 288, 289, 291, 297
Cedar Rapids Gazette, 221
Central Aerological Observatory, Moscow, 216

404

INDEX

Central Intelligence Agency (CIA), 107
Centre National d'Etudes Spatiales (CNES), 17, 18, 20, 21, 22, 24–6, 30, 31, 381
Centre National de Recherche Scientifique (CNRS), 13, 27; Service d'Aernomie, 366, 372
cerebral aneurisms, 64
CFC-11, 63, 164, 209, 263, 319, 347, 348, 350, 353, 363
CFC-12, 63, 164, 209, 319, 350, 363
CFC-113, 351
CFC-114, 351
CFC-115, 351
CFCs (chlorofluorocarbons) 385; absorption of ultraviolet light, 13; in aerosols/spray-can propellants, 1, 5, 56, 62–3; in atmosphere, 62, 63, 68, 69, 324, 362; and blue books, 323; as cause of ozone depletion, 53, 69, 70, 99, 270, 335; chlorine in see under chlorine; and ClO measurements, 98–9; and clean air samples, 252, 362; deceleration, 362–3, 387 and Climate Monitoring and Diagnostic Laboratory, 191; damage to ozone, 61, 164, 384; and greenhouse effect, 208–10; HCFCs as substitutes for, 342, 347–9, 353; as a 'hot topic', 250; and IMOS, 71; increased production, 69, 70, 74; 'lesson' of, 52; Lovelock measures, 62–4, 67; and methane, 135; Midgley creates, 55–7, 61; Nimbus 7 confirms ozone depletion by 147; provision of production data, 306; question of ban/phasing out, 6, 7, 72–3, 98, 175, 188, 238–9, 256, 258, 262, 317–58; and refrigeration 55–6, 72, 300, 333, 338; safe disposal, 357; Sherwood recognises threat posed by, 5, 6, 65, 256; in stratosphere, 5, 6, 7, 13, 67, 68, 69, 135, 263, 265, 387; and Third World, 332, 334, 345, 346; Watson's part in appraising White House of situation, 239
Challenger, 151, 153, 154, 178
Channel 7, KGO, 278
Chapman, Sydney, 45, 47, 48, 49, 58, 161
CHEM-1 (chemistry satellite), 187–8
Chemical Manufacturers Association, 6, 318, 329
chemistry: and dynamics, 224; successes in First World War, 53, 57
Chemistry of the Polar Stratosphere (CHEOPS), 101; CHEOPS I, 26; CHEOPS II, 26; CHEOPS III, 26–7

Cherwell, Lord *see* Lindemann, Frederick
Chicago, University of, 64, 65, 303, 304
Chile, 3, 99, 111, 288–91, 301, 302, 310, 315
Chile, University of, 304
Chilean Agriculture and Farming Service, 278
China, 49, 332, 334, 345, 346
chlorine, 377; and Antarctic vortex, 96; and Arctic vortex, 369–70 and CFCs, 5, 6, 7, 13, 29, 164, 264, 269, 338; and HCFCs, 342, 347; and MLS, 170; and Montreal Protocol, 276; and need for continuous ozone-monitoring system, 344; and neutron accelerator, 67; nitrogen reacts with, 29; and ozone loss, 1, 6, 13–14, 68, 69, 71, 97, 101, 133, 160, 212, 337, 341, 373, 377; peak of loading in atmosphere, 188, 214, 363–4, 366, 376; and Pinatubo aerosol, 368; and potential Arctic ozone hole, 213; on PSCs, 88; Rowland studies, 66; Singer on, 269; at South Pole, 110; in stratosphere, 6, 87, 91, 140, 149, 172, 188, 189, 258, 303, 328, 363, 368, 369, 376
chlorine dioxide (OClO), 89, 170
chlorine monoxide (ClO) ('smoking gun'), 99, 101, 111, 171, 238, 240, 367, 369, 373, 374; CFCs' role clinched, 98–9; and chlorine dioxide, 89; consumption of ozone, 74, 87, 88, 97, 245; ER-2 measures, 98–9; formation of, 13, 90; high levels recorded, 104, 232–5; levels fall, 242; and MLS, 160, 168, 170; and nitric acid, 100; and NOZE, 89–90; and ozone hole, 90, 160, 230; in stratosphere, 74, 75; and UGAMP model, 226; Waters and, 224, 247
chlorine nitrate (ClNO$_3$) 29, 71–2, 86–7, 91, 101, 219, 233, 365, 368, 370, 379
chlorofluorocarbons *see* CFCs
Christchurch, New Zealand, 89, 120, 378, 381, 382
Churchill, Winston Spencer, 122
CIAP *see* Climatic Impact Assessment Program
Circular 175, 329, 330
CLAES *see* Cryogenic Limb Array Etalon Spectrometer
Clarendon Laboratory, Oxford University, 41, 42, 44, 46, 121, 146, 161, 162, 166, 167
Clark Air Force Base, 157
Clean Air Act (US; 1976), 320, 357

climate forecasting *see under* atmospheric modelling
Climate Monitoring and Diagnostics Laboratory, Boulder (NOAA), 191, 215, 362
Climatic Impact Assessment Program (CIAP), 61
Clinton Bill, 185, 355–6, 358
$ClNO^3$ *see* chlorine nitrate
ClO *see* chlorine monoxide
ClOx, 80
clouds: cumulonimbus, 124; ice, 133, 136; and models, 72, 210–11
see also polar stratospheric clouds (PSCs)
CNES *see* Centre National d'Etudes Spatiales
CNRS *see* Centre National de Recherche Scientifique
Coffey, Michael, 266
cold sores, 298
Colorado, 362
Colorado, University of, 191
Colorado Rockies, 130
Columbia University, 220
Committee on the Inadvertent Modification of the Stratosphere (IMOS) 71
Concorde, 58, 61, 376, 381
condoms, and ozone, 36
Condon, Estelle, 104, 113, 114, 276
conferences, scientific, 67
contrails, 122, 123
Convair 990s, 109, 110
Coordinating Committees on the Ozone Layer, 322
Copenhagen (third amendment to Montreal Protocol), 346, 350–51, 354, 358
cosmetic products, 56
Cousins, Norman, 144
Cousteau, Jacques, 81
Covey, Curt, 207
Cox, Tony, 246
Coxwell, Henry 37
Cray Research, 193
Cray supercomputers, 192–4, 197, 204, 205, 212, 224, 226, 228–9, 240, 379
Creighton, John, 154–5, 156
Croatians (in Chile), 274, 280
crops, 313–15, 353–4
Cryogenic Limb Array Etalon Spectrometer (CLAES), 163–4, 171, 374
cryogenic liquids, 163
Cuban missile crisis, 108

Daedalus Project, 137
Daily Mail, 241–2
Daily Telegraph, 241
Darwin, 110, 377
data assimilation technique, 227–8
Davies, Susan, 242
Day the Earth Caught Fire, The (film), 386, 387
DC-8s, 96, 98, 99, 101, 106, 110, 111–14, 131, 132, 233, 240, 367, 368, 382
DDT, 56
de Zafra, Bob, 89–90
Defence Ministry of, 254
denitrification, 101–2
Denver, University of, 19, 21, 31
Deuce, 115
diatomic oxygen, 35
diffuser plates, 173–4, 184
dimer, 88
Discovery (shuttle), 154, 156
DKK Schjerfenstein, 352
DNA (deoxyribonucleic acid), 298, 313
Dobson, Professor Gordon Miller Bourne, 41–7, 49–50, 78, 121–2, 124, 133, 146, 161, 263, 264
Dobson instruments *see* spectrophotometers
Dobson units, 45, 50, 84, 131, 225, 264, 301, 304, 363, 372
Domestic Engineering Company, 54
Dotto, Lydia, 1, 68, 251
Dragon Lady, 115
drones, uncrewed, 137–9
Du Pont, 54, 56, 329, 337
Dukakis, Michael, 240
dynamic hypothesis/dynamicists , 90, 91, 92, 94, 224, 225

Earth: and Gaia theory, 62; magnetic field, 19, 48–9; Nimbus 1 and, 145; public consciousness of Earth as a planet, 144; rotation from west to east, 179
Earth in the Balance (Gore), 356
Earth Observing System (EOS), 185–7, 345; EOS-AM, 187; EOS-PM, 187
Earth Summit, Rio, 212, 240, 324
Earthprobe, 184
EASOE *see* European Arctic Stratospheric Ozone Experiment
ECMWF *see* European Centre for Medium-range Weather Forecasting
Edinburgh University, 160, 240
Edison, Thomas Alva, 53, 54
Edwards Air Force Base, 136
El Agujero, 273, 276
El Chichon volcano, Mexico, 150–51, 219, 222
El Miuage Dry Lake, 136
El Nino, 214–15, 222–3

INDEX

electron-capture detector, 62–4, 64, 251–2
Electronic Numerical Integrator and Computer, 203
electrostatic plates, 32
emission spectrometer, 20
emphysema, 35
energy conservation, 73
Environmental, Department of the, 321, 339, 346
Environmental Affairs, 326–7
environmental movement, 58
Environmental Science Services Administration, 144
environmentalism, 73
EOS *see* Earth Observing System
EPA *see* US Environmental Protection Agency
equator: motion of air from, 90; ozone at, 46, 124, 373, 374, 375; tropopause at, 124
ER-2 aircraft, 126, 283, 378; and AASE, 99; and AASE II, 104, 233, 369; and Ames Research Center, 97; and Arctic stratosphere' behaviour, 367; in Argentina, 94; and ASHOE, 381–3; and ClO, 98–9, 170, 235, 242; and DC-8s, 111–12, 113; equipped with self-contained laboratories, 105; flown over polar regions, 93, 95–8, 103; flying, 115–21; future of, 139–40; and NASA 706, 708 and 709, 115; and nitrogen, 100; and nitrogen oxide, 237; observations constrained by altitude, 137; pre-eminence of, 105, 109–10; sensitive to weather, 112; and SPADE, 377; as TR-2 series, 109; and tracers, 105–6; and volcanic aerosol, 365, 366
erythema, 285, 286, 300, 301
Eskdalemuir weather station, 199
ESRANGE, 12, 15, 17–18, 21–5, 27, 31, 241
ethyl (tetraethyl lead), 54–5
European Arctic Stratospheric Ozone Experiment (EASOE), 11–13, 15, 17, 18, 21, 23, 28, 29, 30, 37, 104, 105, 226, 232, 240, 241, 242, 244, 245, 246, 344, 345, 365–6, 368–71, 378, 379, 380
European Centre for Medium-range Weather Forecasting (ECMWF), 12, 19, 225
European Commission, 245, 321, 328, 378
European Community: and Arctic experiment (1991–92), 15; Benedick and, 326, 332; and Montreal Protocol, 327, 333; Ozone Research Coordinating Unit, 23; production of CFCs, 319, 334, 345
European Council of Ministers, 338–9
European Ozone Research Coordinating Unit, 241
European Parliament, 339
European Space Agency, 123
European Space Research Organisation, 254
European UV-B Monitoring Network, 313
exhausts: car, 35–6; space shuttle, 68; SST, 58
Explorer 1, 48
Exploring the Atmosphere (Dobson), 50, 264
eye problems, 2, 36, 272, 279–81, 288, 289, 297

Fabry, Charles, 39–41, 43, 45
Fahey, David, 237, 243, 369
Fairbanks, Alaska, 104, 234
Falklands Islands and Dependencies Survey, 49
Falklands War, 79, 94
Farman, Joseph Charles, ix, 91, 215, 246, 258, 264, 302, 326, 341, 363, 385–6; and blue books, 323; and Bondi, 254; at British Antarctic Survey, 6, 50, 78, 253, 254; and changes in ozone layer over Antarctica, 78; and chemical mechanisms, 80, 86; and comparatively low Antarctic ozone, 50; delays publication of findings, 79–80; discovery of ozone hole, 6, 78–81, 253, 255, 256, 363, 369, 385; visits Dobson, 46; and HCFCs, 342; and ozone assessments, 359; reactions to the paper, 82–4, 92; Susan Solomon and, 81, 82; TOMS proves him to have been correct, 84; and value of small science, 254, 255
Farmer, Barney, 89, 91
Farnborough, 42, 121, 123, 124, 126
Federal Aviation Authority, 113
Fermi, Enrico, 65
'ferocious fifties', 94
Financial Times, 241, 284
fire-fighting equipment, 5, 319, 325, 341
First World War, chemistry's successes in, 53, 57
Fishman, Jack, 131–2
Fisk, Lennard, 154
Flat Irons, 191
float altitude, 19, 26
Flood, C., 204
fluorine, 55, 66, 67, 164
Flying Fortresses, 122

407

Fokker aircraft, 106
Fondecyt 1143 (National Foundation for Scientific Technological Development in Chile), 304, 305, 307
Food and Drug Administration, 73
Fort Lauderdale, Florida, 67
Frankenstein (Shelley), 217
Frederick, John, 303, 304
free radicals, 59, 74, 75, 87, 97–8, 105, 160
French Academy of Science, 38
Freon, 56
Frieman, Ed, 186
Frigidaire, 56

Gaia theory, 62
Galileo spacecraft, 177
Gang of Four instruments, 152, 153, 155, 158, 161, 163, 166, 170, 186
Garcia, Rolando, 82
Gardiner, Brian G., 79, 80, 82, 302–3, 363, 364
gas chromatography, 105, 126, 362
gas-phase chemistry, 87, 215, 342, 377
Gay-Lussac, Joseph, 37
General Electric, 145, 151, 153, 154
General Motors, 56
Geneva, 331
Geophysical Research Letters, 222
geophysics, 48–9
Geophysika (Russian plane), 381
Germany, and CFC ban, 345; and UV-B levels, 300, 313
Gibbons, Carolina, 281
Gibbons, John, 280, 281
Gibbons, Maria-Paz, 281
Gibbons, Oscar, 280
Gibbons family of Chile, 279
'gigaflop' supercomputer, 229
Gilbert, W.S., 33–4, 101
GISS *see* NASA Goddard Institute of Space Science
Glaisher, James, 37
Global Positioning Satellites, 117–18, 138
global warming, 49, 130, 133, 136, 188, 208–10, 212–14, 215, 239, 256, 277, 352, 353, 357, 358, 359
Goddard Space Flight Center *see* NASA Goddard Space Flight Center
Goldin, Daniel, 187
Goldsmith, Philip, 123, 124
Good Housekeeping, 296
Gorbachev, Mikhail, 178
Gore, Senator Albert, 238, 239, 355–6, 357
GPS receivers *see* Global Positioning Satellites
Graham, William R., 330

Granier, Claire, 134, 135
Gray, Lesley, 225–6
Greece, 303
Green magazine, ix
Green Party, 339
greenhouse effect, 277, 352, 353, 356, 358; and doubled-CO^2, 213; Hansen and, 221; and ozone depletion, 136, 208–10
Greenland, 27
Greenpeace, 7, 246–7, 278, 279, 308, 351–2, 371
Guardian, 136, 241
Gulf War, 118

H-bomb test, first American, 66
Haagen-Smit, Jan-Arie, 35, 36, 60, 128
Hackney, Brian, 278–9
Hadley, George, 204
Hadley Centre for Climate Prediction, Bracknell, 204–5, 211, 212
Haight-Ashbury, 64
Halifaxes, 123
Halley Bay, 51, 78–80, 89, 253, 254, 326
halocarbons, 264, 348, 349
HALOE *see* Halogen Occultation Experiment
Halogen Occultation Experiment (HALOE), 164–5, 171, 172, 374, 383–4,
halons, 325, 350; absorption of ultraviolet light, 13; cut in production, 331–2, 333; 'halon bank', 341–2; release of chlorine, 13; in stratosphere, 5, 13
Hansen, James, 220–24
Harries, Professor John, 153
Harris, Neil, 12, 13, 19, 20, 23, 24, 25, 85, 175, 246,
Harrison, Claire, 342
Hartley, W.N., 40
Hartley bands, 40
Harvard, 76, 97, 377
Harwell Laboratory, 224
Hawaii, 362, 383
HBFCs *see* hydrobromofluorocarbons
HCFC-22, 342, 348
HCFC-123, 348
HCFC-124, 348
HCFC-141b, 348
HCFCs *see* hydrofluorocarbons
HCl *see* hydrogen chloride
headaches, 35, 281
health warnings, 282, 292, 294–5
Heath, Don, 335–6
Heath, Edward, 204
'heliotherapy', 292, 297
helium, 39

INDEX

Helsinki, 340
Hercules transporters, 88, 89, 106, 125–8, 381
Heriot Watt University, 160, 240
Herodotus, 361
herpes, 298
herpes simplex, 298
heterogeneous chemistry 87
HFC-134a, 352
HFCs, 353
hibernation mode, 168
High Altitude Missions Branch *see under* NASA Ames Research Center
High Altitude Research Flight (later Met Research Flight), 122–3
High Cross, Cambridge, 77
High Speed Research Program, *see under* NASA
Hisenrath, Ernest, 176, 177
HIV virus, 36, 298–9
Hodel, Donald 330
Hoffert, Martin, 207
Hoffman, John, 328–30
Hoffmann, David, 86, 89, 90, 219
'Hole Over Kennebunkport Scare', 232, 238, 247
Holes in the Ozone Scare: The Scientific Evidence That The Sky Isn't Falling, The (Moduro and Schauerhammer), 261–5
Holmes, Geoff, 310–14
Horizon (BBC television programme), 278
Horton, Willie, 240
Houghton, John, 47, 146, 147, 161
Houston, 141
Howe, Sir Geoffrey, 340
HOx (oxides of hydrogen), 59, 60
HSRP (high Speed Research Program) *see under* NASA
Hubble Space Telescope, 154, 186
Hurricane Cleo, 145
hydrazine, 180
hydrobromofluorocarbons (HBFCs) 351
hydrocarbons, 35, 127, 352
hydrofluoric acid, 91
hydrofluorocarbons (HCFCs), 342, 346–53
hydrogen, 133, 163–4, 377, 383, 384
oxides of, *see* HOx
hydrogen bomb test, first American, 66
hydrogen chloride (CHl), 29, 30, 86–7, 90, 101, 171, 233, 266, 367, 368, 370
hydrogen fluoride, 265
hydroxyl radical, 97, 129–30, 134
see also OH radical
hygrometers, 37, 122

ICI, 352

IDDEA, 277
IGY *see* International Geophysical Year
immunosuppression, 279, 290, 298
IMOS *see* Committee on the Inadvertent Modification of the Stratosphere
impotence, 36
Improved Stratospheric and Mesospheric Sounder (ISAMS), 143, 144, 161–4, 166–8, 188, 224
Independent 241, 255, 282
India, and CFCs, 332, 334, 345
infectious diseases, 297–8
Infrared Montgolfier (Le Montgolfiér in Frarouge), 380–81
infrared radiometer, 146–7
Innsbruck, University of, 300
insecticides, 56
Insight, 249
Institute for Scientific Information, 250
Institute of Electrical Engineers, 245–6
Intergovernmental Panel on Climate Change (IPCC), 210, 358, 359
Interim Financial Mechanism, 345–6, 354
Intermediate Nuclear Forces Treaty (1987), 178
International Atomic Energy Agency, Vienna, 66, 69
International Civil Aviation Organisation, 377
International Geophysical Year (IGY), 48–50, 78, 188, 268, 385
International Ozone Commission, 46
International Ozone Trends Panel, 336, 337–8, 340
IPCC *see* Intergovernmental Panel on Climate Change
Iran-Contra scandal, 331
ISAMS *see* Improved Stratospheric and Mesospheric Sounder

Jadin, E.A., 216
Japan, and CFCs, 320, 331
Jet Propulsion Laboratory (JPL), Pasadena, 83, 88, 89, 158–9, 162, 244, 317, 318
John Hopkins University, Baltimore, 3, 288–9, 290, 293, 297
Johnson, Clarence 'Kelly', 107
Johnson, Lyndon B., 75
Johnson, Sir Nelson, 121
Johnson Wax, 72–3
Johnston, David, 266–7
Johnston, Harold, 60, 61, 82, 83, 191
Journal of Geophyical Research, 110, 243
Journal of the American Medical Association, 36

JPL *see* Jet Propulsion Laboratory
Junge, Christian, 218
Junge Layer, 218
Jupiter, 177

Kalingrad mission-control center, 181
Karentz, Deneb, 308
Keeling, Charles David, 49
Kennedy, John F., 58, 356
Kennedy Space Center, 154, 157
keratitis, 285
keratocunjunctivitis, 289
Kettering, Thomas, 54, 57
Kiruna, Sweden, 10, 12, 15, 20, 27, 31, 226, 240, 378–81
Kiruna Airport, 24
Kirunavaara, 15
kites, 38
Knorr, 313
Komhyr, Walter, 215, 216
Koshland, Daniel, Jr, 250
Kratatoa, 218
Kripke, Dr Margaret, 298, 330–31
Krueger, Arlin, 147–51, 173, 174, 176, 178–82, 184
krypton-84, 263
Kurylo, Michael, 7, 236, 237
Kuwait, 126

La Rouche, Lyndon, 262
Laboratory for Cytological Analysis, Santiago, 278
Langford, John, 137–9
Langley Research Center *see* NASA Langley Research Center
laser-induced fluorescence, 75, 130
lasers, and OH radical, 75
Lawrence Livermore Laboratory, California, 87, 207
lead in petrol, and Midgley, 53, 54–5
Lee Ray, Dr Dixy, 266–8, 293
Leesburg, Virginia, 328–9, 331
Lemonick, Michael, 238
L.F. Richardson Centre, Bracknell, 204, 205
Libby, Willard, 65
Lidars, 253
life: evolution of, 4–5; and ozone layer, 4, 5
Limb Profile Monitor of the Atmosphere (LPMA), 18–25, 29–31
Limbaugh, Rush, 261, 267
Lindemann, Frederick (Lord Cherwell), 42–3, 45, 121
Lindenberg, 44
Lindzen, Richard, 223
Liu, Shaw, 300
Lockheed, 120

Lockheed Advanced Development Plant, Burbank, 107, 109
Lockheed Palo Alto Research Laboratory, 163
Lodge, David, 203
London Amendment to the Montreal Protocol (1992), 345, 357
London ratification of the Montreal Protocol (1990) 341, 342
Long Term Environmental Research Program, 308
Los Angeles, 247; smog levels in, 35, 132
Lovelock, Chris, 251
Lovelock, Jim, 61–4, 67, 68, 69, 72, 75, 208–9, 251–3
LPMA *see* Limb Profile Monitor of the Atmosphere

McClintick, David, 231
McConnell, Fiona, 326–7
McDonald, James, 58–60, 291
Machiavelli, Niccolo, 194–5
Machta, Lester, 67
McKenna, Dr Danny, 127
MacLean, David, 371
MacLeish, Archibald, 141
McMurdo Sound, 89, 91, 170, 265, 275
McNeal, Robert J. 'Joe', 131, 152
Madronich, Sasha, 134, 135
MAESA *see* Measurements of the Atmospheric Effects of Stratospheric Aircraft
Magallanes region, Chile, 273, 279–84, 288, 302, 309
Magee, John Gillespie, Jr, 103
Magellan, Ferdinand, 273
Major, John, 246, 371
Malaysia, 346
Manchester Guardian 202
Manhattan Project, 64–5
Mankin, William, 266
Marlow, Bill, 67
Mars, 4, 162, 208
Maryland, University of, 314
Massachusetts Institute of Technology (MIT), 137, 159, 203
Mauna Loa, Hawaii, 216
Max Planck Institute for Chemistry, Lindau, 18, 20, 25
Mayo Clinic, Rochester, New York, 287
measles, 298
Measurements of the Atmospheric Effects of Stratospheric Aircraft (MAESA), 377, 378, 383
Medical Research Council, 295
Medium Altitude Missions Branch *see under* NASA Ames Research Center
melanin, 286

INDEX

malanomas, 286, 293, 294–5
Merrick, David, 94
mesosphere, 60, 263
Meteorological Office, 42, 92, 93, 114, 121, 122, 198, 199, 202, 203–4, 211, 212, 213, 321
Meteorological Research Flight (MRF) (previously High Altitude Research Flight), 121, 123–8
meteors, 42, 76
Meteors (weather satellites), 179–83
methane, 90, 133–6, 162, 269, 349, 383
methyl bromide, 350, 353–4, 355
methyl chloride, 55, 265
Mexico, 150, 346
Mexico City, 36
Michigan, University of, 68
Microwave Limb Sounder (MLS), 159–61, 162, 168–72, 188, 224, 226, 236, 240, 245, 247, 373, 374
Microwave, spectroscopy, 146
Midgley, Thomas Edison, 53–7, 61, 63, 329
Miller, Alan, 329
'mini-holes', 27, 28, 101, 372
Minnesota, University of, 148
MIPAS (an emission spectrometer), 20, 25, 29
MLS *see* Microwave Limb Sounder
Moduro, Rogelio A., 262–5, 269
Mohave Desert, 136, 139
Molina, Mario, 5, 68–72, 83, 88, 256, 258, 263
Montgolfier infrarouge, Le (Infrared Montgolfier), 380–81
Montreal Protocol on Substances That Deplete the Ozone Layer, 99, 276, 283, 316, 327, 332–5, 338–41, 345, 349, 350, 355, 362
Moon, as light source, 89
Moon, Reverend Sun Myung, 248
Moreau, Guyton de, 37
Mosquitos, 122, 123, 124
Mount Agung, 218, 221
Mount Augustine, 266, 267
Mount Erebus, 265
Mount Pinatubo, Philippines, 28–9, 156–8, 160, 161, 165, 183, 220–23, 235, 241, 244, 246, 266, 363, 364–6, 368, 371, 372, 373
Mount St Helens, 109, 218–19, 266
Mount Unzen, Japan, 157
MRF *see* Met Research Flight

Nairobi, Kenya, 322
Namias, J., 216
NAS *see* US National Academy of Sciences
NASA, (National Aeronautics and Space Administration), 62, 328, 336, 369; and AASE, 94–6, 276; and AASE II, 104; advice to world governments on state of stratosphere, 10; Airborne Arctic Stratosphere Experiment II, 28; Anderson's grant from, 75; balloon-led assault on the stratosphere, 76; and 'blue books', 322; effect of *Challenger* accident, 178; and CHEOPS, 101; and Congress, 149, 320, 321; and Convair 990s, 109: criticism of, 247–9 337–8; crosschecking drift of space instruments, 177; and Earth Observatory System, 185–7, 345; and Earthprobe-TOMS, 184; and Er-2 campaigns, 109; faith in space shuttle, 151; frequency of large aircraft campaigns, 233; headquarters of, 82, 231, 234, 316, 317, 318; High Speed Research Program, 375–7; lead status, 248; and Meteors, 179; and Perseus aircraft 139; and presentation of aircraft and satellite results on the same stage, 242–3; press conference (3 February, 1992), 235–40, 243, 245, 261, 279; and question of scientific reliability, 250; and research funds, 248, 336; and shot-down U-2 (1960), 107–8; small balloon programme, 105, and sounding rockets, 148; Stratospheric Research Program, 108; on UARS, 143, 162; and uncrewed drones, 137; under close scrutiny, 185
NASA 706, 115, 119
NASA 708, 115
NASA 709, 115
NASA Ames Research Center, California, 97, 98, 108, 109, 110, 112, 115, 119, 120, 131, 377; High Altitude Missions Branch, 119; Medium Altitude Missions Branch, 109
NASA Goddard Institute of Space Science (GISS), 218–19
NASA Goddard Space Flight Center, 83, 84, 141–2, 143, 145, 147, 148, 150, 154, 156, 166, 167, 169, 175, 178, 180–82, 220, 227–9, 335, 375; Atmospheric Chemistry and Dynamics Branch, 174–5, 182; Data Assimilation Office, 227, 228–9; and TOMS, 84–5; World Data Center for Rockets and Satellites A, 188
NASA Langley Research Center, Hampton, Virginia, 164, 383
NASA Wallops, 119
Nash, Nathaniel, 283, 284
National Academy of Sciences, US

411

(NAS), *see* US National Academy of Sciences
National Cancer Institute, 299
National Cash Register, 53–4
National Center for Atmospheric Research, Boulder, 82, 134, 192–7; Atmospheric Chemistry Division, 197; Climate Modelling Section, 194, 196; Computing Center, 192–3
National Foundation for Scientific Technological Development in Chile *see* fondecyt 1143
National Geographic Research and Exploration, 223
National Institute of Medical Research, 62
National Institute of Standards and Technology, 252
National Meteorological Center, Washington DC, 242
National Oceanic and Atmospheric Administration (NOAA), 67, 81, 132, 144–5, 177, 183, 216, 242, 248, 249, 328; Aeronomy Lab, 82, 93, 134, 191, 237, 242, 248, 249, 300, 318, 333, 348, 349, 369, 382, 384; clean air monitoring stations, 132, 215–16, 362, 363; Climate Monitoring and Diagnostics Lab, 191, 362; on weather satellites, 145
National Ozone Expedition (NOZE), 88–92, 93, 97, 191, 275; NOZE II, 96, 98
National Review, 268, 269
National Scientific Balloon Facility (Texas), 75
National Science Foundation, Washington, 91, 248
Natural Environment Research Council (NERC), 63, 253–4, 313
Natural Resources Defence Council, 329
Nature, 41, 64, 68, 70, 81, 91, 210, 212, 213, 219, 243, 303, 314, 326, 323, 348, 363, 373
nausea, and ozone, 35
NCAR *see* National Center for Atmospheric Research
neon, 163
neon-20, 263
NERC *see* Natural Environment Research Council
Netherlands, and methyl bromide, 353
Network for the Detection of Stratospheric Change 253
neutron accelerator, 66, 67
New Scientist, 135
New York State (ban on aerosols), 73
New York Times, 11, 60, 132, 238, 283
New York University, 207

New Yorker, 57, 256
New Zealand, 295, 300, 381
Newton, Isaac, 39
Nimbus satellites, 145, 148, 159, 161, 185, 188
Nimbus 1 satellite, 145
Nimbus 2 satellite, 145
Nimbus 3 satellite, 145
Nimbus 4 satellite, 146, 149, 174, 178, 183
Nimbus 7 satellite, 3, 84, 143, 147, 150–52, 173, 162, 170, 172–6, 183, 184, 215, 234, 301, 336, 370
nitric acid, 35, 100–101, 219, 234, 242, 364
nitrogen, 100–101, 377; reaction with chlorine, 9
nitrogen dioxide, 365, 366, 368–9, 376
nitrogen oxides, 60, 63, 89, 90, 100, 106, 127, 162, 164, 182, 219, 234, 235, 237, 240, 364, 365, 366, 368, 375
see also NOx
nitrogen pentoxide, 365, 375
nitrous dioxide, 368
nitrous oxides, 35, 60, 162, 209
Nixon, Richard, 73
NOAA *see* National Oceanic and Atmospheric Administration
NOAA-9, 176
NOAA-11, 176
NOAA-13, 176
Norburg, Bill, 145
Normand, Sir Charles, 46
North Pole: Barrilleaux and, 104; fear of ozone hole formation over, 8, 13, 161; investigation of ozone layer, 12–32; ozone loss during solar maximum, 149; ozone loss prediction, 102; solar radiation destroys ozone at, 182; varying meteorological conditions, 14; vortex and, 14
northern hemisphere: Bedrich Magas on, 284; and Pinatubo aerosols, 364, 365; planetary waves in, 195; and question of ozone hole, 28, 242; springtime values for ozone, 50, 182; and stripping away of ozone, 6–7, 85, 102, 183, 219, 335, 338, 340, 370, 373, 387
northern lights, 17, 19
Northwestern Airlines, 150
Norway, 27, 99, 126, 303
NOx (nitrogen oxides), 60, 80, 90, 191, 369, 375, 376
NOx/NOy ratio, 369
NOy (odd nitrogen), 369
NOZE *see* National Ozone Expedition

412

INDEX

NOZE II *see under* National Ozone Expedition
nuclear fission, 64
nuclear power plants, 57

O³ Tech, 34
ocean biota, 264, 265
oceans: and El Niño, 214–15; and models, 210, 211
OClO *see* chlorine dioxide
OCTA *see* Oxidising Capacity of the Tropospheric Atmosphere
Ocular and Dermatological Health Effects of Ultraviolet Exposure from the Ozone Hole in Southern Chile: A Pilot Project, 271
odd nitrogen hypothesis, 90, 91
ODPs *see* Ozone Depleting Potentials
Odysseus drone 139
Office of Management and Budget (US), 331
Office of Naval Research, 148
OH radical ('Howard Hughes radical'), 75, 129–30, 135, 349, 377
Operation Deepfreeze, 381
Oregon (first state to ban aerosols), 73
O'Shea, Suzanne, 242
Oxford University, 143, 163
Oxidising Capacity of the Tropospheric Atmosphere (OCTA), 127
oxygen: action of sunlight on, 4; chlorine atoms devour, 13; and creation of ozone layer, 5, 45; and hydroxy radical, 129; liquid, 138; ozone as unstable form of, 5, 32; production of, 5
ozone: absorption of ultraviolet, 40, 41; assessments, 358–9; as a by-product of nitrous oxides and ultraviolet radiation, 35; CFCs damage, 61, 164; chlorine atoms destroy, 1, 6, 13–14, 29, 68, 69, 71, 133, 160; and chlorine monoxide, 74, 99; discovery of, 32; disinfectant properties, 33–4; at the equator, 54–6; first measured from space, 146; and health, 32–6, 44, 128; increased in atmosphere, 126–7;
   measurement of ultraviolet, 40, 41; MLS detection of, 160; NAS predictions of loss, 72; production of, 5, 32, 60; reacts with methane to form water-ice crystals, 135; and the tropics, 377–8, 382–3; tropospheric, 106, 126–32, 134, 209, 300: unexpectedly low levels in Antarctica, 50, 78–9; as unstable form of oxygen, 5, 32; and volcanoes, 150–51; water purification, 34; used in wine-making, 34; world record levels at Spitsbergen, 50
'Ozone' (Gilbert), 33–4
Ozone atmosphérique, L', (Fabry), 40
'ozone backlash', 8, 9, 260–68, 272, 386
ozone depletion: altitude of, 80; and balloons, 105; Bedrich lectures on, 283; and blue books, 323; CFCs cause, 53, 69, 70, 99, 147, 270; and chlorine monoxide, 74, 87, 88, 99, 160; complexities of, 10, 71; and Supercomputer, 194; defining background conditions, 95; economic aspects, 314–15, 330; and El Nino, 214; and global warming, 136, 208–10, 212–14; and halogenated source gases, 90; halons' destructiveness, 341; and health, 3, 7, 59, 271, 272, 277–99, 315, 330; increasing, 69, 70, 77, 98, 102, 343–4, 355, 363; and ISAMS, 143, 168; and Meteors, 179–80; and methane, 133, 134; most yet seen, 6; mostly occurring in lower stratosphere, 209; NAS predictions, 71–4; and North Pole, 14; in northern hemisphere, 6–7, 85 102, 183, 219, 335, 338, 340; and OH radical, 130; and ozone backlash, 8; in a parcel of air, 106; and Pinatubo aerosol, 373; proof of, 5, 6, 172; and PSCs, 95–6; at South Pole, 84, 85, 99; success of backscatter technique, 178; and trace gases' global-warming potential, 130, 359; and UARS, 152, 170; and ultraviolet radiation, 2; and volcanic aerosols, 366
   see also under Antarctica; Arctic
Ozone Depleting Potentials (ODPs), 347, 348
Ozone Diplomacy (Benedick), 316, 326
ozone hole: appearance of, 61, 382, 384; 'best' place for, 307; blames for health problems and weather events, 2, 277–99; and blue books, 323; cause of, 111, 170, 216, 269, 332–3; chemists' v dynamists over, 95; and chlorine monoxide, 90, 232; and CLAES, 164; distortion of ozone layer, 337; and Dobson record, 85; existence confirmed, 143, 151, 174; Farman on, 80, 386; humanity's responsiblity for, 98, 99; increase in, 213, 338; and lack of NASA medium-range aircraft, 109; and measurements in Antarctica, 88, 94, 304; models fail to predict, 198; most depletion yet seen, 6, 362, 363; and phase-outs, 7; and Phytoplankton, 306–7; and

413

plants, 309; Punta Arenas closest to, 273; question of formation over North Pole, 8, 13, 28, 29, 101, 104, 161, 212, 232, 234, 242, 245, 371, 373; Rowland on aerosol use, 258; and skin cancer worries, 190, 288; Susan Solomon on, 82; and TOMS failure, 172; and UV-B, 3; and *Wall Street Journal*, 243–4; Watson on, 84
ozone layer: accelerated stripping away of, 338; and aviation, 58; BUV and, 148; chlorine and, 68, 71, 97, 265, 337; distortion of, 337; ER-2's sample, 126; formed by the action of sunlight upon oxygen, 4, 45; and greenhouse effect, 353; HCFCs and, 348; and hydrogen chloride, 30; investigations over North Pole, 12–32; lack of knowledge of, 369; Lindemann and Dobson's findings, 43; loss in northern hemisphere, 6–7, 85, 183, 219, 335, 338, 370, 387; methane and, 136; and NOZE, 89; proof of existence, 5, 385; protects CFCs in the troposphere, 69; sensitivity to change, 376; soundwaves bounce off, 45; and SSTs, 61; starts to change over Antarctica (1977), 78; stratosphere heated by 43; and sulphate aerosols, 364; 'tattered', 387
ozone sondes, 85, 148
Ozone War, The (Dotto and Schiff), 1, 68

P-80 Shooting Star, 107
Pacific Exploratory Missions, 132–3
Pacific Ocean, 64, 128, 132, 133, 273, 376, 377
Palmer Peninsula, 304, 307
Panama Canal, 273, 282
parallel processing, 194
parameterisation, 210, 211
Paris Air Show (1993), 376
Payload Operations Control Center (POCC), 141, 142, 143, 166, 182
Pearce, Fred, 135
Pegasus (booster), 184
Pei, I.N., 192
Perseus aircraft, 136–9, 377, 378, 381; Perseus B, 139
pesticides, 57, 325
petrol, lead on, 53, 54–5
pharmaceuticals, 325
photochemistry, 197, 377
photodermatoses, 288
photodissociation, 133, 135, 265, 269
photelectric cells, 44
photometers, ultraviolet, 19, 30–31
photometrical measurements, 43

photons, 98
photosynthesis, 5
physics, and Second World War, 57
phytoplankton, 305–9, 315
Pinochet, General Augusto, 283
Piz Buin, 295–6
planetary waves (planetary-scale eddies), 195, 197, 216, 226, 335, 374
planets, formation of, 4
plants: damage to, 2, 272; evolution of, 5; in Punta Arenas, 308–9; research, 310–14; and smog, 35
Plesetsk (Soviet launch centre), 179, 180
Pluto, 158
plutonium, 178
PMR *see* pressure-modulated radiometry
POCC *see* Payload Operations Control Center
Poland, 301
Polar Duke, 306
polar stratospheric clouds (PSCs), 219, 364, 366, 367, 368, 369, 370, 384; and Antarctic ozone hole, 219; and chlorine activation, 88, 91; creating mini-holes, 101; early sightings, 86; first pilots fly through, 96; formation inhibited, 242; and ISAMS, 168; and localised ozone depletion 95–6; and methane, 133, 135; more extensive formation of, 100, 213; and nitric acid, 234; and Perseus, 138, 381; Solomon and, 87–8; and temperature, 170; two types, 100
Polarstern, 275
polymorphus eruptions, 288
Pommereau, Jean-Pierre, 13, 25, 27, 366, 372, 378, 380, 381
Porter, Sir George, 52
Powers, Francis Gary, 107, 108
Prather, Michael, 80
President's Office of Science and Technology Policy, 358
pressure-modulated radiometry, 161–2
Prézelin, Barbara, 306–7
'primeval soup', 4–5
Princeton, 66, 194, 209
prograde orbits, 179
Project Aquatone, 107
propane, 352
PSCs *see* polar stratospheric clouds
Public Act 9595, 320
Puerto Natalas, Chile, 289
Punta Arenas, Chile, 2, 3, 10, 94–5, 98, 110, 111, 117, 120, 121, 236, 248, 271, 272–83, 286–90, 299, 301, 302, 308, 310, 315, 332, 333, 382
Pyle, John, 226–7, 240, 365, 366, 370, 371, 379–80

INDEX

QBO *see* Quasi-Biennial Oscillation
QED (television programme), 296
Quasi-Biennial Oscillation, (QBO), 225–6, 335
Quayle, J. Danforth, 356
Queensland, Australia, 294

radiation, ultraviolet *see* ultraviolet radiation
radiometers, 146, 147, 299, 300, 302
radiometry, thermal emission, 146
RAF Leuchars, 125
RAL *see* Rutherford Appleton Laboratory
Ramaswamy, V., 209–10
Ravishankara, A., 134, 349
Reading University, 146, 209, 212
Reagan, Ronald, 73, 178, 221, 319, 320, 321, 327, 329, 330, 331, 339, 357
Reber, Carl, 154
'red books', 80, 85
refrigeration: alternative, 352; CFCs and, 55–6, 73, 319, 334, 352–3, 357; and Midgley, 55
relative radiative forcing, 359
remote sensing, 108
respiratory problems, 35
Reuters, 238
Revlon, 318–19
Richardson, Lewis Fry, 198–204, 206
Riquelme, Oscar, 277–8
Riverside Airport, 109
'roaring fifties', 93
Robertson-Berger instruments, 300, 301, 302
Roche, Aidan, 163, 164, 166
rockets: 48; sounding, 58, 148, 263
rodent ulcer, 286
Rodriguez, Dr Jose, 367, 368
Rood, Dr Richard ('Ricky'), 227–9
roof insulation, 353
Roosevelt, Franklin D., 122
Rothschild, Lord, 255
Rowland, F. Sherwood, 71, 72, 256–8, 263, 283; address as president of AAAS (1993), 257, 263, 264; attitude to Dobson instruments, 85, 175; background, 65; and chlorine nitrate, 86–7; Fort Lauderdale meeting, 67; investigates CFCs in atmosphere, 67, 68, 69; large number of research papers, 257; as a martyr figure, 256; and methane, 133, 135; Molina joins, 68; presents findings at American Chemical Society, 70; recognises threat posed by CFCs, 56, 65, 256; and stratosphere, 74; and Trends Panel, 336, 337

Rowland-Molina hypothesis, 70–75, 80, 87, 251, 257
Royal Air Force (RAF), 121, 123, 124, 125
Royal Society, 42, 43, 62, 78, 121
Russell, James, III, 164, 165, 172, 383–4
Rutherford, Ernest, 255
Rutherford Appleton Laboratory, Oxfordshire, 153, 160, 163, 224–6, 240, 379

St Petersburg, 18
Salter, Peter, 93–4
Salzburg conference (October, 1970), 66
Samoa 'clean air' monitoring station, 132, 215–16, 362
Santiago, 36, 273, 290, 293
satellites: calibration, 174; and Earth's atmosphere, 173; ESRANGE launches Sweden's first, 12; first dedicated stratospheric chemistry satellite, 187; global perspective, 174, 343; and Goddard, 141–2; GPS, 117–18; Kiruna and, 16; and Mount Pinatubo, 157–8; Nimbus *see* Nimbus; NOAA *see* NOAA; prograde orbits, 179; telecommunications, 142; TIROS, 144; TOMS *see* TOMS; U-2s and, 108; UARS *see* UARS
Sato, Haruo, 87
'Saving the Ozone Layer' conference (London, 1989), 339, 340
SBUV *see* Solar Backscatter Ultraviolet instrument
Schauerhammer, Ralf, 262, 264, 265
Schein, Dr Oliver, 3, 288, 289–90, 297
Schiff, Harold, 1, 68, 251
Schneider, Stephen H., 205–6
Schoerbel, Mark, 95
Schonbein, Christian, 32
Schultz, Ricardo Borrquez, 308–9
Science magazine, 8, 61, 81, 102, 129, 183, 187, 243, 250, 259, 260, 262, 266, 307, 367, 374
Science and Environmental Policy Project, 269
Science Digest, 6
Science funding, 93
Scotto, Joseph 299
Scotto survey, 299–300
Scripps Institution for Oceanography, La Jolla, California, 186
sea surface temperatures: and El Nino, 214, 215; and ozone levels, 216
Second Stratospheric Arctic and Mid-latitude Experiment (SESAME), 378–81, 382
Second World War: first measurements within the stratosphere from Boston

415

bombers, 45; as a triumph for physics, 47, 57
Service d' Aéronomie *see under* Centre National de Recherche Scientifique
SESAME *see* Second Stratospheric Arctic and Mid-latitude Experiment
Shackleton, 62, 63
Shanklin, Jonathan, 79, 80, 303, 363
Shelley, Mary, 217
Shelley, Percy Bysshe, 217
Shine, Keith, 209–10, 212, 213
Shuttle Solar Backscatter Ultraviolet (SSBUV or Shuttle SBUV) instrument, 177, 178
Silent Spring (Carson), 58, 64
Simons, Marlise, 131–2
Singer, S. Fred, 268–9
skin cancers, 2, 7, 59, 60, 190, 272, 279, 284, 286, 287, 288, 291–9, 330, 331, 344
skin problems, 281, 285–6, 289
'Skunk Works' *see* Lockheed Advanced Development Plant, Burbank
Smith, Desmond, 146
Smith, Ray, 306,
smog, 35–6, 60, 128, 132
'Smoking gun' *see* chlorine monoxide
Snoopy (Hercules transporter), 125–8
snow blindness, 285
Snowmass, Colorado, 251–2
Soderman, Paul, 97
sodium, 265
Solar Anomalous and Magnetosphere Explorer, 182
Solar Backscatter Ultraviolet (SBUV) instrument, 149–50, 152, 170, 176–8, 182, 183, 335, 343, 374
solar keratoses, 285
Solar Light Company of Philadelphia, 299
Solomon, Philip, 89–90
Solomon, Susan, 258–60, 373; joins Aeronomy Lab, 81; and Arctic zone variations, 100; and Farman's paper, 81–2; and HCFCs, 348; and NOZE mission, 88–9, 91–2; and NOZE II 96; and PSCs, 87–8; and ratification of Montreal Protocol, 349; as a role model, 259–60; and sulphur-based aerosols, 219; and transitional substances, 347
solvents, 56, 73, 352, 353
SORG *see* Stratospheric Ozone Review Group
soundwaves, and ozone layer, 44–5
South Africa, 132
South America, 94, 315
South Korea, 332
South Pole, 14, 79, 362, 374; ClO levels, 104, 373; DC-8s at, 110; Farman on, 386; Farman's last trip to, 255–6, 386; flow of air around, 383; opening up of hole in ozone layer, 2, 6, 28, 50; ozone dip over, 84, 85, 99
Soviet Union: as perceived threat, 107; released CFC production figures, 329; and space age, 48; TOMS flown on Soviet weather satellite, 178–82; U-2 shot down by, 107–8
space age, launch of, (1957), 48
space shuttle, 89, 136, 142; exhaust, 68; NASA's faith in, 151; and ozone layer, 176; and SSBUV, 177
SPADE *see* Stratospheric Photochemistry, Aerosols and Dynamics Expedition
spectrographs, 44
spectrometers, 40–41, 43, 77, 89, 112, 130, 367
spectrometry, 105
spectrophotometers, Dobson, 43, 44, 49, 50–51, 77, 79, 84, 85, 146, 148, 150, 174, 300, 215–16, 253, 336, 338, 338
spectroradiometers, 302, 304, 305, 306
spectroscopy, 39–40, 74, 146
spectrum, 40, 146, 148, 152, 164, 272, 285, 300, 306
SPF *see* Sun Protection Factor
Spitfires, 122, 125
Spitsbergen, 50
spray-can propellants, 1, 56, 62, 63, 64, 70
see also aerosols
Sputnik 1, 48
squamous-cell cancer, 286, 289, 294
SR-71 Blackbirds, 120, 136
SSBUV *see* Shuttle Solar Backscatter Ultraviolet instruments
SSTs *see* Super Tonic Transport aircraft
Starkey, John, 21, 24
State University of New York, 89
Stavanger, 100, 101, 111, 113, 114, 236, 248, 363
Stolarski, Richard, 174, 175, 176, 183, 375, 376, 377
Straits of Magellan, 2, 94, 272, 273, 274
stratosphere: ability to hold water vapour, 122; air motions within, 39; Antarctic temperatures, 90; and balloons *see* balloons: and Boston bombers, 46; and CFCs, 5, 6, 7, 13, 67, 68, 69, 135, 263, 265; and CHEOPS campaigns 26, 27; chlorine in, 6, 87, 91, 140, 149, 177, 188, 189, 258, 303, 324, 328, 363, 368, 376; circulation of, 193; as dry, 46, 122, 123–4, 127, 133, 195; equatorial, and volcanic aerosols, 219; ER-2s in, 105, 109, 110, 116, 118, 119, 120;

INDEX

first measurements within 45; and greenhouse effect, 213; halons in, 341; heated by ozone layer, 43; and hydrogen, 383–4; and hydrogen chloride, 30; Junge Layer, 218; little consensus on intricacies of, 71; lower, 96; measurement of trace gases, 14–15; methane in, 133; modelling of, 195–6; and North Pole conditions, 8; ozone in, 5, 58, 69, 71, 127, 130, 133, 148, 151, 209, 219, 358;
and planetary waves, 195, 374–5; sampling chlorine monoxide, 74, 75; sensitivity to ozone depletion, 369; south polar, 50; and spectroscopy, 40; and SSTs, 60; and troposphere, 109, 110; U-2s and, 108; and warm surface, 100; weather in, 38–9
Stratospheric, Photochemistry, Aerosols and Dynamics Expedition (SPADE), 377
Stratospheric and Mesospheric Sounder, 162
stratospheric 'jets', 195
Stratospheric modelling 86–7
Stratospheric Ozone Protection Plan (EPA), 328
Stratospheric Ozone Review Group (UNEPSORG), 338, 342, 347
Stratospheric Research Program (NASA), 108
STS-34, 177
Sulphate aerosols, role of, 375
sulphur dioxide, 28–9, 55, 150–51, 160, 364
sulphuric acid, 218, 219
Sun: and air motions within the stratosphere, 39; in Arctic, 28; and catalytic cycles, 90; and chlorine monoxide, 101; and chlorine nitrate, 72; formation of, 4; interaction of magnetic forces with those of Earth, 48–9; and LPMA gondola, 18–21, 30–31; movement from east to west in sky, 179; and ozone stripping, 96, 149; satellites' solar panels track, 168–9: solar maximums, 48, 90, 149; sunspots, 149
Sun Protection Factor (SPF), 295–6
sunburn, 2, 272, 281, 286, 290, 293, 295, 300, 301, 302
sunlight: and formation of ozone layer, 4, 45; ozone stripped away with return of, 79; and smog, 35, 128; and UARS, 152
sunspots (solar keratoses), 285
Sununu, John 240
Super Sonic Transport (SST) aircraft, 58–61

supercomputers, 192–4, 208, 210, 212, 227, 228–30
Superconducting Supercollider, 185
supersonic air travel, 376
Sutherland, Dr John, 314
Sverdlovsk, 108
Synthesis Report, 340

Taiwan, 49
Tambora volcano, 217–18
Tasmania, 294
Taylor, Fredric, 161–3, 167–8
TECHNOPS (French balloon campaign 1989), 26
Teisserenc de Bort, Léon Philippe, 37–9, 42
telemetry, 25
Television and Infra-Red Observation Satellite (TIROS), 144
teraflop supercomputers, 229
Teramura, Allan, 314
tetraethyl lead *see* ethyl
Tevini, Dr Manfred, 313
TEWI *see* Total Equivalent Warning Impact
Texas, University of, 298, 331
Thatcher, Margaret, Baroness, 79, 92–3, 204–5, 321, 332, 339–40
thermal emission radiometry, 146, 147
thermal radiation, 21
thermodynamics, 42
Theseus drone, 139
Third World 327, 340; and CFCs, 322, 334, 345, 346; funding for, 354
Thompson, Anne, 132
Thomson, J.J., 199
Tickell, Sir Crispin, 361
Tierra del Fuego, 3, 273, 274, 277, 278, 279, 284, 303,
Time, 238
Times, The, 1
TIROS *see* Television and Infra-Red Observation Satellite
TOGA-CORE, 128
Tokyo, 36
Tolba, Moustafa, 322, 325, 331, 334, 335, 343, 350, 351, 355
TOMS *see* Total Ozone Mapping Spectrometer
Total Equivalent Warning Impact (TEWI) 353
Total Ozone Mapping Spectrometer (TOMS), 84–5, 98, 149–52 157, 170, 172–6, 178–84, 188, 238, 268, 301, 304, 335, 337, 343, 345, 370, 375
Towyn, 204
TR-2 series, 109
trace gases, 14–15, 18–19, 58–9, 62, 63, 67–8, 76, 106, 120, 124, 128–31,

133, 226, 323, 359; detection of, 146, 159, 161–2; models, 196; and replacements for CFCs, 353
TRACE-A (Transport and Chemistry near the Equator in the Atlantic), 131–2, 133
tracers, 62, 106, 123, 152, 171
Transalls aircraft, 106
'transitioned substances', 347, 352, 355
Transport Department of (US), 61
Transport and Chemistry near the Equator in the Atlantic see TRACE-A
Trappes, 38
Trashing the Planet (Lee Ray), 266
trichlorethane, 325, 333, 341, 350
Triogen, 34
Trippier, David, 345, 346
tritium, 66
tropopause, 38, 42, 46, 105, 106, 111, 123, 124, 132, 150, 195, 375
tropopause folds, 225
troposphere, 38, 69, 90, 106, 109, 110, 126–35, 196, 212, 218, 349, 350
tuberculosis, 292, 298
Tuck, Adrian, 92–6, 99, 234, 321, 382–4
21st Century Science and Technology (magazine), 262

U-2 spy plane, 93, 107–9, 115, 117, 121, 140
UARS see Upper Atmosphere Research Satellite
UFOs, 59–60, 257
UGAMP see Universities' Global Atmospheric Modelling Project
Ulloa, Walter, 281
ultra-violet radiation 35; and chlorine nitrate, 71; and clouds, 72; and creation of ozone layer, 5; danger of, 59, 271–2, 278, 282, 284–5, 289–90; halons and CFCs absorb, 13; health warnings, 282, 292, 294–5; and ozone depletion, 2, 134; ozone layer absorbs, 42; and photometers 30–31 and 'primeval soup', 4–5; reflected from clouds back up to space, 71–2; satellite measure, 148; in Swiss Alps, 44; and trace molecules, 146; and UARS, 152; UV-A, 284, 285, 296, 305; UV-B, 3, 284, 285, 289, 290, 291, 296, 298–307, 310–15; UV-C, 284, 285
UNEPSORG see Stratospheric Ozone Review Group
United Nations Environment Programme (UNEP); 210, 239, 291, 297, 315, 321, 322, 324, 325, 328, 340,

342, 344, 348, 349, 353, 354, 355, 358, 377
United States: and CFC ban, 319–58; and space age, 48; and U-2s' role, 140
US Air Force, 108, 109
Us Atomic Energy Commission (AEC), 65, 67, 68, 266
US Bureau of Standards, 252
US Department of Commerce, 60
US Department of Defense, 142
US Department of Transport, 61
US Environmental Protection Agency (EPA), 3, 6, 73, 75, 268, 288, 289, 290, 291, 293, 313, 328, 329, 357
US National Academy of Sciences, 5–6, 71–4, 251, 252, 258, 320
US National Science Foundation, 259, 308
US National Weather Bureau, 144, 204, 216
US Navy, 381, 382
Universidad de Magallanes, Punta Arenas, 275
Université Pierre et Marie Curie, Paris, 18, 29
Universities' Global Atmospheric Modelling Project (UGAMP), 224–6
University College London, 92
Upper Atmosphere Research Satellite (UARS), 142–4, 151–65, 168–71, 184, 185, 187, 224, 226, 235, 243, 244, 373, 374, 382, 383
Halogen Occultation Experiment, 383–4
Urey, Harold, 65
urocanic acid, 298
Ursa Major, 47
Ushuaia, 303
UV-A/B/C see under ultra-violet radiation

vaccinations, 298
Vaghjiani, G., 134, 349
Van Allen, James, 48, 268
Van Allen radiation belts, 48
Vandenberg Air Force base, Northern Carolina, 145, 147
Vega, Suzanne, 220
vegetation, and parameterisation, 211
Venus, 4, 162
Verrieres-le-Buisson, 372, 380
verrucae, 289–90
Vienna Convention (1985), 323–6, 328, 331
Vietnam, 108, 346
volcanoes, 28–9, 150–51, 156–8, 160, 161, 165–6, 183, 217–24, 264–7, 349, 363, 364–6, 368, 371, 372, 373
Vortex, polar, 114, 139, 370, 381; at Antarctica, 12–13, 50, 90, 91, 93,

INDEX

95–8, 110, 120, 170–72, 383, 384; at Arctic, 12–13, 15, 16, 19, 27–8, 99, 111, 117, 233, 367 369–70, 372, 379, 380; and chlorine monoxide, 226; deactivation of chlorine, 227; and hydrogen chloride, 30; and ozone depletion, 95–6, 105, 226

Wall Street Journal, 243–4, 284
Wallops Island, Virginia, 148, 183
Walshaw, Desmond, 47
Wank Mountain, Bavarian Alps, 128
Washington Post, 8, 238
Washington Times, 247–8
Watch Hill, near Shotover, 47
Watergate, 70, 73
Waters, Joseph W., 159, 162, 172, 188–9, 226, 231–2, 235–8, 244, 245, 246, 249, 373–4, 375
Watson, Elizabeth (nee Gormley), 318
Watson, Robert, 83–5, 88, 93, 94, 97, 99, 110, 114–15, 191, 210, 212, 234, 236, 239, 243, 276, 316–18, 320–23, 326, 327, 328, 330, 333, 336–8, 340, 343–5, 349, 358–60
Watt, James G., Jr, 73
Wavelength pair justification technique, 175, 343
Way Things Ought to Be, The (Limbaugh), 261
weather forecasting, 42, 199–208
Weather Prediction by Numerical Process (Richardson), 199–203

Weddell Sea, 61
Wesoky, Howard, 376, 377, 378
Whitecross, Dr Malcolm, 314
Wigley, Tom, 136
Williams, Ron, 95
Williamsburg, Virginia, 216, 267
Wilson, Harold, 58
Windhoek, 132
WMO *see* World Meteorological Organisation
Woolwich Arsenal, 45
World Health Organisation, 35–6
World Meteorological Organisation (WMO), 239, 321, 322, 328, 340, 342, 344, 348, 349, 353, 358, 362, 372, 377
World Ocean Circulation Experiment, 211
World Ozone Data Center, Toronto, 80
World Weather Watch, 144
Wuebbles, Don, 87
Wyoming, University of, 86

X-ray fluorescence technique, 75

Yugoslavia, 274

Zabarenko, Deborah, 238
Zaire, 132
Zambia, 132
Zenith (balloon), 37
Zhigansk, Siberia, 380
Zugspitzer, 128